# Radiation Protection at Light Water Reactors

Robert Prince

# Radiation Protection at Light Water Reactors

 Springer

Robert Prince
US Nuclear Regulatory Commission
Marquis One Tower
245 Peachtree Center Avenue
NE, Suite 1200
Atlanta
GA 30303-1257
USA

ISBN 978-3-642-44761-7          ISBN 978-3-642-28388-8 (eBook)
DOI 10.1007/978-3-642-28388-8
Springer Heidelberg New York Dordrecht London



*Photo credit:* The two-unit Koeberg Nuclear Power Station is owned and operated by the power utility ESKOM in South Africa. The Koeberg station consists of two PWR units rated at 900 MW(e) each and is located on the west coast of South Africa approximately 20 kilometers north of Cape Town (courtesy of ESKOM and Bjorn Rudner).

Printed on acid-free paper

Springer is part of Springer Science+Business Media (www.springer.com)

# Preface

Numerous excellent Health Physics textbooks have been published over the years with many of these texts now into their third or fourth editions. These texts cover a wide range of health physics discipline areas. General introductory health physics texts cover such subject matter as the interaction of radiation with matter, definition of radiation dose quantities and units, measurement of radiation dose, principles of detection, controls associated with the handling of radioactive materials and basic radiation safety principles. Many of these texts cover specialty health physics discipline areas including medical or environmental health physics or the detection and measurement of radiation for example. Radiation protection aspects associated with the operation of the worlds' current fleet of light water reactors has been marginally presented in existing health physics texts. Many texts that cover various health physics discipline areas often do not contain any discussion relating to light water reactor health physics. This text is meant to fill that void and is targeted to those health physicists currently employed in the light water reactor industry and to serve as the basis for a course of study for students entering the light water reactor radiation protection field. The text does not attempt to cover basic health physics topics in detail such as the principles of interactions of radiation with matter, the design of nuclear instrumentation, theory associated with the design of personnel dosimetry devices, calibration of survey equipment or detailed discussions associated with radiation quantities and units. All these subject areas are covered in much greater detail by others who are recognized as subject matter experts in their field. No useful purpose would be served by the author in attempting to cover these areas for which the author is only a novice. Texts related to the topics covered in this text that present supporting subject matter in greater detail are listed at the end of each chapter. Many of these texts provide extensive coverage of a given topic and often serve as the basis for an in-depth course.

This text was written for those individuals who wish to gain an understanding of radiation protection aspects associated with the operation and maintenance of commercial light water reactors in operation today. The author has attempted to focus each chapter on those topic areas directly related to radiation protection program activities required to support the operation and maintenance of light water reactors. Chapters 2 and 3 provide an overview of pressurized and boiling water reactor systems of radiological concern along with an overview of the purpose and function of those systems. A discussion of the parameters that effect radiological conditions of the various systems is also presented along with the radiological environment associated with these systems. Chapter 4 discusses the radiological source terms at LWR facilities and those parameters that influence the magnitude of source terms. Chapter 5 defines the various radiological zone classifications and the requirements associated with the posting of radiological areas. Chapter 6 presents the elements of those activities associated with a LWR radiation protection program at the functional level. Radiological surveillance activities, radiological signposting, RCA access control measures, job coverage aspects, use and purpose of radiation work permits, departmental interfaces and work control activities are presented. Chapter 7 presents those elements associated with the planning, scheduling and implementation of radiological work activities and techniques and methods employed to minimize personnel exposures. Chapter 8 describes contamination and radiation source control measures and techniques to minimize the presence and spread of radioactive contamination. Various measures to minimize and control the production of contamination source terms and the affects of water chemistry on source terms are presented. The various types of protective clothing and their use, equipment and supplies commonly used to control the spread of radioactive material, and elements of a respiratory protection program are presented in Chap. 9. Chapter 10 describes the elements of a LWR personnel dosimetry program. The utilization of various dosimetry devices for whole-body and extremity monitoring are discussed along with those elements that comprise a LWR bioassay program. An overview of the instrumentation requirements to support a LWR radiation protection program is presented in Chap. 11. Instrumentation commonly used along with the purpose and function of various types of fixed and portable survey equipment is described.

Problems and exercises have been provided that encompass those issues most likely to pose radiological incidents at a LWR. Many of the problems present a unique situation whereby a health physicist is challenged to evaluate a given issue in sufficient detail to ensure that an appropriate radiological assessment of the situation has been performed. The problems are also designed to encourage students to identify root causes and what actions would they take to minimize future radiological incidents or to prevent recurrence. Radiological incidents at LWRs seldom occur as a result of a technical issue requiring detailed calculations to determine doses received by those involved in the incident. Primary dosimeters along with whole-body counts and bioassay data are typically sufficient to support an adequate dose assessment. Oftentimes it is the non-technical and human performance aspects that contribute to a radiological incident that require attention and an adequate evaluation in order to improve radiological safety performance of a LWR radiation protection program. Consequently the problems and exercises are meant to enforce these aspects of radiological incident investigations. For many problems there is no "one right answer" rather the focus is to have students look at the "bigger picture" and utilize the skills and practical knowledge that a LWR health physicist should possess to ensure that corrective actions associated with radiological incidents are identified and thoroughly investigated. Any errors found in the text or problem solutions should be forwarded to the author.

# Contents

Contents

# Chapter 1
# Introduction

Radiation protection programs at nuclear power plants have developed and matured as experience in operating and maintaining nuclear power plants has been gained. Initial programs grew quickly in both size and complexity with the number and size of nuclear units entering commercial operation. Operational radiation protection programs evolved to face various challenges confronted by the nuclear power industry resulting in an overall increase in the effectiveness of radiological safety measures. Industry improvements in radiological safety have resulted in a significant decrease in the annual collective exposures over the past decades.

A well-planned, organized and managed radiation protection program is an essential element in ensuring the safe operation of nuclear power plants. The successful implementation of those discipline areas associated with a radiological safety program support efficient plant operations. Efficient use of plant resources, effective planning, prudent use of technological innovations, minimization of the time required to perform a given activity and ensuring that tasks are performed correctly the first time are cornerstones of an effective radiological safety program. A strong radiation protection program is integral in ensuring that activities conducted in radiologically controlled areas of nuclear power plants are performed safely and efficiently to minimize collective radiation exposure of plant personnel.

## 1.1 The Early Years

Light water reactor (LWR) radiation protection (RP) programs have evolved in concert with the growth of the American nuclear power industry over the last half-century. The units built under the "Eisenhower Atoms for Peace Program" and the early demonstration plants of the mid to late 1950s gave birth to the first generation of nuclear power plants. These early plants, such as the 150 MW(e)-Shippingport unit, which entered commercial operation in December of 1957, had generation capacities typically of 50 MW(e) to 200 MW(e). Larger capacity units

R. Prince, *Radiation Protection at Light Water Reactors,*
DOI: 10.1007/978-3-642-28388-8_1, © Springer-Verlag Berlin Heidelberg 2012

of several hundred megawatts soon followed in the 1960s. By the early 1970s the height of the nuclear power plant construction boom was in full swing with the majority of units under construction sized with capacities in excess of 1000 MW(e). By 1976 there were 25 units in commercial operation in the USA and 100 units by 1987. The world-wide number of units in operation and under construction also increased significantly during this period with France, Great Britain, Japan and Russia all having active nuclear power programs. Essentially in a period of 10 years, light water reactor radiation protection had grown from programs supporting relatively small capacity, single-unit sites, with small operating staffs to programs supporting multi-unit sites, with large units and operating staffs of 1,000 or more workers. It was a time of unprecedented growth. It was a challenging time for a program still in its infancy.

Health physics (HP) programs developed in support of the first generation of nuclear power plants were characterized by small staff sizes and in some cases were an outgrowth of chemistry departments. In fact during the 1960s and 1970s it was common practice in the USA to have combined radiation protection and chemistry programs (i.e., Rad-Chem) with technicians responsible for performing both chemistry and radiation protection functions. This approach was suitable when dealing with relatively small-sized, single-unit plants, when radiation protection programs were still being developed and in some respects not yet fully defined. Chemistry workloads are typically highest when nuclear units are at power, while RP workloads reach a peak during outages and other non-operating periods. Thus the dual roles were in some respects complimentary, maximizing the utilization of resources. In time it soon became apparent that the complexity of LWR radiological safety programs would demand greater numbers of specialized technicians and health physicists.

The early years of the nuclear power industry were marked by low capacity factors and numerous unplanned shutdowns. Planning and scheduling of radiological work activities was weakly executed and extended refueling and maintenance outages were the norm. Additionally source control efforts were limited, plant chemistry controls were relatively weak, and a host of other operational parameters that impacted radiological performance were not fully understood. A common attitude that prevailed during the early days was that worker exposures were more an indicator of who was working and who was not. Those with the higher exposures were obviously the "real workers" and higher doses were almost looked upon as a "rite of passage" into the nuclear power plant work force. The LWR health physics profession, as with any new field of endeavor, has grown from an infant program, has experienced the growing pains of adolescence and has matured as the nuclear power industry garnered hundreds of reactor operating years of experience.

The early days of nuclear power plant radiation protection could be characterized as a period when programs were being developed on-the-job. Programs were highly independent with limited exchange of ideas or lessons-learned within the industry. New equipment, processes, procedures and radiological safety techniques and practices had to be developed and tested. First generation power plant health physicists and radiation workers were being recruited and trained. There was no wealth of experience to fall back on to anticipate and prepare for the radiological challenges associated with the operation of large-scale LWR units. In this environment radiation protection programs typically had sole responsibility for all things to do with radiological safety. Workers were not directly responsible for their own radiological safety and the prevailing attitude was that this responsibility rested with the RP group.[1] This fostered an environment in which the skills and knowledge of craft personnel were not fully utilized when implementing radiological safety program improvements. Workers were content to leave this little understood field with its strange terms of roentgens, rads and DAC-hours and anything to do with radiological safety in general, strictly within the domain of the radiation protection group.

Typically radiological safety and job coverage planning was initiated when the work crew showed up at the Radiological Control Area (RCA) access control point. A radiation protection technician would be assigned to the task and responsible for establishing radiological controls for the job. There were no extensive, computer-based, job history files or radiation work permit databases to assist with the planning and performance of the task. Workers probably had not even given any thought to the radiological aspects of the task until a member of the radiation protection department was assigned job coverage responsibilities and even then their involvement may have been limited. Consider the radiological challenges this would pose when confronted with unplanned shutdowns, component malfunctions, unplanned operational events and other issues that pose radiological safety challenges?

Even though the vast majority of personnel exposures are received during refueling outages, pre-planning of these activities in the early days was virtually non-existent. Planning for outage activities would begin when the "breaker" was opened. Outage durations of 60–90 days and often much longer were the norm. The amount of maintenance and testing rework was excessive and there was little incentive in a regulated industry to shorten outage durations. No wonder that annual collective exposures averaging several hundred person-rem and annual collective exposures in excess of 1000 person-rem, were not uncommon, during the early years of the nuclear power industry. Figure 1.1 depicts historical annual collective radiation exposures for the LWR industry in the USA.

A "good" day in the life of the power plant health physicist was one in which the unit was on-line and there were no workers alarming the personnel contamination monitors at the RCA exit point as a result of some "gas" leak. However, these early experiences would serve as cornerstones in identifying needed program improvements. These improvements included such items as the design of more sensitive and improved radiation monitoring equipment, the development and

---

[1] Some countries such as Canada and France have established programs whereby workers receive more extensive training in radiological safety and surveillance techniques and have a greater degree of responsibility in implementing radiological safety measures.

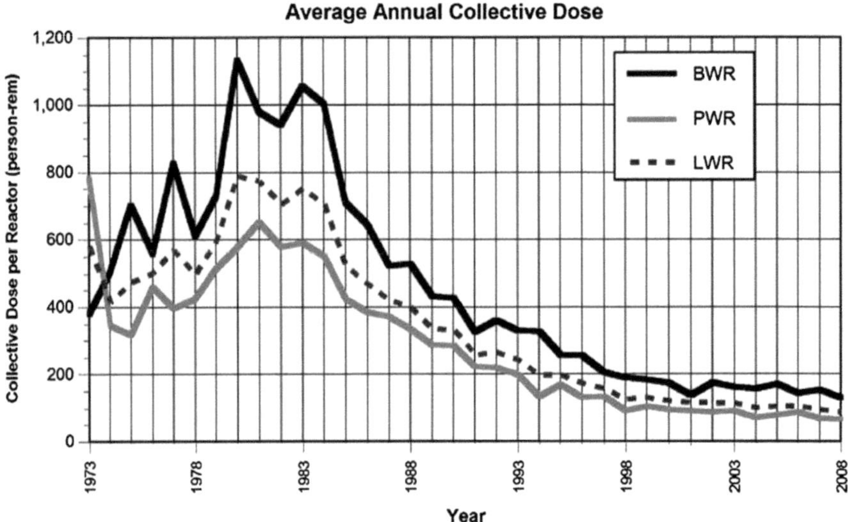



**Fig. 1.1** Average annual collective doses for BWR and PWR reactors from 1973 through 2006. (Source: US NRC NUREG-0713, Volume 28, www.nrc.gov)

implementation of innovative radiological control measures, improved radiation worker training and forging the development of effective ALARA programs among others. Additionally, the support of industry-sponsored research to improve radiological work practices and reduce worker exposures garnered momentum as radiation protection programs matured. Then on March 28, 1979 the accident at Three Mile Island, unit 2, occurred which would forever change the industry. This one event underscored the need to ensure operational, and by extension radiological safety, excellence in the daily operation of nuclear units. The sharing of lessons-learned and operational experiences and the need to proactively establish and pursue programmatic improvements would become industry cornerstones. Stemming from this event important cultural changes were beginning to emerge that would result in a broad-based interdepartmental approach towards radiological safety, resulting in improved performance. The days when RCA access control points were housed in converted locker rooms or restroom facilities with RP organizations of 10–15 people were coming to an end.

Some key aspects associated with current LWR radiation protection programs are introduced here. These discipline areas are integral to establishing and maintaining an effective LWR radiation protection program. Various attributes associated with these topic areas are intertwined and expanded upon in related chapters of this text. Though the topics discussed below are not an exhaustive list, and some subject areas are not necessarily unique to RP, these areas must be successfully addressed to ensure the effective implementation and continued development of LWR radiation protection programs.

## 1.2 Source Control and Reduction

Industry efforts associated with source reduction have dramatically reduced personnel exposures and greatly improved contamination control aspects of radiological safety at nuclear plants over the years. These efforts have been targeted at reducing and controlling the inventory of activation products, most notably cobalt-60, improved fuel cladding performance and controlling contamination at the "source". In the early days little if any thought was given to cobalt reduction or controlling contamination at the source. Oftentimes a large area or entire room was "posted" more for convenience regardless of the impact on work activities, even if the source of the contamination was an isolated leaking valve or component. Aggressive control and minimization of leaking components has resulted in large reductions in the amount of contaminated square footage within radiological control areas. It was not uncommon to have upwards of 10% of the RCA floor area posted as contaminated. This complicated entry into areas for maintenance work and routine inspection and operator rounds. Time to perform tasks in contaminated areas was increased, coupled with ambient radiation levels, resulting in additional exposures to workers. Secondary impacts included the generation of increased radioactive waste volumes and the concomitant handling and disposal costs. Today contaminated square footage is routinely maintained below 1% of the RCA area and oftentimes much less.

Though results have been impressive it is essential that the industry continue to implement measures to further reduce source terms and minimize the extent, duration and magnitude of contaminated areas. The deregulated marketplace will continue to exert pressure on the industry to reduce costs and staff sizes, or otherwise find ways to improve worker performance and productivity. A radiological environment in which individuals are not encumbered with needless use of protective clothing and equipment can play an important role in achieving these objectives. Control of contamination at the source versus the "room", improved chemistry controls, continued cobalt reduction efforts, and utilization of various system decontamination processes and other techniques will all play an important role in reaching the next level of radiological excellence within the industry. It is essential that the industry not become complacent in this regard. Even though the magnitude of additional improvements may not be as dramatic as those already achieved they will be no less important. It is essential that industry working-groups continue their efforts to identify new processes and take advantage of new technologies that will help to improve radiological safety performance.

## 1.3 Outage Management

Though it is not the sole domain of radiation protection, radiological safety performance is uniquely entwined with outage performance. Since outage periods represent the most challenging radiological opportunities it is essential that continued focus be placed on improving outage performance. This is an area that has been significantly impacted by deregulation. One obvious way to increase plant capacity factors, and hence the generation of megawatt-hours, is by shortening outage durations. Shorter-duration outages have a significant impact on radiological safety. Outages that are well planned and that incorporate measures to improve worker productivity (e.g., use of automated equipment, better tooling and work methods) result in minimal re-work and overall are shorter in length. These factors typically result in reducing outage exposures. The dramatic reduction in outage durations that began in the mid 1980s or thereabouts, from 40 to 60 days, to a current average of less than 30 days, with many plants routinely completing outages in less than 25 days, has been a key factor in lowering annual industry exposures.

Even though major improvements in outage performance have been achieved this is an area in which continued improvement will be required to maintain and improve upon the current level of radiological safety. Radiological preparations and support of outage activities play an important role in the successful execution of outage tasks. Oftentimes effective radiological preparation and planning not only result in improved radiological safety for employees but also increases worker productivity and efficiency. Radiation protection departments have become fully integrated with outage planning and scheduling. This integration has resulted in many of the noted improvements. However, as in other areas, it is vital that efforts continue in this regard and that new and innovative techniques with potential radiological safety benefits continue to be evaluated and implemented to support improved outage performance.

## 1.4 Benchmarking and Lessons-Learned

Many painful and important lessons-learned relating to radiological safety have been experienced by the industry since its inception. The nuclear power industry exists in an environment that is intolerant of mistakes. There is an unconditional mandate to protect the health and safety of the public and to maintain public trust and confidence. To achieve these objectives nuclear power must be held to a high standard. The TMI and Chernobyl accidents serve as vivid reminders of the impact an event at one nuclear plant can have on the entire industry. This point is particularly well illustrated in Joseph Rees' 1996 book entitled "Hostages of Each Other" the title aptly captures the situation of the nuclear power industry. The consequences of industry events such as unplanned worker exposures, contamination events involving several or more individuals, inadvertent release of radioactive material, and other events involving radiological safety concerns are not limited to one plant and often end up on the front pages of local newspapers. To prevent recurrence of these and similar radiological safety events and to promote continued improvement in LWR radiation protection programs, it is essential that the industry continue efforts to communicate lessons-learned in a timely and effective manner.

Benchmarking efforts should be considered an integral part of any LWR radiation protection program. Participation in internal and external RP program assessments and evaluations is an effective way to exchange ideas, to strengthen existing programs and to promote implementation of new techniques. Benchmarking efforts may involve actual visits to other nuclear units, participation in workshops and symposiums, industry surveys or simply phoning or contacting colleagues to inquire about specific topics. As noted previously the impact of deregulation and economic issues in addition to the absolute need to maintain excellence in radiological safety are issues that weigh heavily on future prospects of the industry. Diligent use of lessons-learned and aggressive benchmarking efforts play an important role in this regard. Program improvements must be predicated on proactive measures and should not rely solely on lessons-learned stemming from after-the-fact evaluations. Industry efforts in this area have been instrumental in reducing both the frequency and severity of radiological incidents experienced by the industry. The continued pursuit of benchmarking and sharing of lessons-learned will be no less important in the years to come.

## 1.5 Innovation and Use of New Technology

We are living in an age of explosive growth and innovation with new and more versatile and powerful tools and products introduced to society almost on a daily basis. Creative use of these products has resulted in significant improvement in the performance of LWR radiation protection programs. Many RCA access control facilities now resemble miniature control rooms. Computer terminals provide live-time display of in-plant area radiation monitors and continuous air monitor readings. Video monitors and a system of cameras may be available to provide live-time coverage of key RCA areas. Capabilities may also include the availability and use of numerous teledosimeters, and other wireless technology, placed throughout the RCA and in the vicinity of key components to monitor ambient radiation levels. Such remote monitoring systems may preclude the need for RP technicians to enter certain areas of the plant, such as high radiation areas, to perform radiological survey functions. The use of remote or wireless technology for personnel dosimetry purposes and remote job coverage has resulted in both significant dose and cost savings. State-of-the-art communication equipment from cell phones to cordless headsets and video and camera equipment have had a tremendous impact on improving the productivity and efficiency of tasks conducted in radiologically significant areas. Personnel providing job coverage support activities, various inspections, observation type tasks among others can now perform these functions at remote locations, oftentimes outside the RCA itself. Plant documents and procedures can now be accessed by a worker inside the RCA via a handheld or pocket device, displaying the information on a small display screen attached to workers safety glasses. Improved chemistry controls and processes have greatly reduced source term inventories and have minimized the long-term buildup of source material, which has been a major factor in lowering personnel exposures. The diligent application of new technologies for radiological safety purposes will be an important ingredient in fostering continued improvements in LWR radiation protection programs.

Possible areas where new technology and processes could have significant impact on radiological safety would include applications associated with source reduction, reducing outage durations, products that improve the productivity, efficiency and reliability of maintenance and operational activities among others. Radiation protection personnel should stay abreast of new technological innovations and products in order to take full advantage of these opportunities. The business climate will continue to demand that the industry do more with less. How many nuclear plants or RP programs have increased staff levels over the last 10–15 years, probably not many? A trained health physicist may be the first to recognize the potential radiological safety applications of a new product. This could be something as simple as the introduction of masslinn clothes or Velcro straps in lieu of tape that are now in common use, to the use of power tools and automated equipment for routine maintenance work or hand-held electronic devices to reduce the time required to read and record instrument readings located in radiological areas. If the recent past is any indication of the potential impact that future technological innovations may have in the area of radiological safety than RP programs must be positioned to take full advantage of these opportunities.

## 1.6 Training the Next Generation of Health Physicists

The first generation of LWR health physicists with their vast knowledge and operational experience are reaching retirement age at the same time that the nuclear power industry may be on a threshold of experiencing resurgence. Some countries with a long dormant nuclear power program (e.g., USA) are reconsidering the need for an increase in nuclear power generation in order to meet growing energy needs in an environmentally sound manner. In addition, countries with improving economies and high economic growth rates, such as China and India, are relying heavily on nuclear power to support their future energy requirements. Economic viability is promising and many plants in the USA have already applied for and obtained 20-year license extensions. The industry in general and RP in particular, is confronted with the need to hire and train the next generation of LWR health physicists. The availability of trained and qualified health physicists and RP technicians will be important in sustaining a viable nuclear option. Where will these people come from and will sufficient numbers of suitably qualified professionals be available to support the future needs of a resurgent nuclear power industry?

The number of colleges and institutions offering degree programs in radiation protection and technical institutions offering radiation protection-related certifications has dramatically declined over the last 10–15 years. Industry and academia are confronted with the task of recruiting the next generation of health physicists for the nuclear power industry. The perception on the part of many college-age people, that nuclear power is a "dying" industry makes this a difficult task. It is essential that suitably qualified individuals be recruited to assume responsible positions in the coming years as the current generation of health physicists retire or otherwise leave the industry. To ensure continuity of existing programs, while at the same time sustaining ongoing programmatic improvements, recruitment must be ongoing. The current generation of health physicists represents an extensive amount of experience gained in the school of "hard knocks" that must be imparted to the next generation in an orderly and systematic fashion. It is essential that this knowledge be retained and expanded to ensure that past lessons-learned are not prone to be repeated.

As operating licenses are extended and plant lifetimes become longer there will be new radiological challenges facing the industry. These challenges will have to be addressed in a safe, timely and economical manner. Consider the recent situation whereby many PWR facilities were confronted with the need to replace reactor vessel heads. This challenging radiological issue arose in the late 1990s with many units confronted with the need to replace reactor vessel heads starting in the early 2000s. Those facilities initially confronted with this emerging issue were on the "leading edge" of this learning curve and had to devote significant RP resources to address the radiological safety challenges associated with this major task. This is a significant radiological task that has been addressed remarkably well and one in which experienced health physicists drew upon their experience to mitigate the radiological safety aspects associated with projects of this nature. What other challenges are on the horizon? Whatever challenges may lie ahead it will be essential that the industry have experienced health physicists who possess the necessary skills to address emergent radiological issues.

Some utilities in recent years have initiated training programs for new RP technicians. However, these efforts will need to be expanded, perhaps with utilities partnering with local colleges, to ensure that sufficient numbers of degreed health physicists are available. The expansion of scholarship programs and recruitment at the high school-level will play an important role in this effort. Though the industry is uniquely aware of the potential shortage of health physicists it is imperative that present efforts continue and be expanded to ensure that industry resource needs are met.

## 1.7 Deregulation

Deregulation has presented new challenges and opportunities for the nuclear power industry to say the least. Twenty years ago most power plant health physicists probably never worried about such esoteric issues as earnings per share, generation costs, outage durations, the budgeting of resources for capital projects or economic aspects of plant operations in general. How many health physicists are wondering if all these recently introduced training programs are in an attempt to make them financial gurus? In fact many health physicists may find they are spending more time dealing with financial issues than radiological safety issues on a given day. The movement toward a deregulated, competitive industry will play a vital role in the planning and execution of RP programs. Consider the impact that shorter outage durations have had on the number of contractor RP technicians available for short-term assignments. The pool of qualified RP technicians as well as other radiation protection professionals available to support outages has drastically declined over the last decade. Radiation protection programs had to adapt to this environment and implement creative measures to ensure the continued success of radiological safety programs. The continuing impact of deregulation and the overall need to control generation costs will have to be anticipated and appropriate measures implemented to minimize any potential impact on the effectiveness of radiation protection programs.

## 1.8  Summary

Light water reactor radiation protection programs have improved greatly since the initial conception of the nuclear power industry. The industry has gone through its glory days, has been on the "ropes" fighting for its economic survival, has withstood many challenges and is now at a crossroad. The continued safe and economic operations of today's nuclear plants are vital if the nuclear industry is to play a larger role in meeting America's future energy needs as well as that of other countries. Radiation protection programs must continue to explore and implement innovative measures to ensure continual improvements in performance. Many radiological safety improvements often result in secondary benefits that increase worker productivity and efficiency. In other words, sound radiological work practices go hand-in-hand with improved economic performance, a fact that was not fully appreciated in the early days of the industry when ALARA resources were often scarce and ALARA programs under funded. Radiation protection programs will continue to play a vital role in ensuring excellence in nuclear and radiological safety, while supporting industry efforts to successfully face economic and other challenges that lay ahead.

## Bibliography

1. Barley W., and Hiatt J., Acceptable Experience and Training for HP Technicians as Nuclear Power Plants, Radiation Protection Management, 5:70–74; 1988
2. Blevins M., and Andersen R., Radiation Protection at U.S. Nuclear Power Plants—Today and Tomorrow, Health Physics, 100(1):35–38, 2011
3. Brooks M.E., San Onofre's Work Controls Planning Program, Radiation Protection Management, 8:40–50; 1991

4. International Atomic Energy Agency, Best Practices for Identifying, Reporting and Screening Operating Experience at Nuclear Power Plants, IAEA-TECDOC-1581, Vienna, 2007
5. International Atomic Energy Agency, Best Practices in the Management of an Operating Experience Programme at Nuclear Power Plants, IAEA-TECDOC-1653, Vienna, 2010
6. International Atomic Energy Agency, Best Practices in the Organization, Management and Conduct of an Effective Investigation of Events at Nuclear Power Plants, IAEA-TECDOC-1600, Vienna, 2008
7. Leclercq J., The Nuclear Age, Hachette, France, Euro American Consulting & Service; 1986
8. Lish K., Nuclear Power Plant Systems and Equipment, New York, Industrial Press, Inc.; 1972
9. Millsap W.J, Zitevitz L.T., Glennon P.T., Sejvar J., Health Physics Aspects of Reactor Lower Internals Transfer at Salem Unit 1, Radiation Protection Management 7:57–68; 1990
10. Mothena P., RTD Bypass Systems Elimination of Virgil C. Summer Station, Radiation Protection Management 9:59–70; 1992
11. National Council on Radiation Protection and Measurements, NCRP Report No. 120, Dose Control at Nuclear Power Plants, Methesda, MD: 1994
12. National Council on Radiation Protection and Measurements, Self Assessment of Radiation Safety Programs: Recommendations of the National Council on Radiation Protection and Measurement, NCRP Report No. 162, Bethesda, MD, 2009
13. Rahn F., Adamantiades A.G., Kenton J.E., and Braun C., A guide to Nuclear Power Technology—A Resource for Decision Making, New York: Wiley & Sons; 1984
14. Rees J., Hostages of Each Other: The Transformation of Nuclear Safety Since Three Mile Island, The University of Chicago Press, Chicago; 1994
15. United States Nuclear Regulatory Commission, Guidance for Performance-Based Regulation, NUREG/BR-0303, Washington, D.C., 2002
16. United States Nuclear Regulatory Commission, Occupational Radiation Exposure at Commercial Nuclear Power Reactors and Other Facilities 2001, Thirty-Ninth Annual Report, NUREG-0713, Washington, D.C., 2006
17. United States Nuclear Regulatory Commission, Qualification and Training of Personnel for Nuclear Power Plants, Regulatory Guide 1.8, Washington, D.C., 2002
18. United States Nuclear Regulatory Commission, Reactor License Renewal—Preparing for Tomorrow's Safety Today, NUREG/BR-0291, Washington, D.C., 2002
19. United States Nuclear Regulatory Commission, Reactor Oversight Process, Revision 3, NUREG-1649, Washington, D.C., 2000

# Chapter 2
# Radiological Aspects of PWR Systems

## 2.1 Overview

Light water reactors are characterized by the fact that water serves as both the coolant and moderator. Two major types of reactors dominate the LWR industry, the pressurized water reactor (PWR) and boiling water reactor (BWR). This chapter describes those PWR systems of concern to radiation protection personnel while Chap. 3 provides an overview of BWR systems. The primary objective is to present those aspects of system design and interrelationships that impact plant radiological conditions. The function and purpose of various systems are presented along with their associated radiological hazards. System descriptions are provided in sufficient detail to allow radiation protection personnel to assess radiological conditions associated with various plant operating conditions.

Radiation protection personnel should have a basic understanding of various plant systems in order to evaluate actual and potential radiological hazards associated with the operation of LWR's. It is not necessary for radiation protection personnel to have an in-depth working knowledge concerning all aspects of system operational-related parameters as required of plant operators. Consequently, the intricate details of system design and functions comparable to the level of knowledge required of plant operators are not covered. However, it is essential that they have sufficient knowledge of plant systems to adequately address the radiological requirements for activities performed either on or in the vicinity of plant systems.

Pressurized water reactors currently operating in the United States have been designed by the Westinghouse Electric Corporation, Combustion Engineering, Inc., and the Babcock and Wilcox Company (now Framatome). Other suppliers include Framatome (France), Kraftwerk Union (Germany) and Toshiba and Mitsubishi Heavy Industries (Japan) among others.

Several PWR systems are of direct concern from a radiological aspect. These include the reactor coolant (or primary) system and various auxiliary systems. The auxiliary systems of most concern include the chemical and volume control

R. Prince, *Radiation Protection at Light Water Reactors,*
DOI: 10.1007/978-3-642-28388-8_2, © Springer-Verlag Berlin Heidelberg 2012

system, residual heat removal system, reactor cavity purification, spent fuel pool cooling and purification, safety injection system, containment spray system, plant ventilation systems, radioactive waste handling and processing and radiochemistry sampling systems.[1] These systems and those having a potential of becoming contaminated under certain conditions are described.


## 2.2  Plant Layout

A PWR facility consists of three or four distinct buildings in addition to those that are required to support site operations such as administrative office buildings, security access facilities, and warehouses among others. Major buildings commonly associated with PWR stations include the containment building (or reactor building), the fuel building, the auxiliary building and the control building (Fig. 2.1).

The containment building is a large reinforced concrete cylindrical structure which houses the primary system components, components of emergency core cooling equipment, and air handling and ventilation equipment. Depending upon the design and size of a particular PWR unit various components (e.g., RHR system) may be located in either the containment building or the auxiliary building. The fuel building contains the spent fuel storage pool, building ventilation and cooling equipment, spent fuel pool cooling and purification components, and new fuel storage facilities. Systems and components associated with the chemical and volume control system, safety injection system pumps and heat exchangers, perhaps residual heat removal system components, various storage and hold-up tanks, radioactive waste processing facilities, air handling equipment, filter and demineralizer compartments, and associated electrical equipment, piping and valves are located within the auxiliary building. Additionally numerous components and equipment associated with the secondary side are located within areas of the auxiliary building. The main control room is located in the control building. Typically the control building will also include the battery rooms, motor control centers, electrical cable and relay rooms, and emergency ventilation equipment. Figure 2.2 depicts a typical PWR containment building and steam flow to the turbine generator and secondary side systems.

---

[1] System nomenclature of the various reactor vendors differs to some degree. For instance residual heat removal and decay heat are synonymous terms as are the makeup and chemical and volume control systems. Readers may want to refer to the specific nomenclature used at their facilities. Terms used in this text are descriptive in nature and may differ somewhat from site-specific terminology.

**Fig. 2.1**  Photograph of a two-unit PWR unit (Courtesy of Luminant)

## 2.3  Primary System

As the name implies a PWR maintains the primary circuit at an elevated pressure, approximately 15.5 Mpa (2,200–2,300 psi) with an operating temperature of about 332°C (629°F). The primary or reactor coolant system (RCS) contains the reactor core. The reactor coolant system provides cooling for the reactor core and transfers the heat to the secondary side via steam generators, producing steam to drive the turbine-generator. The primary system consists of 2–4 loops. Each loop contains a reactor coolant pump (or pumps), steam generator and associated piping. In addition to the reactor vessel, the pressurizer and pressurizer relief tank are the other major components associated with the primary system. Figure 2.3 depicts the basic PWR primary system components.

The reactor vessel contains the fuel assemblies, core support structures, control rods, thermal shield, incore guide tubes and related components. The reactor vessel contains the heat generated by the core, provides a flow path for the moderator-coolant through the core, allows access to the fuel during refueling operations and provides penetrations to allow the control rods and incore instrumentation to access the core. Figure 2.4 depicts a reactor vessel along with some of its' major components.

The major components of the RCS are located within the biological shield wall area of the containment building. Depending upon the number of loops and specific design, loops may be equipped with individual shield walls. Access to major components such as steam generators, reactor coolant pumps and the pressurizer are

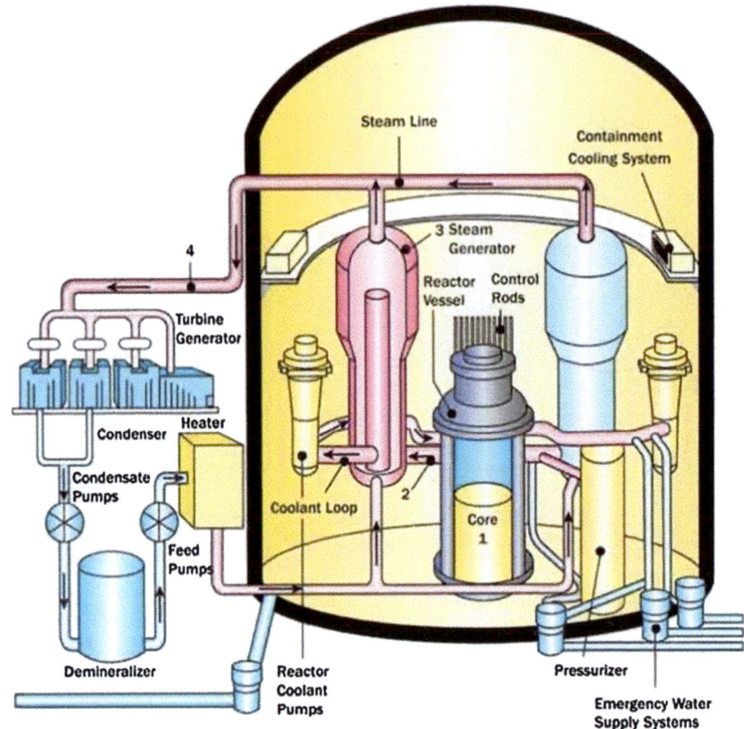



**Fig. 2.2** Simplified PWR plant layout showing the major components of the containment building (www.nrc.gov/reactors/pwrs)

strictly limited during periods of operation. When a unit is at 100% power dose rates inside the primary shield wall, enclosing the loop rooms is typically on the order of 100–250 mSv/h (10–25 rem/h). Contributors to these radiation fields include the presence of short-lived radionuclides, primarily N-16, in addition to activation and corrosion and fission product radionuclides that are present in the coolant. Obviously neutron radiation levels may be significant, the magnitude of which increases with reactor power. Neutron radiation levels inside loop rooms at 100% power could easily be in the range of a few tens of mSv/h (several rem/h). Consequently access to reactor coolant pumps, steam generators and pressurizer areas is strictly limited while at power. Under certain circumstances (e.g., <10% reactor power) and depending upon the specific plant design, access for short periods of time, on the order of minutes, may be possible for emergency type entries to investigate trouble alarms or equipment problems. Any such entries must be properly planned and strictly controlled.

Similar conditions and reasoning applies to areas in direct line of site of the reactor vessel head and reactor cavity area. Dose rates on top of the reactor head structure; in the cavity area and in close proximity to the edge of the reactor cavity

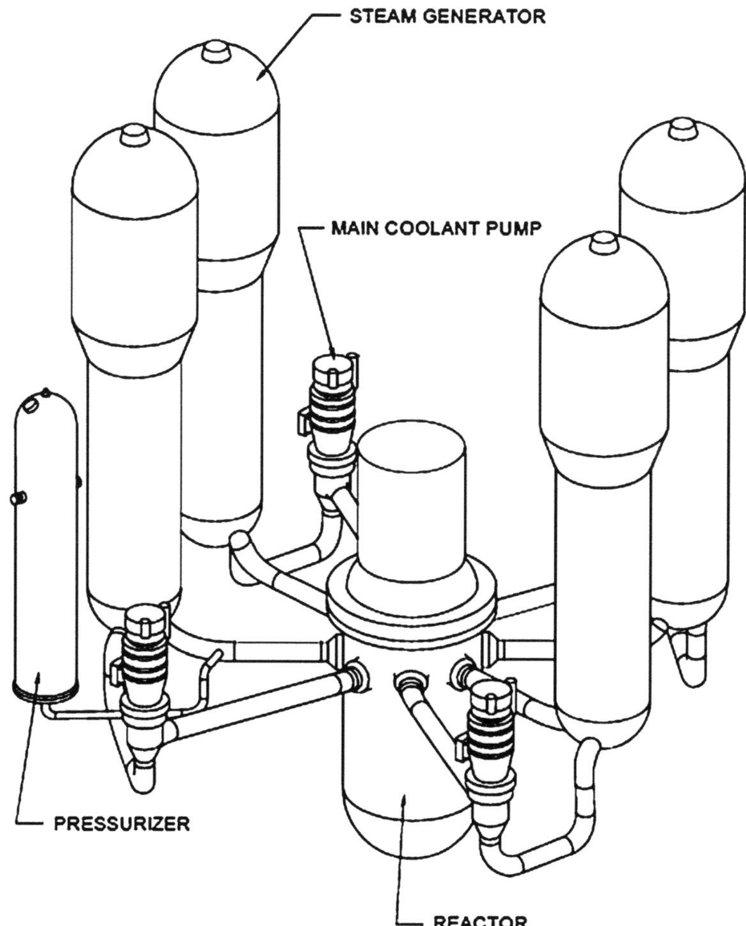



**Fig. 2.3** Major PWR primary system components for a four-loop PWR (adopted from www.nrc.gov/reading-rm/basic-ref/teachers)

usually prevent entry to these areas at power due to high radiation levels, which include a significant neutron radiation component. General area radiation levels in the range of 100–200 mSv/h (10–20 rem/h) on top of the reactor vessel head and in the vicinity of control rod drive mechanisms are not uncommon.

Each primary system loop contains a U-tube or once-through steam generator (SG). Primary system water flows through the tube bundle and transfers its heat to the feed water circulating on the shell side of the steam generator. Located directly above the tube bundle is the steam drum section. The steam drum section extracts moisture from the steam returning it to the feed water stream and dries the steam before it leaves the steam generator. This ensures a high quality steam supply to the turbine-generator. Manways and handways (or hand holes) are provided at

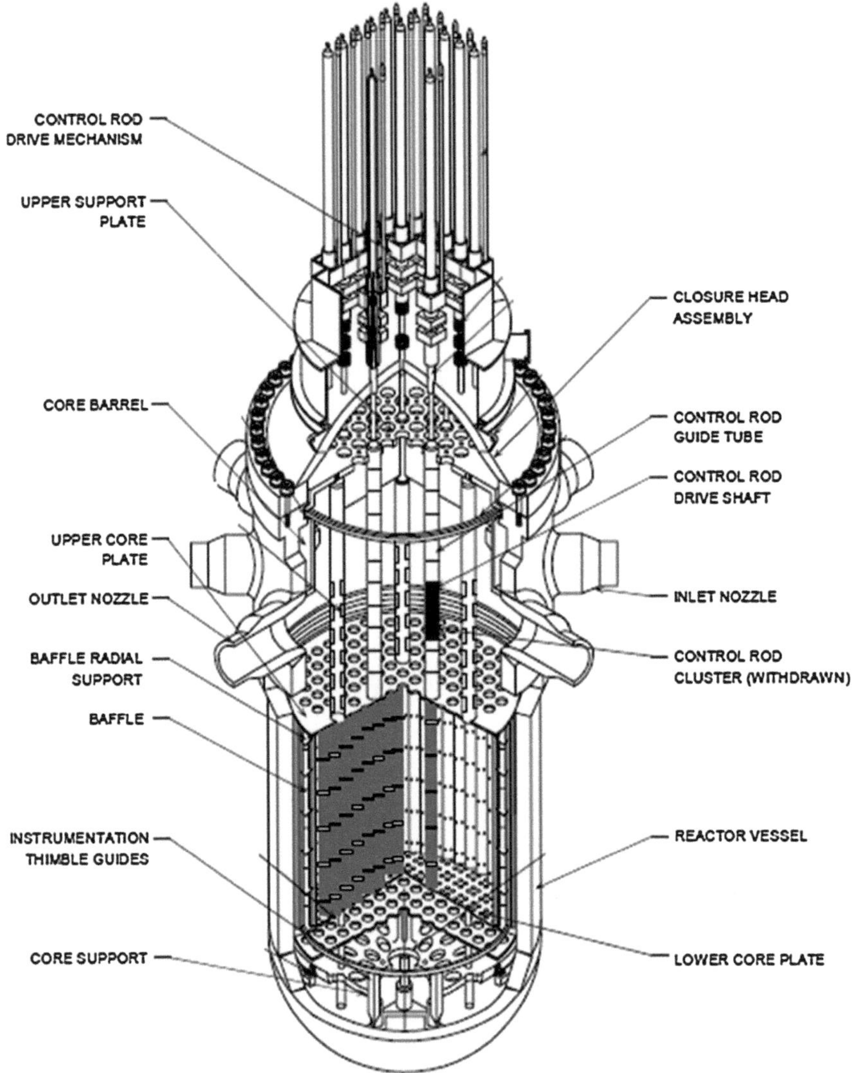



**Fig. 2.4** PWR reactor vessel and major components (adopted from www.nrc.gov/reading-rm/basic-ref/teachers)

strategic locations on each steam generator to allow access for inspection and maintenance activities during outage periods. Each steam generator contains thousands of individual tubes to provide the necessary surface area to afford sufficient heat transfer to produce the steam required in order to generate the large number of megawatts typical of nuclear power plants. Steam generators must be maintained in good operational condition to ensure operational efficiency of a nuclear unit. Degraded conditions impacting the quality or amount of steam produced by each steam generator can result in reduced megawatt output, negatively impacting economical operation of the unit. Steam generators are subject to comprehensive inspections and maintenance activities during outages. These activities can comprise a significant portion of outage exposures at PWR units.

Access in the vicinity of steam generators is not typically performed while at power due to high radiation levels. Additionally, even if access were possible the scope of inspections would be limited. The most likely reason to inspect a steam generator while at power would be in the event of a suspect manway leak or hand hole inspection port leak. Assuming these areas are readily accessible and could be easily observed entries for emergency investigative purposes could be warranted under certain circumstances.

Radiation levels associated with steam generators are of primary concern during outage periods when inspection and maintenance activities can be performed. Once the primary manway channel heads are removed dose rates of several mSv/h (several hundred mrem/h) may be encountered in the vicinity of the open manways. Radiation levels up to tens of mSv/h (several rem/h) or higher on contact to the tube sheet, inside the channel head, are not uncommon. Steam generator channel head dose rates may vary significantly from one unit to the next and are highly dependent on operating history of the plant, maintenance of good operational chemistry controls, shut down chemistry methods employed and integrity of fuel cladding. Figure 2.5 depicts the major components of a steam generator; notice the location of the manways and the tube sheet area of the steam generator.

Reactor coolant pumps (RCP) circulate reactor coolant through the reactor vessel taking suction from the steam generators. Depending upon the reactor supplier there may be either one or two coolant pumps for each steam generator. Each pump is composed of a hydraulic section, seal section and motor package. General area radiation levels in the vicinity of reactor coolant pumps during shutdown conditions may be as high as few mSv/h (a few hundred mrem/h), especially in close proximity to the seal section of the pump. Depending upon the design and specific location of the motor section, dose rates in the vicinity of the motor section are usually significantly lower, perhaps less than hundreds of μSv/h (ten's of mrem/h) or lower. Figure 2.6 depicts a reactor coolant pump.

The last major component of the primary system is the pressurizer. The pressurizer is a large vessel, maintained partially filled with RCS water and a cover gas maintained in the upper portion of the vessel. The pressurizer maintains the reactor coolant system pressure within prescribed limits. Electrical heaters located internally to the pressurizer are switched on when RCS pressure must be increased. When system pressure must be decreased, cold water is sprayed into the pressurizer void space via an internal spray nozzle located at the top of the pressurizer. Discharges from the pressurizer are routed to the pressurizer relief tank (Fig. 2.7).

As with other RCS components, radiation levels at power severely restrict entry to the pressurizer. Depending upon the design and compartmental layout it may be possible to access the lower regions of the pressurizer while at power. General area

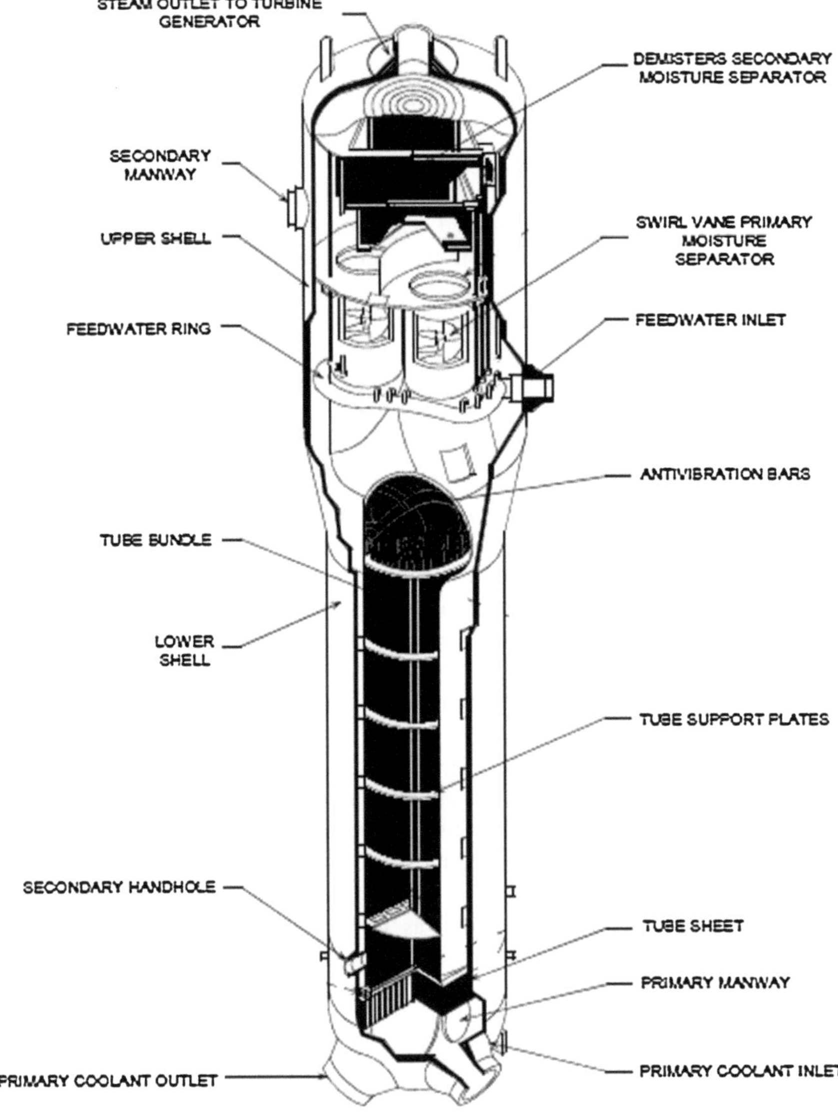


**Fig. 2.5** Steam generator and its major components (adopted from www.nrc.gov/reading-rm/basic-ref/teachers)

radiation levels in the vicinity of the pressurizer at power could range from a few mSv/h (few hundred mrem/h) to tens of mSv/h (several rem/h), again depending on the unique plant layout. Consideration must also be given to environmental conditions in the pressurizer compartment while at power. Ambient temperatures,

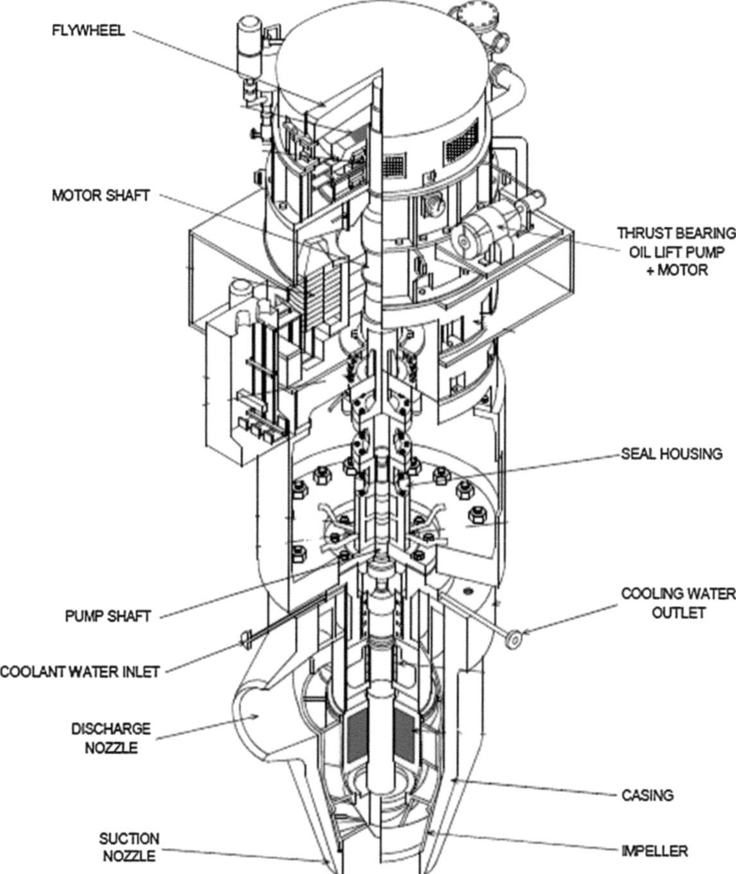



**Fig. 2.6** A reactor coolant pump and its major components (adopted from www.nrc.gov/reading-rm/basic-ref/teachers)

especially towards the top of the pressurizer, may be excessively high and could also be a factor in limiting access during periods of operation. The most likely situation requiring access to the pressurizer while at power would probably be associated with the need to inspect the pressurizer safeties for possible leakage. The pressurizer safeties are relief valves that provide over-pressure protection for the RCS. If the pressurizer safeties are accessible for visual observation it could be possible to allow entry for a short period of time. This would assume that environmental conditions allow entry and that the estimated dose to the individuals is acceptable based upon the urgency and benefits to be gained.

Radiation levels associated with the pressurizer are of primary concern during outage conditions since this is the period of time that maintenance and inspection activities are performed. The pressurizer heaters require routine maintenance and inspection. Radiation levels associated with these heaters could be in the range of

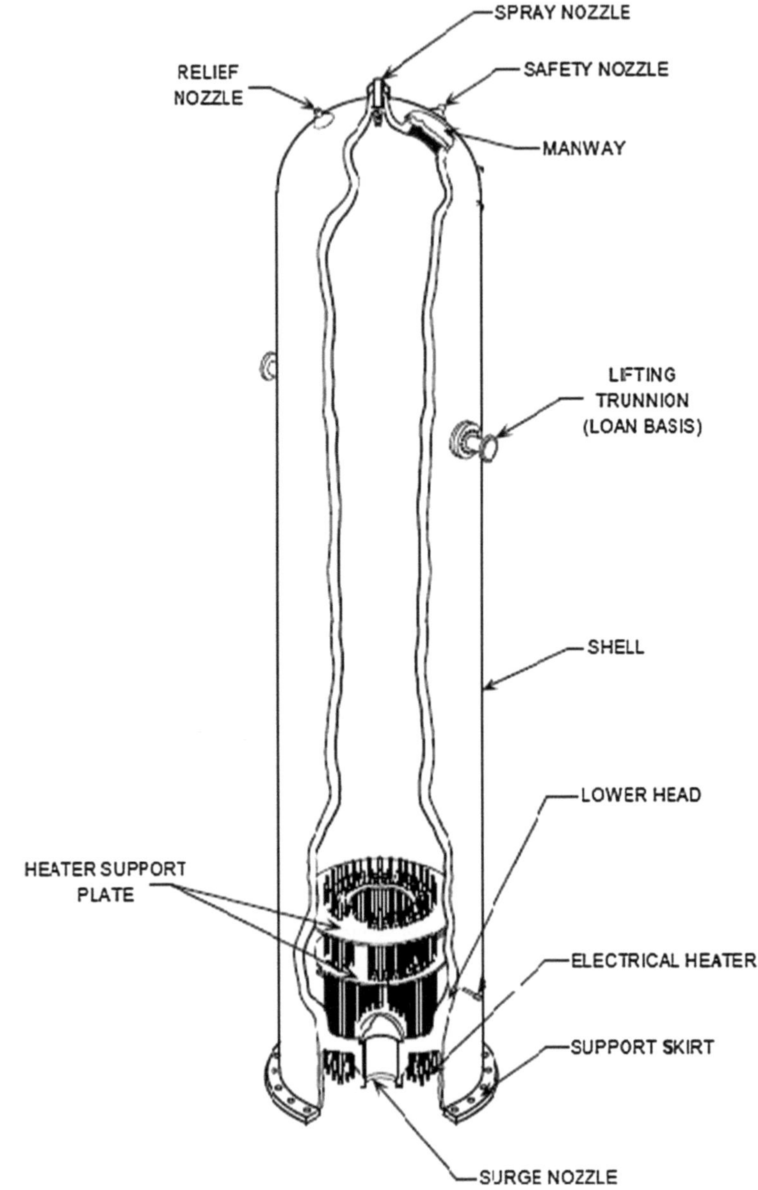


**Fig. 2.7** Pressurizer and its major components (adopted from www.nrc.gov/reading-rm/basic-ref/teachers)

several mSv/h to a few hundred mSv/h (several hundred mrem/h to perhaps a few rem/h), and will be highly dependent upon operational history pertaining to fuel integrity and RCS chemistry conditions. The pressurizer safeties also require routine maintenance and testing. These valves are located on the top of the pressurizer, oftentimes in very close proximity to the pressurizer and in a relatively confined area due to the presence of related piping, pipe supports and related components. Radiation levels on contact to the safeties and adjacent piping may be on the order of a few tens of mSv/h (a few rem/h) during outage periods with general area dose rates of a couple of mSv/h (a few hundred mrem/h) not uncommon.

## 2.4   Chemical and Volume Control System

The chemical and volume column system (CVCS) purifies the reactor coolant by removing fission and activation products by filtration and demineralization, maintains reactor coolant system inventory, adjusts the boron concentration in the reactor coolant and serves as an integral component of the emergency core cooling system. Major components of the CVCS system include pumps, heat exchangers, a volume control tank, purification filters and demineralizer beds in addition to associated valves and piping. Another common name for this system is the makeup system.

Reactor coolant is discharged to the CVCS system (i.e., letdown flow) and flows through the shell side of the regenerative heat exchanger where the temperature of the RCS letdown is reduced. The coolant is reduced in pressure and next flows through the tube side of the letdown heat exchanger. The letdown flow next passes through a mixed bed demineralizer and reactor coolant filter and enters the volume control tank via a spray nozzle located at the top of the tank. The purified and chemically treated flow (i.e., charging flow) is returned to the RCS via charging pumps. Most of the charging flow is directed to the reactor coolant system through the tube side of the regenerative heat exchanger to reduce the temperature of the letdown flow. The remaining portion of the charging flow is routed to the reactor coolant pump seals and returns to the CVCS through the seal water filter and seal water heat exchanger. If the normal letdown path is not available, reactor coolant may be returned to the volume control tank (VCT) via the letdown heat exchanger. Figure 2.8 displays the major components of the CVCS system. Obviously due to the function of the CVCS system and the fact that it contains letdown from the RCS it represents a system of significant radiological concern to radiation protection personnel. A more detailed description of key CVCS components follows in order to provide a foundation for evaluating and understanding potential radiological conditions associated with this system.

The regenerative heat exchanger recovers heat from the letdown flow by reheating the charging flow. This reduces the reactivity affects resulting from the insertion of relatively colder water into the core and reduces thermal shock to reactor coolant system piping. The regenerative heat exchanger is the first component that the RCS letdown flow enters. Consequently it is essentially an "RCS component" from a radiological perspective. The regenerative heat

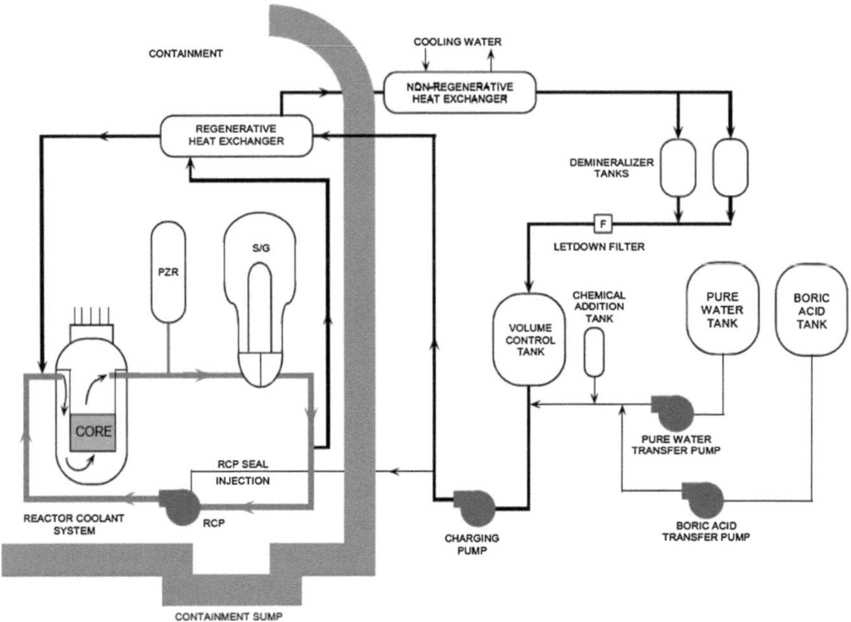


**Fig. 2.8** Schematic of CVCS system and its major components (www.nrc.gov/reading-rm/basic-ref/teachers)

exchanger is typically located within the containment building in a shielded compartment and considered not accessible while the plant is at power. Contact radiation levels tens of mSv/h (several rem/h) are typical for this component, while general area radiation levels of several mSv/h (several hundred mrem/h) could be common.

The letdown heat exchanger cools the letdown to ensure that the demineralizer resins are not damaged and that the water routed to the reactor coolant pump seals is at the proper temperature. This component as well as all the other major CVCS components, with the exception of the regenerative heat exchanger discussed above, is typically located in the auxiliary building. Dose rates in the vicinity of the letdown heat exchanger may be on the order of tens of mSv/h (a few rem/h) while the plant is operating.

The volume control tank (VCT), or make up tank (nomenclature varies among reactor vendors) provides a means for introducing hydrogen into the RCS coolant and is used for degassing the reactor coolant during shutdown. The hydrogen gas serves to scavenge excess oxygen that may be present in the RCS that is an important corrosion control function. Fission gases are vented to the waste gas handling system. The VCT also provides excess surge capacity for the reactor coolant. During periods of operation the radiation levels in the VCT room may fluctuate rapidly. Typically the volume control tank (another common name is the RCS bleed tank) is located within a shielded room or compartment. Associated valves and gauges are usually located in a pipe chase or valve alley outside the tank room itself, negating the need for individuals to physically enter the VCT room on a routine basis. Consequently the need to enter the VCT room is infrequent. General area radiation levels in the VCT room are typically on the order of a couple mSv/h (a few hundred mrem/h) to tens of mSv/h (a few rem/h) and subject to fluctuation.

The CVCS system purification loop typically consists of a mixed-bed and cation demineralizer to remove ionic species and a reactor coolant filter that collects resin fines and suspended particulate matter from the letdown stream. The demineralizers are usually sized to process a maximum letdown flow. As the inventory of activated corrosion and fission products accumulate on the resin beds and reactor filters significant dose rates will be encountered. During outage periods or when the resin beds are exposed to significant quantities of crud (e.g., during crud bursts) dose rates in excess of a few Sv/h (a few hundred rem/h) are not uncommon in the vicinity of the resin tanks. These components are located behind heavily shielded vaults.

A portion of the CVCS charging flow is routed to the RCP seals and returns via the seal water heat exchanger. The seal water heat exchanger reduces the temperature of the returning seal water to the operating temperature of the volume control tank. The seal water heat exchanger is cooled by component cooling water that flows through the shell side of the heat exchanger. Since the seal water flow has been purified dose rates associated with this heat exchanger are typically much lower than those for the letdown heat exchanger. Dose rates in the vicinity of the seal water heat exchanger should not typically exceed a few mSv/h (couple hundred mrem/h).

The pumps that provide the motive force for the CVCS system are the charging (or make-up) pumps. The charging pumps take suction from the volume control tank and route flow back to the reactor coolant system as noted above. During normal operation there is usually only one charging pump in service and it is not uncommon to have as many as three charging pumps per unit. Charging pumps also serve a dual role as part of the safety injection system and provide high-head injection to the RCS in the event of an accident involving loss of coolant. Under accident conditions charging pumps take suction from the refueling water storage tank or other suitable supply of emergency core cooling water. Dose rates associated with these pumps and immediate piping are highly dependent on the activity concentration of the RCS. Plants' with good chemistry controls and operating with little or no fuel defects may experience dose rates in the hundreds of μSv/h (tens of mrem/h) range or lower, on the train in service. If RCS source terms are higher than dose rates approaching 1 mSv/h (100 mrem/h) on contact to the charging pump in operation may be encountered. Unlike other CVCS components noted above the charging pumps are required to be accessible on a daily basis for inspection and to monitor their performance and consequently it is important to maintain good chemistry and crud controls to minimize worker exposures resulting from these routine tasks.

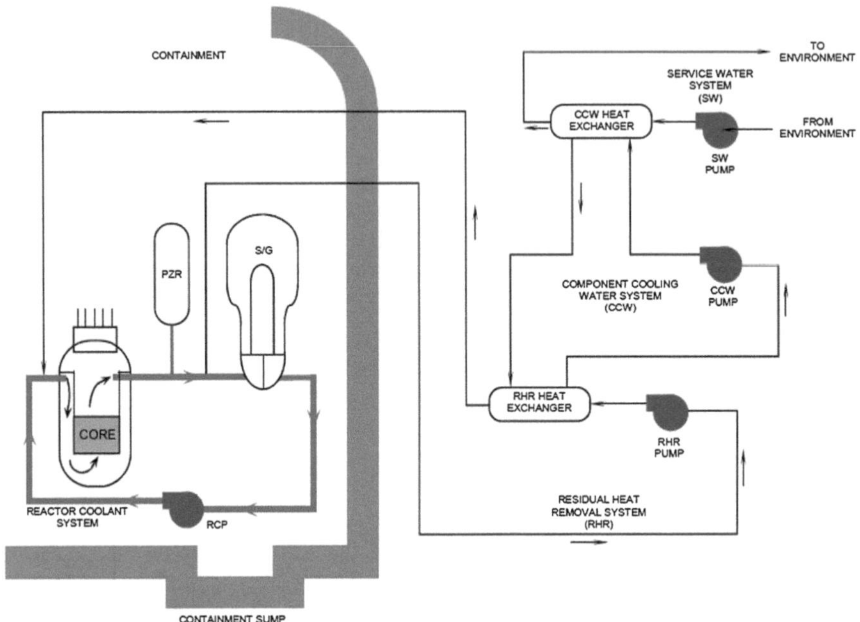


**Fig. 2.9** Schematic of RHR system and its major components (www.nrc.gov/reading-rm/basic-ref/teachers)

## 2.5  Residual Heat Removal System

The primary function of the residual heat removal system is to remove decay heat energy from the core during plant cool down and initial stages of refueling periods. The RHR system (or shutdown cooling or decay heat system) is also utilized to transfer refueling water between the refueling water storage tank and reactor cavity during refueling operations. The RHR system consists of two parallel trains each containing a pump and heat exchanger together with associated piping, valves and control instrumentation. Reactor coolant flows from the RCS via suction from an RHR pump through the tube side of an RHR heat exchanger and transfers heat to the component cooling water flowing through the shell side of the heat exchanger. The pumps are sized to deliver reactor coolant flow through the RHR heat exchangers to meet plant cool down requirements. Each train can provide 100% of shutdown core cooling requirements. Figure 2.9 depicts the basic components of the RHR system.

Depending upon the size of the unit or the manufacturer RHR system components may be located within the containment building or auxiliary building. If RHR major components (e.g., heat exchangers and pumps) are located within the containment building then they will be less accessible while the unit is at power. During normal plant operation the RHR system is maintained in standby and may serve a dual purpose as part of the emergency core cooling system. During these periods dose rates associated with RHR system components are typically on the order of hundreds of μSv/h (tens of mrem/h) or less. Once the plant enters a shutdown mode and goes onto RHR for cool down radiological conditions will change significantly for the train that is in service. During these periods fresh RCS coolant is flowing through RHR system piping. Dose rates in the vicinity of the RHR pump and heat exchanger that are in service could be on the order of a couple of mSv/h (100 mrem/h). Assuming no significant fuel failures are present these dose rates will decrease rapidly several days following shutdown.

## 2.6 Safety Injection System

The safety injection system (SIS) provides emergency core cooling and shutdown margin in the event of a loss of coolant accident. The SIS typically consists of high, low and intermediate pressure safety injection trains. The high-pressure safety injection system (HPSI) is capable of injecting borated water into the primary system while the RCS is at high pressure. The major components include the refueling water storage tank, safety injection pumps, the boron injection tank (BIT) and associated headers and valves. The charging pumps serve as the high pressure SIS pumps thus serving a dual function. Depending upon the design and engineering basis for a given plant a boron injection tank may not be present. The headers inject into both the cold and hot legs of the RCS. If a boron injection tank is utilized it is incorporated into the cold leg header.

The intermediate-pressure safety injection system injects borated water into the primary system from a set of safety injection accumulators (or core flood tanks). The accumulators are water storage tanks typically with a capacity of a few thousand liters that allows an individual accumulator to provide enough borated water to flood the reactor core. The accumulators automatically discharge when the primary system pressure falls below the pressure of the accumulators. Make-up to the accumulators is provided from the refueling water storage tank.

The low-pressure safety injection system injects borated water into the primary system at low pressure and also serves to increase the suction pressure of the HPSI pumps to prevent cavitations. The RHR system pumps usually serve as the low-pressure safety injection system pumps. The refueling water storage tank (RWST) also serves as the source of water for this system. When the RWST is exhausted then the RHR pumps take suction from the containment sump operating in a closed loop. The capacity of the RWST is typically on the order of 350,000–450,000 l.

The radiological conditions associated with major safety injection components were noted above for the case of the RHR system and CVCS system components that serve dual purposes. Boron injection tanks, for those plant designs that require a BIT tank, may have dose rates on the order of hundreds of µSv/h (tens of mrem/h). Dose rates in the vicinity of the BIT tank could also be influenced by source terms from nearby components. Accumulators are located inside the containment building and typically have relatively low dose rates since they contain clean water that has not been mixed with fresh primary coolant. Often-times the primary influence of radiological conditions in the vicinity of the accumulators is more dependent upon their location within the containment building and their proximity to primary system components or other major components of radiological concern.

## 2.7  Containment Spray System

The containment spray system (CSS) reduces the airborne contamination levels inside the containment and reduces the containment building pressure and temperature to maintain containment integrity after a loss-of-coolant accident. The CSS consists of two independent trains each capable of performing CSS functions. Each train consists of a spray pump, a heat exchanger (i.e., via the RHR system), a chemical additive injector and spray headers. Make-up is usually supplied from the RWST or the condensate storage tank (Fig. 2.10).

Upon actuation of a spray signal the CSS spray pumps take suction from the RWST and pump borated water to the CSS headers which are attached to the inside of the containment dome, high above the refueling floor. Multiple spray nozzles are attached to the header. Upon receipt of a low level signal in the RWST, the containment spray system enters a recirculation mode drawing water from the containment sump. Water spraying out from the CSS nozzles condenses steam in the containment building reducing the pressure inside the building. The CSS heat exchangers serve to cool the spray water as it passes through the tube side of the heat exchanger. Component cooling water passes through the shell side of these once-through heat exchangers.

A sodium hydroxide solution may be injected into the CSS system via a chemical additive tank. The purpose of this solution is to reduce the amount of iodine in the containment atmosphere that may be present during accident con-ditions. The sodium hydroxide mixture increases the pH of the spray to mitigate corrosion concerns.

The CSS system is maintained in standby and filled with clean water that may have low amounts of radioactive contaminants. Radiation levels in the vicinity of CSS components during normal plant operation usually do not pose a radiological concern and may be in the range of tens of µSv/h (a few mrem/h). Obviously during accident conditions when the CSS is drawing highly contaminated water from the containment building sump these components will have significant radiological source terms associated with their operation and access to which may be impractical due to high radiation levels that could be in excess of several Sv/h (hundreds of rem/h), or higher.

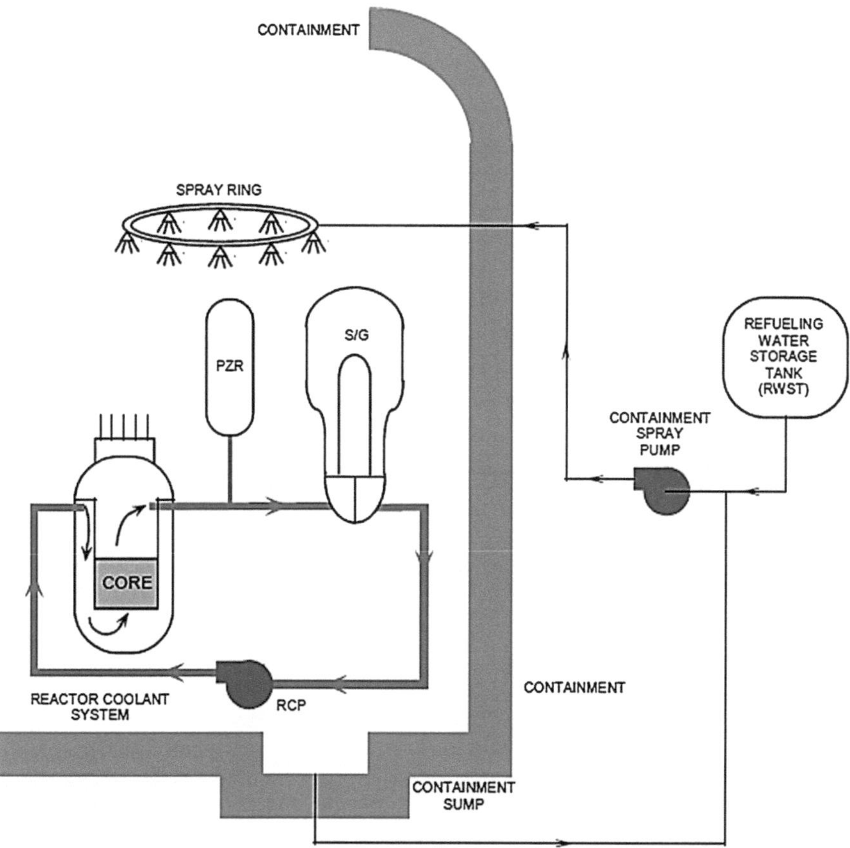


**Fig. 2.10** Schematic of containment spray system and its major components (www.nrc.gov/reading-rm/basic-ref/teachers)

## 2.8 Spent Fuel Pool Cooling and Purification

The spent fuel pool cooling and purification system serves to remove heat generated by spent fuel assemblies stored in the spent fuel pool. Filtering and purification serves to maintain the clarity of the spent fuel pool water and to reduce radionuclide activity concentrations. The concentration of fission and corrosion products is minimized to reduce personnel exposures associated with fuel handling operations.

The spent fuel pool at PWR facilities is maintained in a separate building referred to as the fuel building. In addition to housing the spent fuel storage pool this building has facilities for the storage of new fuel, a truck bay for receiving new fuel and shipping of spent fuel along with ventilation and other

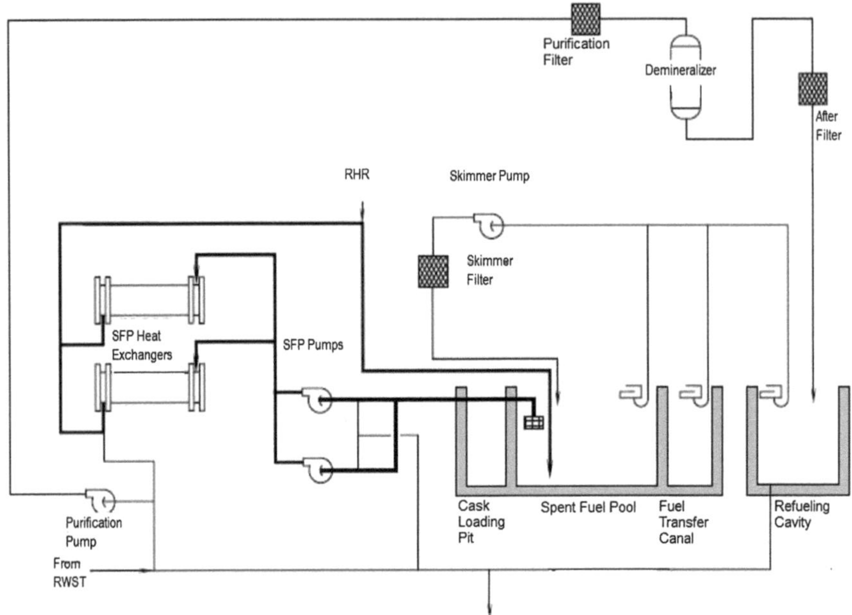


**Fig. 2.11** Schematic of spent fuel pool cooling and filtration and reactor cavity filtration systems (adopted from www.nrc.gov)

support systems. The major components of the spent fuel pool cooling and purification system are also located within the fuel building. The spent fuel pool itself contains a spent fuel storage area; an area to load spent fuel assemblies into a shipping cask, and associated transfer compartments. A fuel transfer tube connects the spent fuel pool to the reactor cavity inside the containment building.

The spent fuel pool cooling circuit consists of two 100% capacity trains each equipped with a pump, heat exchanger and associated piping and valves. The pumps take suction from a header below the spent fuel pool surface and route the water through the heat exchangers for cooling. The water is returned to the spent fuel pool via an orifice distribution system. The heat exchangers are cooled by the component cooling water system. The water may also be filtered and passed through a demineralizer to remove impurities and maintain radioactivity concentrations within acceptable levels. A skimmer box draws water from the surface of the pool for filtration and to maintain surface clarity to support fuel-handling activities (Fig. 2.11).

Radiological conditions associated with the spent fuel pool cooling system components are primarily determined by the inventory of spent fuel elements in the spent fuel pool and the integrity of spent fuel rod cladding. One objective is to minimize dose rates to operating personnel located on the refueling bridge during

**Fig. 2.12** Spent fuel handling operations in the fuel building of a PWR unit (Courtesy of Luminant)

the movement of spent fuel. The operation of the purification loop is typically optimized during refueling periods to maintain dose rates emanating from the surface of the spent fuel pool as low as possible. Radiation levels of tens of μSv/h (a few mrem/h) are probably acceptable during these periods. Dose rates much higher than this (e.g., approaching several tens of μSv/h) should be cause for concern. Dose rates in the vicinity of spent fuel pool cooling pumps and heat exchangers could be in the hundreds of μSv/h range if no significant cladding damage is associated with the elements in storage. Significant cladding damage could be defined as any leaks exceeding technical specification limits. Conversely if fuel cladding damage is present SFP heat exchangers, pumps, and associated piping could easily have radiation levels in excess of 1 mSv/h (100 mrem/h) with localized hot spots of perhaps 10 mSv/h (one rem/h) or more. Due to the parameters that influence the source inventory of radionuclides that may be circulating in the SFP cooling system at any given moment, the range of dose rates that may be encountered are subject to variation. The capacity of the purification loop, age of the plant and operating history, effectiveness of plant chemistry controls, and whether or not damaged fuel elements have been containerized are some of the variables that will impact radiological conditions associated with this system (Fig. 2.12).

## 2.9  Reactor Cavity Filtration

The reactor cavity provides sufficient volume to afford the necessary water depth for adequate shielding to personnel during spent fuel handling operations. The reactor cavity may have a capacity to hold up to 2.5 million liters (several hundred thousand gallons) of water. The cavity contains the reactor vessel positioned below the flange-level. The cavity is sized to allow storage of vessel internals during refueling and maintenance periods. During refueling operations the cavity is filled with borated water to a depth sufficient to provide adequate shielding to workers on the refuel floor and other plant areas that may be impacted from radiation levels emanating from an exposed reactor vessel. The refueling water storage tank (RWST) serves as the source of borated water for the reactor cavity during refueling periods. As previously noted the RWST also serves as a water supply to the HPSI, LPSI and CSS systems.

Dose rates in the vicinity of the reactor cavity are primarily influenced by the radionuclide concentration of the cavity water. The objective of the reactor cavity filtration system is to minimize exposures to operating personnel. The system is placed into operation during periods when the reactor cavity is flooded. The system consists of pumps and filtration units and possibly demineralizers. Pumps circulate the water through a filter network, returning the purified stream back to the cavity (see Fig. 2.10). Skimming and filtration of the water surface is also provided. Skimming of impurities and debris from the water surface improves visibility to support fuel movement and core alteration activities. Dose rates emanating from the surface of the reactor cavity water should be maintained in the range of several tens of µSv/h or less during refueling periods to minimize exposures to operators on the refuel bridge and support personnel in close proximity to the reactor cavity. Dose rates higher than these levels may be indicative of excessive fuel cladding defects or otherwise higher than anticipated radionuclide concentrations present in the water source.

## 2.10  Radioactive Waste Treatment Systems

The purpose of radioactive waste treatment systems is to reduce radioactivity concentration levels in plant effluents to acceptable levels. The systems are designed to maintain activity releases as low as possible. Radioactive wastes may include liquid, gaseous and solid radioactive material. Effluent batch releases are sampled and analyzed prior to release. Continuous effluent release streams (e.g., stack and liquid discharges) are routinely monitored when discharges are in progress to supplement grab sample analysis results and to ensure that releases are maintained within applicable limits. Solid wastes are packaged and shipped for offsite disposal in accordance with appropriate regulations and other applicable requirements (e.g., NRC, DOT and IAEA).

Various reactor suppliers as well as several engineering firms design radioactive waste treatment systems. The following discussion will of necessity be generic in nature due to the differences in both the design offered by various suppliers and the terms used to designate components and systems. Since the source of the radioactive waste stream is the primary determinant when considering the radiological conditions associated with a particular radioactive waste treatment system component; primary consideration should be given to understanding the source and constituents of a given waste stream. Dose rates emanating from primary system leak collection tanks versus waste water holdup tanks containing processed water will differ significantly.

## 2.10.1 Liquid Waste Treatment System

Liquid waste treatment systems provide for the storage, monitoring and processing of liquid wastes and effluent streams. Liquid wastes typically originate from valve and component leakage, demineralizer regeneration and resin replacement, chemistry sampling and laboratory activities, laundry facilities, equipment and floor drains, building sumps, and decontamination activities. In addition, miscellaneous sources of liquid waste could arise from plant operational occurrences involving thousands of liters of wastewater. Liquid wastes are routed to various collection tanks of varying storage capacities for collection and processing. Liquid collection tanks may have capacities ranging from a few thousand liters for the collection of laboratory wastewater, to several hundred thousand liter holdup tanks for the batch processing of liquid waste discharges. The various waste collection and holdup tanks are often designated by the source of the liquid waste and could include such terms as floor drain collection tanks, chemistry drain tanks or laundry waste water collection tanks for example. The number, type and designation of liquid waste storage tanks may differ depending upon the reactor size and type and the specific design considerations. Liquid wastes are usually collected and processed based upon the level of radioactivity present in the waste stream. Low activity wastewater is usually segregated from high-activity wastewater to simplify processing and to minimize costs. Additionally, it is oftentimes advantageous to segregate clean wastewater (i.e., primary system and auxiliary system leakages) from dirty wastewater to minimize the volume of wastewater requiring extensive processing and treatment. Dirty wastes may include those liquids that are contaminated with detergents or chemicals (e.g., from decontamination activities) or otherwise contain relatively high concentrations of impurities (e.g., auxiliary building sumps and reactor cavity drain tank after a refueling outage). The descriptions that follow are more generic in nature. The designations and capacities of the various liquid waste collection and processing tanks may differ from that encountered at a particular nuclear plant.

Floor drain collection tanks are used for the storage and holdup of non-reactor grade wastewater streams. This water may originate from decontamination processes, waste water from janitorial and housekeeping activities within clean areas of the RCA, non-recoverable leaks, and other miscellaneous sources. This water is sampled and analyzed, and if within applicable discharge limits, may be released directly to the environment with no further processing. If concentration levels exceed established release values the liquid waste may be filtered and routed through a demineralizer bed to reduce radioactivity concentrations prior to discharge. The treated wastewater would then be routed to a liquid waste holdup tank to prevent cross-contamination. Liquid waste holdup tanks may have capacities of 200,000 l or more. Several collection tanks could serve as feed to a given holdup tank. Many nuclear units were equipped with evaporators to reduce the volume of liquid waste required to be processed and discharged. The performance of evaporators has been mixed with the advantages associated with the reduction achieved in liquid waste volumes often overshadowed by operational and maintenance costs to maintain evaporator system components. Significant personnel exposures could result due to the handling and processing of evaporator "bottoms", containing high concentrations of radionuclides, in addition to routine maintenance activities. Dose rates in the vicinity of floor drain collection tanks are typically in the range of hundreds of μSv/h (tens of mrem/h). These tanks may or may not be enclosed behind walls and it is not uncommon for these tanks to be situated in open floor areas.

Waste holdup tanks collect wastewater originating from equipment drains and demineralizer regeneration and resin replacement (i.e., sluicing) operations. These wastes are treated and processed prior to release. If evaporators are available for use at a given facility these wastes could constitute the major feed to the waste evaporators. Waste holdup tanks are usually located within a separate room surrounded by concrete walls, not necessarily designed as shield walls. Dose rates in the vicinity of waste holdup tanks could be on the order of several hundred μSv/h (several hundred mrem/h) depending on the source of the water. In addition dose rates could be higher during outage periods when higher-activity wastewater is typically generated in support of outage activities.

Sites equipped with laundry facilities may have separate laundry holdup tanks or service effluent tanks to collect wastewater from laundry facilities and other sources such as shower facilities. Activity concentrations in this waste stream are typically low. Dose rates in the vicinity of laundry wastewater storage tanks could be on the order of hundreds of μSv/h (tens of mrem/h). Typically these wastes can be filtered and discharged without any further processing. The storage capacity allotted for these tanks, particularly for laundry wastewater, has been chronically undersized in the industry. Laundry facility modifications to provide additional wastewater storage capacity or the introduction of dry cleaners have often been necessitated to alleviate this problem.

Storage facilities may be available for the collection and storage of chemical liquid wastes, originating primarily from chemistry laboratories. The major source of influent to this tank is from the primary sample room and perhaps the chemistry laboratory. Wastes from laboratory drains, chemistry-sampling stations, and perhaps from various decontamination facilities could be routed to this tank. This tank, often referred to as the chemical drain tank, may have a capacity of a couple thousand liters. Radiation levels in the vicinity of this tank could be on the order of tens of μSv/h (tens of mrem/h). Radiation levels could be significantly higher in the event of failed fuel or if abnormal amounts of primary system water is allowed to be introduced to the tank during sampling activities. Chemical drain tank waste may be filtered and treated by demineralization or evaporation prior to release.

Liquid waste from different waste streams may be processed and routed to a common collection tank prior to discharge. This tank serves as an intermediate storage, or holdup, tank and facilitates batch processing and release of radioactive effluent. Collecting smaller volumes of processed liquid waste in one large common collection tank reduces the number of batch releases performed over a given time period. This "monitor" tank may have a capacity of 200,000 l (50,000 gallons) or more. A batch release typically requires the contents of the waste tank to be mixed for a period of time. This requires time to perform the necessary valve lineups and to place necessary equipment (e.g., pumps) in service. Following mixing of the tank contents samples are obtained and analyzed. Sample results are evaluated to ensure compliance with both radiological and non-radiological release limits. Depending upon the facility, arrangements must be made to ensure sufficient dilution flow is available while the batch release is in progress. This may entail placing into service various discharge pumps to maintain the necessary dilution flow. The radioactivity level of batch releases is usually monitored while the release is in progress. The alarm set point of the radiation monitor on the discharge line must be confirmed and adjusted to the appropriate value. These and other activities required to support a batch release may be included in an "effluent release package" of some kind. The package must be completed and reviewed by such departments as chemistry and operations. The availability of a waste monitor tank reduces the number of batch releases, the operational time required to support a batch release, and the administrative burden associated with the preparation of effluent release packages.

Liquid wastes may be treated by a combination of filtration, distillation, demineralization or holdup to minimize the volume and activity of radioactive waste discharged to the environment. The radioactive constituents may be concentrated and solidified prior to shipment for offsite disposal. Resin beds used in the treatment of liquid wastes may concentrate radionuclides by many orders of magnitude. Resin beds influent and effluent streams are routinely sampled or monitored to measure the depletion of the resin bed. Decontamination factors (i.e., the reduction in activity concentrations of the effluent stream compared to that of the influent stream) of 100 or higher are considered to be indicative of a normally functioning, or non-depleted resin bed. The processing of hundreds of thousands of gallons of wastewater with radionuclide concentrations in the 20,000–40,000 Bq/ml range (μCi/ml range) can result in resin beds reading greater than several Sv/h (several hundred rem/h). Resin beds used in the treatment of radioactive waste streams are often entombed in a shielded compartment or as a minimum, located within a labyrinth configuration behind shield walls and entrances equipped with multiple locked-door barriers. A similar arrangement is associated with filters used in the treatment of liquid waste streams; however, the magnitude of radiation levels may be on the order of a few Sv/h and often will not reach the higher levels encountered in the vicinity of demineralizers. Nevertheless certain filter housings (in such systems as the CVCS, RWCU, reactor cavity filtration and various other purification systems) are also located within shielded vaults or compartments equipped with strict access design controls.

Usually there are at least two waste evaporators arranged in parallel to process wastes from liquid waste holdup tanks. Evaporators concentrate liquid wastes by boiling off the water in the process stream. The vapor (i.e., condensate) produced in the evaporator may be demineralized and filtered and routed to a waste evaporator condensate tank and used for makeup or other purposes. The concentrate, commonly referred to as evaporator bottoms, is discharged to the solid waste packaging and drumming facility for solidification and packaging.

## 2.10.2 *Gaseous Waste Treatment System*

The gaseous waste treatment system provides for holdup, filtration and dilution of gaseous waste produced during plant operations. Since in a PWR the primary system is closed, gas volumes produced are relatively small. Gases collected from the primary system are compressed and stored in waste gas decay tanks. Each PWR unit may be equipped with several waste gas storage tanks. Each tank having a capacity to store several weeks or months of waste gas generated during normal plant operation. Waste gas originates from the CVCS system via gases stripped from the volume control tank, the boron recycle system and gases vented from various liquid waste storage tanks. Waste gas streams may contain hydrogen and nitrogen in addition to fission gases.

Two waste gas compressor trains are usually provided. One train supports normal operations while the second train serves as a backup and supplies additional capacity for peak load periods encountered during refueling when the reactor coolant system is degassed. The waste gas is then pumped through a recombiner where oxygen is added to minimize potential explosive gas concentrations. The oxygen combines with hydrogen to produce water vapor that is removed from the process stream. The compressed gases are routed to a waste gas storage tank and allowed to decay for a period of time sufficient to allow short-lived fission gases to decay to insignificant levels prior to release. Depending upon the number of waste gas storage tanks (or decay tanks) available, storage times of 30–60 days are common. For those PWR units with several waste gas storage tanks, storage periods of several months may be achievable.

Waste gas storage tanks are located within a shielded room or area, often entombed to prevent access. Radiation levels in the vicinity of these tanks will fluctuate over a large range depending on the age of the gas contained within a given storage tank. Radiation levels in the vicinity of the gas storage tank in use may approach hundreds of mSv/h (tens of rem/h). Once a tank is full to capacity (based on pressure readings) it is removed from service, isolated, and fission gases allowed to decay. By the time of release radiation levels may have decayed by orders of magnitude. The contents of waste gas storage tanks are released at a subsequent date when additional storage capacity is required, based on operational needs, or if the tank must be placed into service. These tanks are released on a batch basis, similar to the process noted above for liquid batch releases. The contents are sampled and radiation levels of the batch release monitored while the contents of the tank are being released to the plant stack.

The volumes of waste gas produced fluctuate based upon reactor power level and plant operating conditions. A buffer tank may be provided to allow temporary holdup of waste gases so that an even gas flow can be provided for the waste gas process stream. The buffer tank helps to eliminate transient pressure spikes due to changing gas volumes.

## *2.10.3  Solid Waste Treatment System*

The solid waste treatment system provides for the handling, compacting, solidi-fication and packaging of solid wastes. Solid wastes include spent ion exchange resins, evaporator bottoms, used filter materials and miscellaneous solid wastes. Miscellaneous wastes include consumables and a host of materials expended in the normal operation and maintenance of the nuclear unit. These items consist of worn protective clothing, covering and enclosure materials used for contamination control, solid waste generated during decontamination activities (e.g., mop heads and rags), worn or unusable contaminated tools and equipment.

There is no "one-size-fits-all" solid waste treatment system. In fact the history of nuclear power plants has been plagued with the lack of a dependable and efficient onsite system for processing solid waste. Many of the original systems underwent extensive modifications only to be abandoned at a later date. This situation has been compounded, at least in the case of the USA, for the need to dramatically reduce the volume of generated solid waste due to the high cost of disposal and limited access to disposal facilities. These factors were not a concern when solid waste treatment systems were first designed for many of the nuclear plants now operating. Essentially solid wastes were to undergo minimal volume reduction via compaction into 55-gallon drums or other suitable containers. Consequently, many nuclear power plants in the USA rely on radioactive waste processing firms. Current practice now is to package dry active waste in sealand containers, or other suitable packaging, for shipment to offsite processors for volume reduction and disposal. These firms specialize in volume reduction and segregation processing. The capabilities of these offsite processes include super-compaction, metal-melt, incineration and other pro-cesses that are targeted towards minimizing the volume of solid radioactive waste ultimately requiring disposal.

# Bibliography


1. Lish K., Nuclear Power Plant Systems and Equipment, Industrial Press, New York, NY, 1972
2. Neeb, Karl-Heinz, The Radiochemistry of Nuclear Power Plants with Light Water Reactors, Walter de Gruyter & Co., Berlin, 1997
3. Rahn F.J., Adamantiades A.G., Kenton J.E., and Braun C., A guide to Nuclear Power Technology – A Resource for Decision Making, New York: Wiley & Sons; 1984
4. Whicker F.W., and Schultz V., Radioecology: Nuclear Energy and the Environment, Volume 1, CRC Press, Boca Raton, Florida, 1982
5. US Nuclear Regulatory Commission, Reactor Concepts Manual, USNRC Technical Training Center

# Chapter 3
# Radiological Aspects of BWR Systems

## 3.1 Overview

The other major LWR design currently in widespread use throughout the world is the boiling water reactor (BWR). The distinguishing design feature of the BWR is that water is allowed to boil in the reactor vessel in contrast to the PWR design described in Chap. 2, in which water in the primary system is maintained under high pressure to preclude boiling. Many of the BWR units currently in operation are of the General Electric (GE) design. The first GE unit to enter commercial operation was the Humboldt Bay plant near Eureka, California. The Humboldt Bay plant has subsequently been decommissioned. In addition to GE, ASEA-Atom (Sweden), Toshiba and Hitachi (Japan), among others, are other suppliers of BWR-designed reactor types.

Various generations or model types of the GE BWR are in operation. These model types are signified by various classifications such as the BWR-6 series. Series classifications are dominated by the BWR-2 through the BWR-6 series. The earlier generations of BWR units are represented by the BWR-2 and 3 series, progressing to the more recent designed series represented by the BWR-6. Examples of the various BWR designs are provided below.

- BWR-1 Big Rock Point and Dresden-1
- BWR-2 Oyster Creek
- BWR-3 Monticello
- BWR-4 Hatch, Susquehanna and Limerick
- BWR-5 LaSalle, Columbia and Nine Mile Point
- BWR-6 Perry, River Bend and Grand Gulf

System descriptions provided in this chapter are generic in nature and the descriptions may not necessarily reflect actual system configurations encountered at a given BWR facility. The system descriptions are primarily based upon the GE series of BWR units.

R. Prince, *Radiation Protection at Light Water Reactors*,
DOI: 10.1007/978-3-642-28388-8_3, © Springer-Verlag Berlin Heidelberg 2012

The BWR is a direct cycle steam generator system in which steam is produced directly in the reactor core by allowing the water coolant to boil. The reactor coolant loop system pressure is maintained at approximately 7 MPa (1040 psi) under these conditions water will boil at 285°C (545°F). As water flows through the core it is heated causing some of the water to boil. The water-steam mixture then rises upward to the steam separator where the liquid is extracted from the steam. The water is recirculated to the reactor vessel and the steam is passed through a steam dryer assembly and then routed to the turbine-generator. Consequently, the main steam system of a BWR is radioactively contaminated. Unlike in a PWR in which various chemicals may be added to the RCS for reactivity control and other purposes, no such chemical additions are made to the reactor coolant system of a BWR. Since the reactor coolant in a BWR is converted to steam, chemical additions are not utilized, in order to produce high-purity steam and to minimize chemical corrosion and fouling concerns in secondary side systems.

The systems of radiological concern at a BWR facility include the reactor water recirculation system and auxiliary systems including the reactor water cleanup system, fuel pool cooling and cleanup system, and the residual heat removal system. In addition the main steam system and other secondary side systems located within the turbine building will also pose radiological concerns due to the presence of radioactive species carried over in the steam and the presence of short-lived activation products. Other systems of radiological concern include the radioactive waste treatment, radiochemistry sampling, and plant ventilation systems. Figure 3.1 depicts a typical BWR reactor building and steam flow to the turbine generator.

## 3.2  Plant Layout

The distinctive buildings associated with a BWR facility include the reactor building, auxiliary building, the turbine building, the radioactive waste processing building and the control complex. As with a PWR, other facilities required to support plant operations such as administrative office buildings, security access facilities, and receiving and storage warehouses are also present.

The reactor building is a large square or circular, concrete reinforced structure that houses the spent fuel pool, various components associated with emergency core cooling systems, the reactor water cleanup system, auxiliary systems and ventilation systems. Located within the reactor building is the drywell (i.e., the containment structure) that houses the reactor vessel and associated coolant recirculation equipment and components. The radioactive waste process building contains radioactive waste handling and treatment systems and associated ventilation equipment while the control building houses the main control room, motor control centers, battery rooms, electrical cable and relay rooms and emergency plant ventilation equipment (e.g., air intake fans, blowers and dampers).

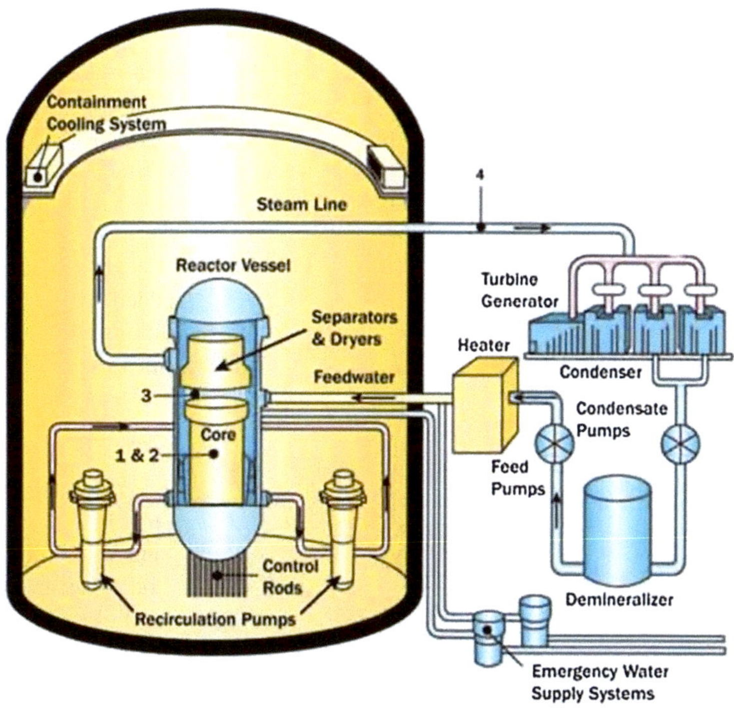


**Fig. 3.1** Typical BWR plant layout showing the major components of the reactor building (www.nrc.gov/reactors/bwrs)

## 3.3 Containment Systems

Containment systems for BWR plants consist of a primary and secondary containment. The primary containment consists of the drywell which encloses the pressure suppression chamber and the reactor vessel. The secondary containment includes the reactor building, which encloses the primary containment. As noted above the various BWR product lines differ in design. Product lines BWR-2 and BWR-3 and some earlier model BWR-4 designs are equipped with the Mark I containment design. These designs consist of a drywell shaped like an inverted light bulb (see Fig. 3.2). A suppression chamber is located below the drywell. The drywell forms part of the primary pressure suppression system and directs steam to the pressure suppression chamber in the event of an accident. The reactor vessel and recirculation system are housed within the drywell. The suppression chamber has the shape of a doughnut and maintained partially filled with water. The suppression chamber serves as a heat sink to cool and condense steam that may be routed to the torus during plant upset conditions.

The BWR-6 series incorporates the Mark III containment design (see Fig. 3.3). This design has a larger containment that in many respects looks similar to a PWR

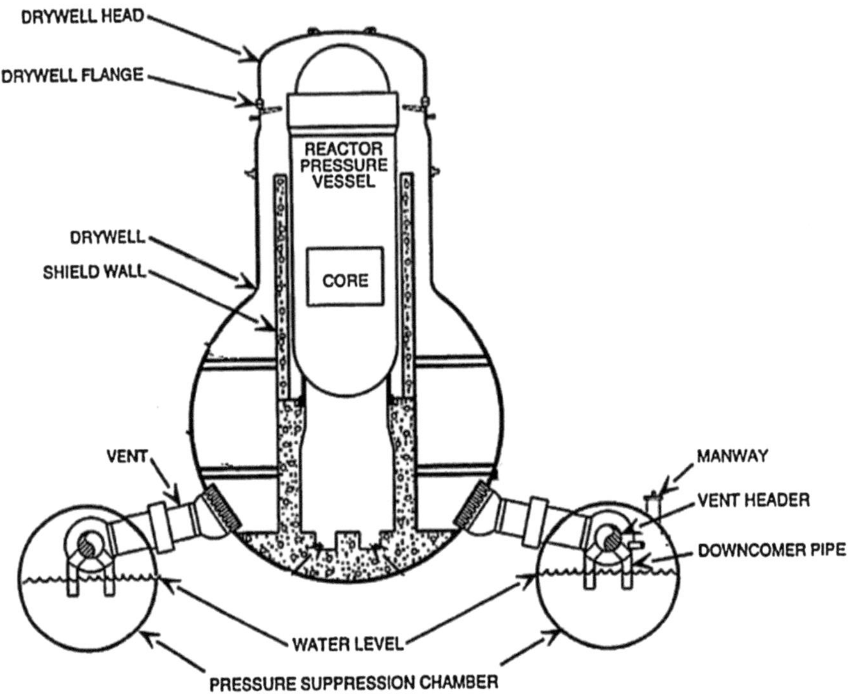


**Fig. 3.2** Figure depicting the earlier Mark I BWR containment design with the distinctive "inverted light bulb" shape (adopted from www.nrc.gov)

containment building. Mark III containment encloses the drywell and suppression pool. The Mark III containment is accessible during all modes of operation whereas the Mark I and II containments maintain an inerted atmosphere during power operations.

The primary containment serves to contain fission products and radioactive materials that may be released as a result of a LOCA. The primary containment serves a vital function in minimizing off site radiological consequences during an accident.

## 3.4  Reactor Vessel

The reactor vessel serves many purposes in addition to the obvious function which is to house the reactor core. Other functions of the reactor vessel are to support and align fuel assemblies and control rods to maintain proper configuration of the core, to provide a flow path for the circulation of coolant through the fuel, to remove moisture from the steam and to serve as part of the reactor coolant boundary. Steam is produced directly within the reactor vessel assembly to drive the turbine generator.

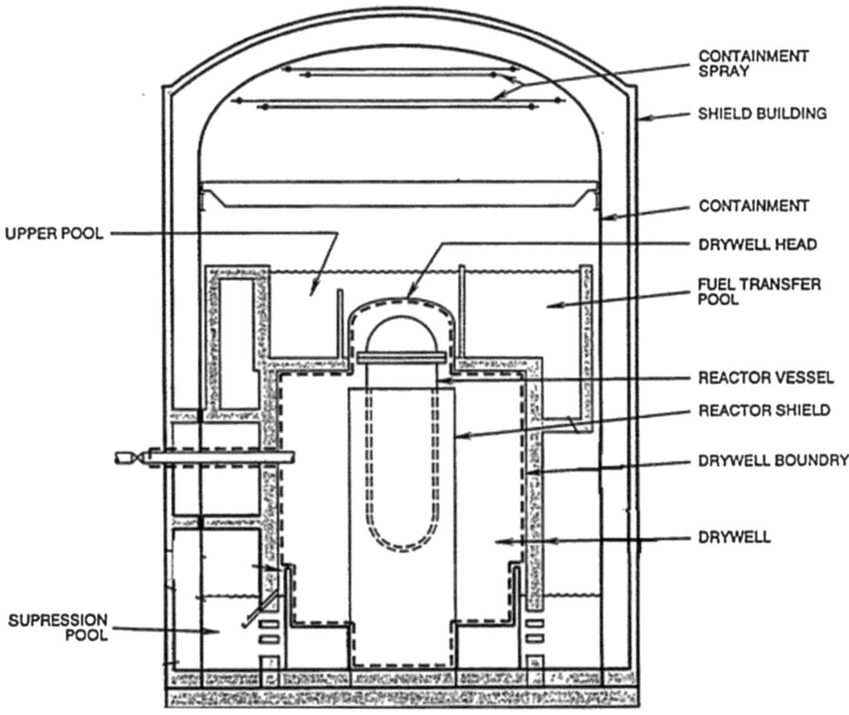



**Fig. 3.3** Figure depicting the Mark III BWR containment design (adopted from www.nrc.gov)

The reactor vessel is hemispherical-shaped with a removable top head to gain access to the core for refueling. A BWR reactor vessel is larger than that of a PWR since it contains the steam separator/steam dryer assembly, in addition to the core, and can have an overall height of approximately 20 or more meters, whereas a PWR reactor vessel is approximately 12 m in height. Figure 3.4 depicts the major components associated with a BWR reactor vessel.

The reactor vessel is located within the drywell and has a thickness of approximately 10–15 cm and is classified as an ASME code pressure vessel. The reactor vessel is of alloy steel construction with an inner stainless steel cladding. The overall dimensions of the reactor vessel are on the order of 20 m, as noted above, in height with an inner diameter of about 6 m. The vessel head is secured to the reactor vessel with studs. Control rods enter the reactor vessel via penetrations in the bottom of the vessel. A drain line located at the vessel bottom directs flow to the reactor water cleanup system. Feedwater is supplied to the reactor vessel via feedwater inlet nozzles while steam outlet nozzles, located above the moisture separator and steam dryer segments provide a flow path for steam to the turbine generator. The steam production process is described in more detail below.

A steam-water mixture is produced inside a BWR reactor vessel. The steam-water mixture exits from the top of the core, flows through standpipes connected to

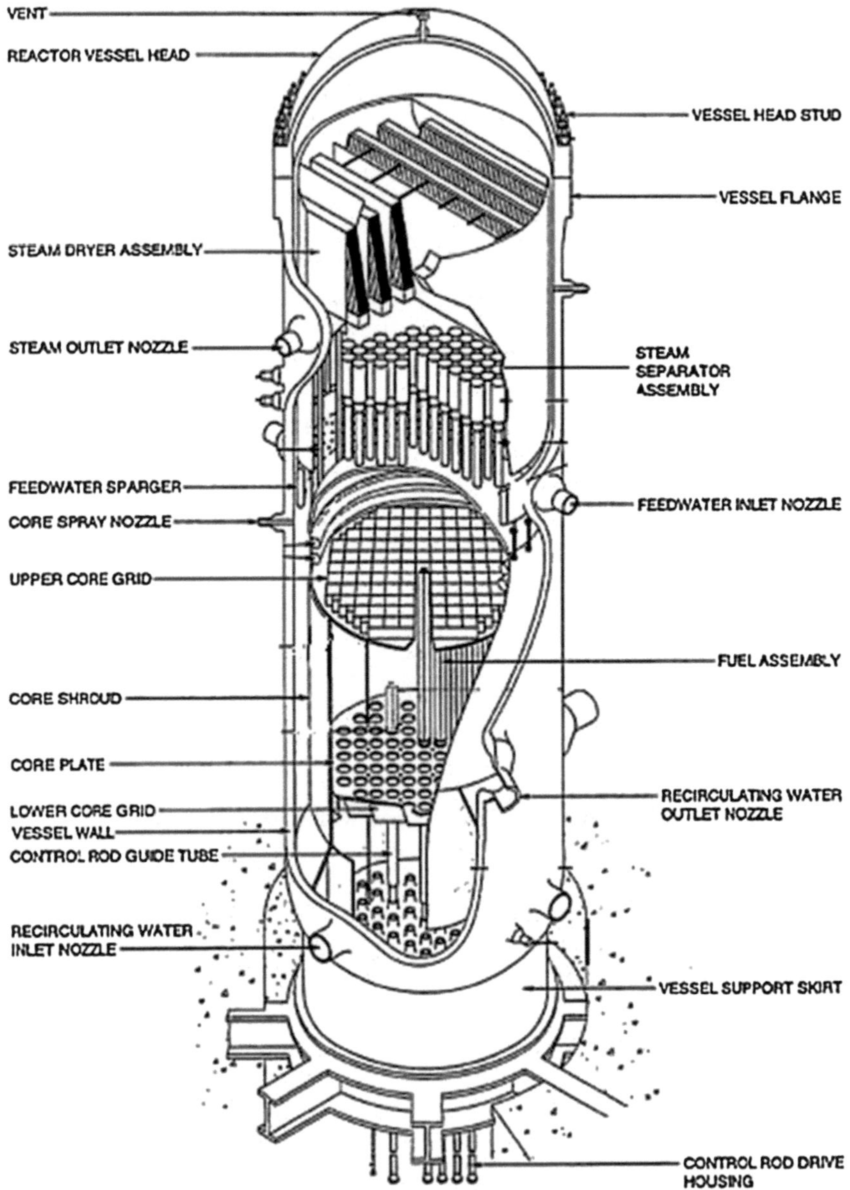


**Fig. 3.4** BWR reactor vessel depicting the major components of the vessel (www.nrc.gov/reading-rm/basic-ref/teachers)

the reactor vessel shroud, and then enter the steam separator assembly. The steam separator assembly is composed of cyclone type separators. Depending upon the rated capacity of a given BWR, the steam separator assembly may consist of 200 or more, cyclone separators. Essentially the cyclone separators consist of fixed turning vanes that cause the steam to separate from the water. The water is recirculated and the steam enters the steam dryer assembly. The separators are welded to the standpipes as an integral component.

Located above the steam separator assembly is the steam dryer assembly. Steam exiting from the steam separator assembly is essentially "wet" steam that must be dried to provide high quality steam to the main steam lines. The steam dryer assembly is configured with internal panels that force the wet steam to make a series of directional changes in flow path. These directional flow changes cause moisture to be removed from the steam by impinging in the dryer vanes or panels. The moisture is collected and routed back to the reactor vessel annulus region. Dry steam is routed to the main steam outlet penetrations (nozzles), welded to the upper reactor vessel shell area. Moisture from the steam dryer is collected as water and recirculated by jet pumps back to the core. Part of the recirculation flow is routed through the external recirculation loops and returned to the reactor vessel.

## 3.5 Recirculation System

The purpose of the recirculation system is to provide forced circulation of water through the reactor core. The amount of water flow through the core determines the power level achievable by a given BWR unit. The recirculation system for the various GE product lines has their own unique design and configuration. The recirculation system is located within the primary containment (i.e., the drywell). The recirculation system for the BWR 5/6 product lines consists of two independent trains. Each recirculation train consists of a recirculation pump, a number of jet pumps, a flow control valve and suction and discharge isolation valves and associate piping and instrumentation.

Recirculation pumps take suction from the annulus area of the vessel and discharge into the riser manifold. The recirculation pumps provide water to the inlet riser of each jet pump assembly via reactor vessel recirculation inlet penetrations. The risers in turn provide driving flow to the jet pumps. The jet pump assemblies are located between the core shroud and the reactor vessel wall. Each assembly consists of an inlet riser that supplies flow to a pair of jet pumps. Each individual jet pump consists of an inlet nozzle, a mixing section and a diffuser.

A typical arrangement for a BWR-6 may consist of a ring header supplying five jet pump risers. The major components of the recirculation system are located within the drywell, a concrete re-enforced structure. Figure 3.5 depicts a simplified diagram of a BWR recirculation system and flow through the reactor vessel. In addition to providing sufficient coolant flow to the core to maintain the core temperature within allowable limits, recirculation pump flow rates also control the power level in BWR units. As noted above the recirculation loop provides flow to the jet pumps. The jet pumps allow for an increase core flow, while minimizing the recirculation flow, required to provide a given flow through the core (see Fig. 3.6).

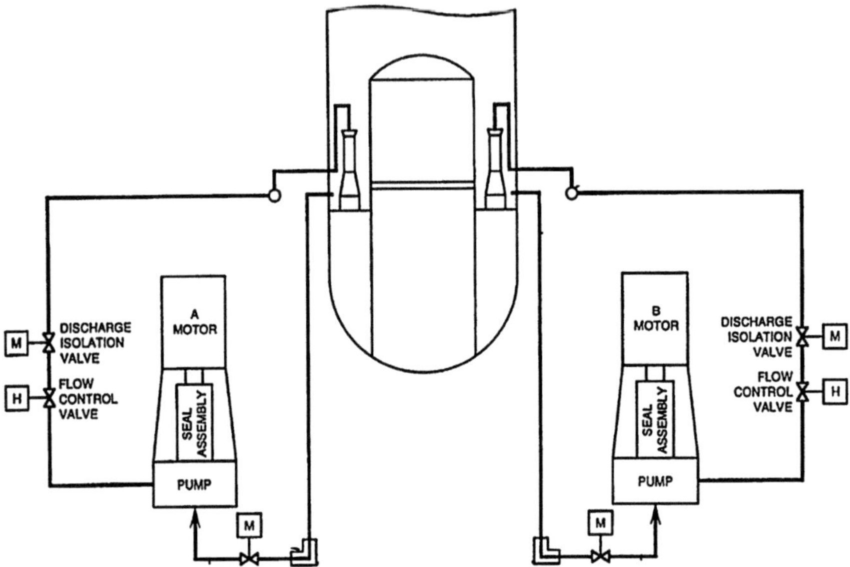


**Fig. 3.5** Schematic of BWR recirculation system for a BWR 5 and 6 series (www.nrc.gov/reading-rm/basic-ref/teachers)

During full-power operation, jet pumps may provide approximately two-thirds of the core flow while the recirculation pumps supply approximately one-third of the core flow. First generation BWR units relied upon natural recirculation through the core. The limited flow rates provided by natural circulation severely limited the upper power level achievable for the early BWR units.

Access to the recirculation system at power is limited due to its close proximity to the core. Additionally, depending upon the design of a specific BWR unit, access may be further limited based on heat stress and operational-related access restrictions. General area radiation levels in the vicinity of recirculation pumps during shutdown conditions may be as high as few mSv/h (a few hundred mrem/h). Depending upon the design and specific location of the motor section, dose rates in the vicinity of the motor section are usually significantly lower, perhaps less than hundreds of μSv/h (tens of mrem/h) or lower.

## 3.6  Reactor Water Cleanup System

The reactor water cleanup system (RWCU) purifies the reactor water by removing fission and activation products and impurities by filtration and demineralization, and promotes the circulation of water in the reactor vessel bottom head region to minimize thermal stratification. The RWCU system is similar in function and design to that of the CVCS system in PWR units. Major components of the RWCU

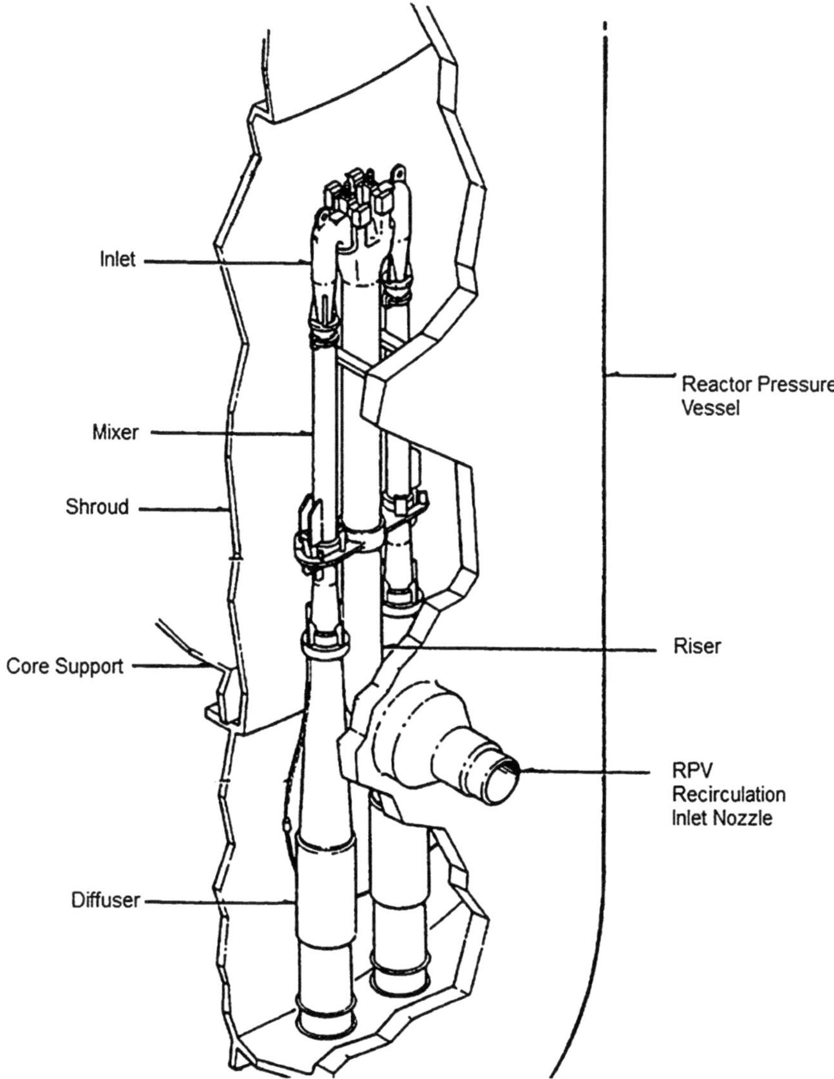



**Fig. 3.6** BWR jet pump assembly (www.nrc.gov/reading-rm/basic-ref/teachers)

system include pumps, heat exchangers, purification filters and demineralizer beds along with associated valves and piping.

The RWCU system treats that water component of the primary system that circulates between the reactor vessel and the recirculation loops. Much of the water-steam mixture exiting the top of the core contains a significant liquid component that is recirculated and not routed to the turbine. Reactor coolant is discharged from the recirculation system loops and the reactor vessel bottom head

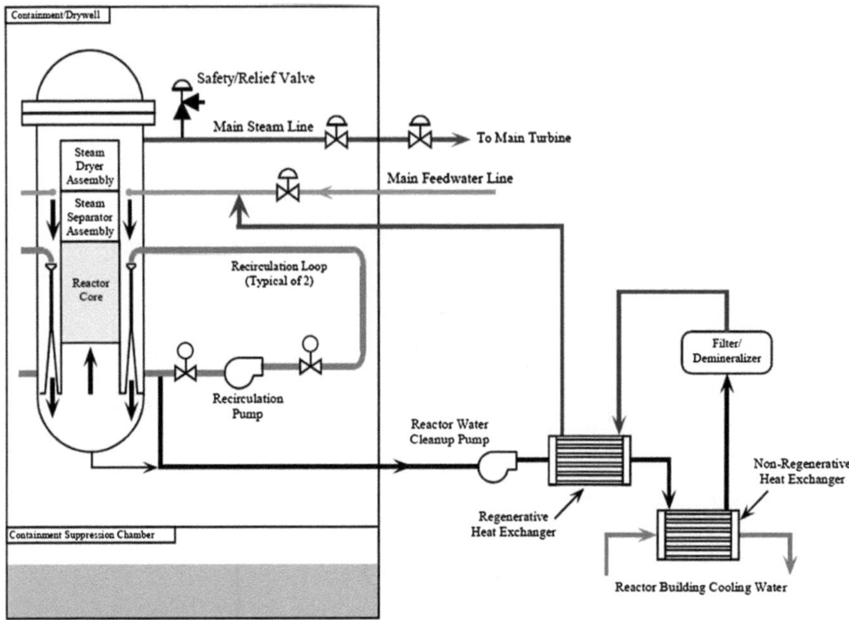


**Fig. 3.7** The reactor water cleanup system and its major components (www.nrc.gov/reading-rm/basic-ref/teachers)

to a common discharge header. This discharge is pumped through the shell side of the regenerative heat exchangers where the discharge flow is reduced in pressure and temperature. The flow is then directed to the tube side of the non-regenerative heat exchangers before passing through the filter demineralizers. The purified water is returned to the reactor vessel via the feedwater piping. Figure 3.7 displays the major components of the RWCU system and their inter-relationship.

Since the RWCU system must be operable to support plant operations the system is provided with redundant components to ensure the availability of at least one fully functional RWCU train at all times. This is vitally important since the RWCU system also serves a safety-related function. The number of components may differ somewhat depending upon the rated capacity and generation design of a given BWR. The function of the RWCU and the fact that it handles and processes water directly discharged from the reactor vessel and core makes this a system of significant radiological concern. A more detailed description of key RWCU system components and their radiological conditions is provided below.

The RWCU pumps draw suction from the recirculation loops and direct flow to the regenerative heat exchangers. Each pump is sized to provide 100% system flow requirements. During normal operation there is usually one RWCU pump in service and it is not uncommon to have as many as three RWCU pumps per unit. Since these pumps handle fresh reactor water discharge flow with minimal decay time since exiting the reactor core dose rates associated with these pumps and immediate piping will be excessive. Dose rates of 1–10 mSv/h (100–1,000 mrem/h) on contact with the operating RWCU pump may be encountered. Consequently these pumps are located in shielded compartments. Dose rates associated with these pumps and the interconnecting piping and valves, are highly influenced by the activity concentration of the discharge flow. The maintenance of sound chemistry controls and operating with no fuel defects are instrumental in minimizing radiation levels in the vicinity of the RWCU pumps.

The RWCU pumps direct the discharge flow to the regenerative heat exchangers. The regenerative heat exchanger reduces the temperature of the reactor water discharge by transferring the heat to the feedwater flow, thus minimizing thermal stress on system components and piping. Contact dose rates on the regenerative heat exchangers could be in the range of 10–40 mSv/h (1,000–4,000 mrem/h), or higher.

RWCU flow is next routed to the non-regenerative heat exchanger from the regenerative heat exchangers. The non-regenerative heat exchanger cools the discharge to ensure that the filter demineralizer resins are not damaged. The RWCU water is on the tube side of the non-regenerative heat exchanger that is cooled by CCW water on the shell side. Dose rates in the vicinity of the non-regenerative heat exchanger may be on the order of tens of mSv/h (a few rem/h) while the plant is operating.

The RWCU system purification loop consists of a filter demineralizer unit to remove soluble and insoluble species from the reactor water. A filter demineralizer unit is composed of a pressure vessel that contains screen type filter tubes, a holding pump, and associated valves and piping to support operation and servicing of each unit. The filter tubes are coated with a filter medium and powdered ion exchange resin. The filtering materials are held in place by the differential pressure created by water flowing through the pressure vessel. These materials serve as the filter media and ion exchange sites. As the inventory of activated corrosion and fission products accumulate on the powdered resin and filter medium significant dose rates will result. Depending upon the service time of a particular unit and the presence of any fuel cladding failures dose rates in excess of a few Sv/h (a few hundred rem/h) are not uncommon in the vicinity of the filter demineralizer unit. These vessels are located behind heavily shielded vaults. When the filter demineralizer unit is depleted the unit is removed from service and backwashed. The filter tubes are precoated with fresh filter and ion exchange resin and returned to service.


## 3.7  Residual Heat Removal System

The residual heat removal (RHR) system of a BWR is similar to that of the RHR system for a PWR in that it serves several functions. In addition to removing decay heat from the reactor core during reactor shutdown, the RHR system serves as the low pressure coolant injection (LPCI) system in the event of a loss of coolant

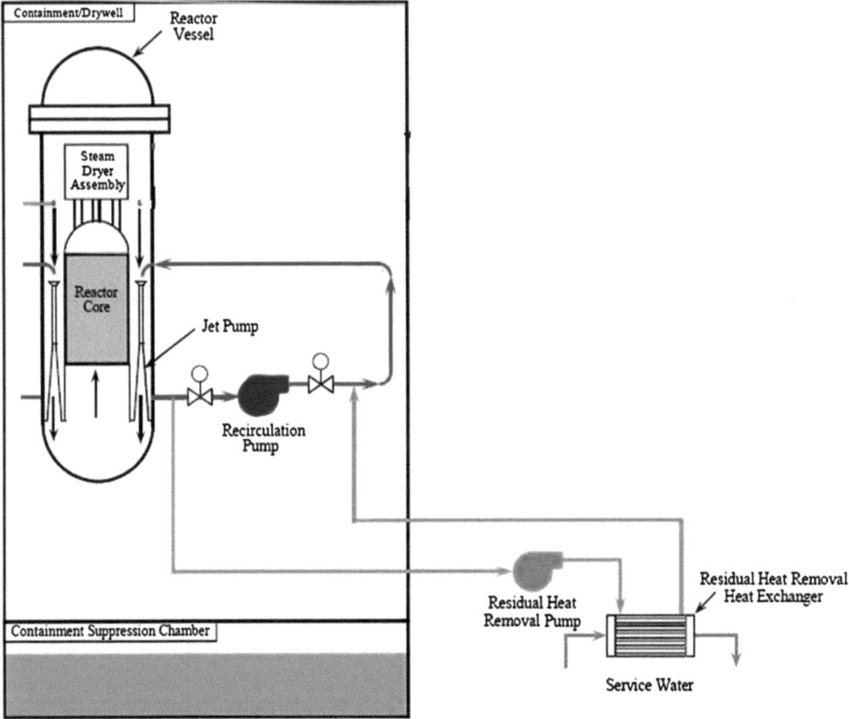


**Fig. 3.8** Simplified schematic of RHR system (www.nrc.gov/reading-rm/basic-ref/teachers)

accident. The system also supports the containment spray system used to reduce primary containment pressure and temperature in the event of an accident. The RHR system also removes heat from the suppression pool, among other functions. In the LPCI mode, the RHR pumps take suction from the suppression pool. In the containment spray mode the RHR pumps feed suppression pool water to the containment spray spargers.

During normal plant operation the RHR system is maintained in standby. The system also serves a dual purpose as part of the emergency core cooling system. The radiological conditions associated with the RHR system of a BWR are very similar to those described for the RHR system of a PWR. Dose rates for the train in standby are typically on the order of hundreds of μSv/h (tens of mrem/h) or less. As before radiological conditions will change significantly for the train that is in service providing shutdown cooling due to the presence of fresh RCS coolant flowing through the system. Dose rates in the vicinity of the RHR components for the train in service could be on the order of a couple of mSv/h (100–200 mrem/h). Assuming no significant fuel failures are present, these dose rates will decrease rapidly several days following shutdown (Fig. 3.8).

## 3.8 Fuel Pool Cooling and Cleanup System

The fuel pool cooling and cleanup system removes decay heat generated by spent fuel, purifies the fuel pool cooling water, and maintains a sufficient water level in the fuel pool to maintain adequate cooling of spent fuel assemblies. Spent fuel assemblies, control rods and miscellaneous components may be stored in the fuel pool at any given time. Unlike a PWR, whereby the spent fuel pool is located in a separate building from that of the reactor vessel, the spent fuel pool (SFP) is located immediately adjacent to the reactor cavity area. A cask set down area is provided in the SFP in a location that would allow placement of casks or other heavy objects in the SFP without the need to move these heavy objects over spent fuel storage racks. The spent fuel pool is connected to the reactor cavity by means of a transfer canal that may be only a few meters in length. The transfer canal is isolated from the reactor cavity area during power operation by means of a gate and shield blocks. Adjacent to the reactor cavity, opposite to that of the SFP, is the steam dryer and moisture separator storage pool. Figure 3.9 depicts the physical arrangement of the SFP, reactor cavity and steam dryer storage pool for a BWR.

The fuel pool cooling circuit consists of two 100% capacity trains, each containing a pump and heat exchanger. Water is pumped from the SFP through a heat exchanger and a filter-demineralizer unit. The SFP heat exchangers are cooled by the reactor building closed cooling water system that provides water to the shell side of the heat exchangers. After cooling and purification the water is returned to the SFP. Skimmer units maintain surface clarity by draining water from the surface of the fuel pool and reactor cavity (when the reactor cavity area is flooded during outage periods). Water from the skimmer units is filtered in a closed loop configuration.

Radiation levels in the vicinity of fuel pool cooling and cleanup system components are highly dependent upon the condition of spent fuel stored in the SFP. Filters and demineralizers could read hundreds of mSv/h to Sv/h (tens to hundreds of rem/h) depending upon the length of service and activity concentrations in the SFP water. These components are located in shielded cubicles. The SFP heat exchangers could have radiation levels on the order of a few mSv/h (a few hundred mrem/h) and subject to localized hot spots exceeding several tens of mSv/h (several rem/h). Radiation levels in the vicinity of the spent fuel pool cooling and cleanup system pumps could be on the order of hundreds of $\mu$Sv/h (tens of mrem/h).

Depending upon the system layout and configuration, spent fuel pool cooling and cleanup system pumps and heat exchangers may be located in readily assessable locations. Under these circumstances these component could pose a significant radiation source to individuals working in nearby areas. If the unit has been plagued with a history of fuel cladding defects it may be necessary to provide some sort of shielding around the heat exchangers and pumps to reduce ambient radiation levels near these components.

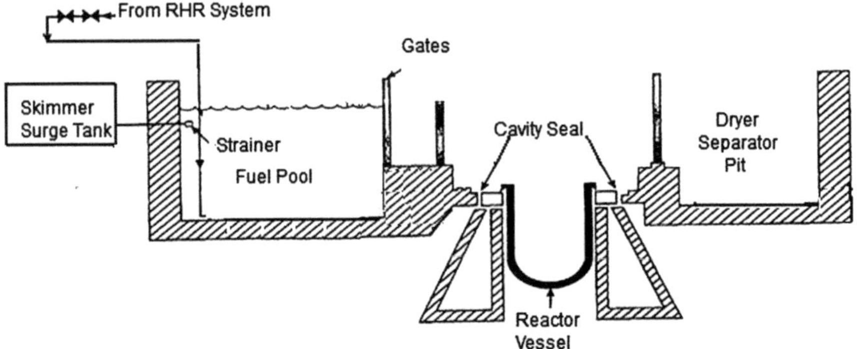



**Fig. 3.9** Reactor cavity, spent fuel pool and dryer-separator storage pool layout for BWR (Source: NUREG-1275, 1997)

## 3.9 Reactor Core Isolation Cooling System

The reactor core isolation cooling (RCIC) system provides makeup water to the core to maintain core cooling. The RCIC is required when the main steam lines are isolated or the condensate and feedwater system is unavailable. Both situations result in less than adequate core cooling capability. The RCIC system initiates automatically upon detection of low reactor vessel water level. The RCIC system consists of a steam driven turbine pump and associated valves and piping to deliver water to the reactor vessel when called upon. The turbine steam supply is provided by decay heat and the turbine exhaust is routed to the suppression pool. A steam driven turbine drives the RCIC pump which takes suction from the condensate storage tank. The RCIC pump discharge flow is via a feedwater line that provides cooling water to the reactor vessel by the feedwater spargers.

The RCIC system is maintained in standby and filled with clean water that may have low amounts of radioactive contaminants. Radiation levels in the vicinity of RCIC components and piping during normal plant operation usually do not pose a radiological concern and may be in the range of tens of µSv/h (a few mrem/h). Obviously during accident conditions when the RCIC is drawing highly contaminated water from the suppression pool these components will have significant radiological source terms associated with their operation and access to which may be impractical due to high radiation levels that could be in excess of several Sv/h (hundreds of rem/h), or higher (Fig. 3.10).

## 3.10 Reactor Building Closed Loop Cooling Water System

The Reactor Building Closed Loop Cooling Water System (RBCLCW) or simply the closed cooling water system (CCW) provides cooling to the major components

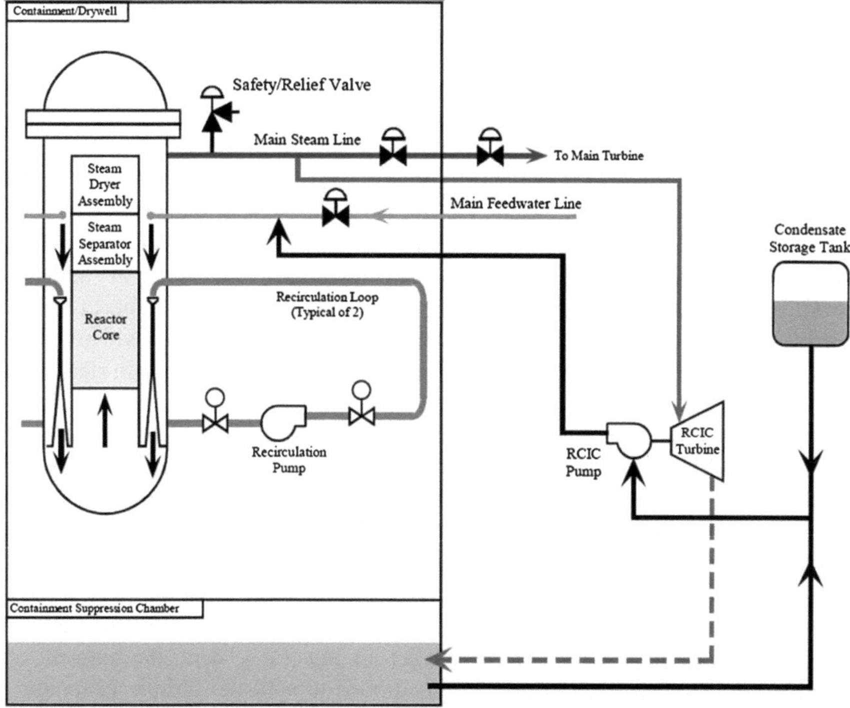


**Fig. 3.10** Simplified schematic of reactor core isolation cooling system (www.nrc.gov/reading-rm/basic-ref/teachers)

of several systems, both safety and non-safety related. Major components cooled by the CCW system include the recirculation system pumps and motors, RHR pumps, fuel pool heat exchangers, drywell air coolers, control rod drive pump coolers, RWCU pumps and heat exchangers among others. The system typically consists of two safety-related loops and two non-safety related loops. Due to the important function of the CCW system each loop is sized to provide 100% of the coolant flow to the components served by a given train. Redundant head tanks serve as the source of water for the system. The CCW system pumps are each sized at 100% capacity. This allows for one of the pumps to be out of service for maintenance. Cooling flow is directed to the various components served by the CCW system.

The CCW system contains high purity demineralized water and under normal conditions is non-contaminated. Obviously any leakage into the CCW system from the various components cooled by the CCW could introduce contamination into the system. Radiation levels in the vicinity of CCW system components are primarily influenced by the location of a given CCW component versus the concentration of radioactive contaminants within the CCW system.

## 3.11 Main Steam System

Unlike PWR units, the main steam system of a BWR poses radiological concerns. The primary purpose of the main steam system is to direct steam from the reactor vessel to the main turbine. Since the steam is produced from water that has flowed through the reactor vessel it contains various activation products. Principle coolant activation products listed in Table 4.2 will all be present in the steam to varying degrees. The activation product of primary concern is N-16. The amount of N-16 carried over in steam is influenced by the use of hydrogen water chemistry and noble metal injection as part of BWR feedwater chemistry controls. The impact of these feedwater chemistry controls on N-16 concentrations in steam is discussed in Chap. 8.

The presence of N-16 and other coolant activation products in main steam lines and associated systems and piping will result in high radiation areas in the vicinity of these components. Dose rates emanating from the main steam lines exiting the reactor vessel are sufficiently high to require steam lines to be routed through a shielded steam tunnel. Due to the short half-life of N-16, radiation levels will decrease as the process steam makes its way to the turbine generator and processed through various systems and components. Principal components impacted by N-16 radiation levels may include, in addition to the main steam lines, the high-pressure turbine, low-pressure turbine stages, feedwater heaters and steam extraction components, moisture separators, steam jet air injector system, the hotwell, and intermediate piping. Shield walls are positioned around the turbine generator to reduce dose rates on the turbine deck. Contact dose rates on the turbine housing may approach a few mSv/h (a few hundred mrem/h) during power operation.

## 3.12 Radioactive Waste Treatment Systems

The discussion for PWR radioactive waste treatment systems also generally applies to BWR units. The purpose and overall function of these systems is the same with the exception of various terms and nomenclature. The secondary side of BWR units will also contribute to the various radioactive waste streams. The radiological parameters and approximate dose rates associated with various BWR liquid waste collection tanks, chemical drain tanks, hold-up tanks and laundry collection tanks is essentially analogous as that described in Chap. 2 for PWR waste treatment systems.

### *3.12.1 Offgas System*

A major function of a BWR offgas system, from a radiological perspective, is to delay the release of radioactive species contained in the gaseous waste stream. Vacuum pumps remove non condensable gases from the condenser, while steam jet air ejectors route the gas stream to the offgas system. The non condensable gases are comprised of fission gases, water activation gases and radiolytic gases. The primary fission product gases include xenon and krypton. The halogen iodine is the other primary radioactive species in the offgas. The water activation products are discussed in detail in Chap. 4. What follows is a simplified description of the offgas processing stages. Emphasis is placed on those system components that impact the radiological aspects of the offgas system.

Prior to processing and holdup the offgas stream is heated and routed through a hydrogen recombiner to reduce hydrogen gas concentrations in the gas stream to reduce hydrogen gas concentrations to preclude possible explosions. The temperature of the gas stream increases as it passes through the hydrogen recombiner. The temperature of the gas stream discharged from the recombiner then passes through a condenser where it is cooled. This condenser reduces the flow rate of the offgas stream thus affording additional holdup time that allows short-lived fission gases and their daughter products to further decay.

The offgas stream is eventuality routed though one of two sacrificial charcoal decay beds located in the radioactive waste processing building. One bed is in service while the plant is in operation and the other bed maintained in standby. The charcoal beds delay the flow of xenon and krypton through the beds and retain the short-lived daughter products of the fission gases. The offgas stream is dried to decrease the moisture content of the gas before routing to a charcoal adsorber tank. Decreasing the moisture content in the process stream increases the efficiency of the charcoal adsorber tanks. The charcoal adsorber tanks provide sufficient holdup time to allow the short-lived xenon and krypton radionuclides to decay into their particulate daughter products. Any iodine that may be present in the offgas stream at this stage is adsorbed on the charcoal. There are two trains of charcoal adsorber tanks each train consisting of several tanks. The waste stream is routed through HEPA filters prior to being discharged to the exhaust stack.

### 3.12.2 Liquid Waste Treatment System

The collection and processing of liquid wastes is essentially similar to that as for PWR units. Liquid wastes are collected and processed based on their purity level to minimize processing costs. Various waste collection and holdup tanks, chemical drain tanks, and floor drain collection tanks comprise the liquid waste treatment system. Due to the larger volumes of liquid waste processed by BWR units a waste evaporator is typically available to reduce the volume of liquid waste required to be processed and released.

### 3.12.3 Solid Waste Treatment System

The solid waste treatment system provides for the handling, compacting, solidification and packaging of solid wastes. Solid wastes include spent ion exchange resins, evaporator bottoms, used filter materials and miscellaneous solid wastes. Again the handling of these wastes and other miscellaneous wastes described in Chap. 2 for PWR units applies to the treatment of solid waste at BWR units. This also includes the reliance on offsite radioactive waste handling, processing, and disposal firms.

The use of powdered ion exchange resin in various systems, most notably the reactor water cleanup system produces a high-activity solid waste stream. Spent powdered ion exchange resin is stored in phase separator tanks (i.e., spent resin storage tanks). There may be as many as three phase separator tanks. These tanks are located within shielded vault rooms due to the high radiation levels associated with RWCU system spent resin. Lower-activity spent condensate demineralizer resins are stored in separate phase separator tanks prior to processing.


## Bibliography


1. Lish K., Nuclear Power Plant Systems and Equipment, Industrial Press, New York, NY, 1972
2. Neeb, Karl-Heinz, The Radiochemistry of Nuclear Power Plants with Light Water Reactors, Walter de Gruyter & Co., Berlin, 1997
3. Rahn F.J., Adamantiades A.G., Kenton J.E., and Braun C., A guide to Nuclear Power Technology – A Resource for Decision Making, New York: Wiley & Sons; 1984
4. Whicker F.W., and Schultz V., Radioecology: Nuclear Energy and the Environment, Volume 1, CRC Press, Boca Raton, Florida, 1982
5. US Nuclear Regulatory Commission, Reactor Concepts Manual, USNRC Technical Training Center

# Chapter 4
# Sources of Occupational Radiation Exposure

## 4.1 Radiation Sources

The reactor core is the primary source of the radionuclides encountered at LWR's whether they are fission products or activation products. Radiation is produced when the fissile nuclides U-235 and later in core life, Pu-239 fission. The fission process produces neutrons, beta particles and gamma radiations directly along with radioactive fission products. Along with the two fission products, two to three neutrons are also emitted in addition to the energy given off during the process. The fission products consist of numerous radionuclides many of which emit beta and gamma radiations. In addition to neutrons posing a direct radiation hazard when the reactor is critical, they also produce radionuclides by means of neutron activation.

Obviously neutron radiation is present as long as the fission process is ongoing while the reactor is critical. After shutdown, neutron radiation fields are no longer a radiological concern. Less than 1% of all neutrons originate from decay of fission products. Neutrons emitted by fission products are referred to as delayed neutrons. These delayed neutrons have half-lives on the order of minutes or less, and consequently do not represent any significant exposure concerns during outages or maintenance periods.

Gamma and neutron radiations may escape the core region and penetrate the biological shielding surrounding the reactor vessel. Beta particles originating from the core will be attenuated by both the coolant and reactor vessel and do not contribute significantly to general area radiation levels external to plant systems and components. Dose rates in the vicinity of the reactor vessel and primary system components will be a function of reactor power, increasing with reactor power. Since the number of fissions is directly proportional to the reactor power level, neutron radiation levels are also a function of reactor power level. Reactor power levels should be noted when performing surveys in those plant areas affected by reactor power level so that survey results may be properly interpreted. This chapter summarizes the sources and the production mechanisms for those

R. Prince, *Radiation Protection at Light Water Reactors*,
DOI: 10.1007/978-3-642-28388-8_4, © Springer-Verlag Berlin Heidelberg 2012

radionulcides that are the significant contributors to ex-core radiation levels and personnel radiation exposures.


## 4.2 Neutron Activation Source Terms

Stable nuclides introduced into the core and subsequently exposed to the intense neutron flux may become radioactive as a result of neutron activation. Many of these activated radionuclides result due to the presence of corrosion products. Neutron interactions with the primary coolant also result in the production of radionuclides. The type and quantity of corrosion product activity is dependent upon several factors. Coolant chemistry, reactor type, construction materials, age and operating history of a given plant are all parameters that influence corrosion product inventories. Additionally, fuel cladding and plant components in close proximity to the core become activated in situ contributing to local radiation fields. As these components corrode they may release activated corrosion products into the coolant.

The major corrosion products produced depend primarily on the metal alloys present in the primary circuit and to a lesser degree, components and systems in direct communication with the primary system. One of the more significant activated corrosion products is cobalt-60. Cobalt is present in structural materials such as various types of stainless steel, and originates via an n-$\gamma$ reaction with cobalt-59, the natural constituent of cobalt. Cobalt is found in PWRs in stainless steels used for steam generator tubing, fuel assembly grid spacers and control rod cladding, while in BWRs cobalt is utilized for control blades and in recirculation lines. Cobalt-59 is also found in trace quantities in Zircaloy and Inconel materials used in the fabrication of various reactor coolant system components. Significant quantities of cobalt and nickel may be present in those components subject to mechanical wear where they are utilized in hard facing alloys such as stellite. Corrosion of valves in various auxiliary systems that communicate with the RCS may be a significant source of cobalt. This subject is discussed further in Chap. 8.

Another corrosion product of concern is cobalt-58 which is produced from nickel-58 via an n-p reaction. Nickel is present as an impurity in many high quality alloys. Both Co-58 and Co-60 contribute significantly to in-plant radiation fields. Cobalt-58 may comprise a larger percentage of the radiation source term early in plant life, especially for PWR facilities, while other radioactive corrosion product species, such as Co-60, have yet to reach equilibrium values. Obviously considering the maturity of today's LWR industry the plants currently operating have long since reached Co-60 equilibrium values.

Some of the more common activation corrosion products along with their production processes and principle radiations are provided in Table 4.1.

Activation of the primary coolant also produces several radioactive activation products. The most abundant isotope of oxygen is oxygen-16. Neutron activation of oxygen-16 produces nitrogen-16, while proton activation of oxygen-16

**Table 4.1** Common activation corrosion products

| Radionuclide | Production processes | Half-life | Major radiations |
|---|---|---|---|
| Co-60 | $^{59}Co(n,\gamma)^{60}Co$ | 5.2 y | $\beta$-$\gamma$ |
| Co-58 | $^{58}Ni(n,p)^{58}Co$ | 71.4 d | $\beta$-$\gamma$ |
| Fe-55 | $^{54}Fe(n,\gamma)^{55}Fe$ | 2.6 y | $\gamma$ |
| Fe-59 | $^{58}Fe(n,\gamma)^{59}Fe$ | 45.1 d | $\beta$-$\gamma$ |
| Cr-51 | $^{50}Cr(n,\gamma)^{51}Cr$ | 27.8 d | $\gamma$ |
| Mn-54 | $^{54}Fe(n,p)^{54}Mn$ | 300 d | $\gamma$ |
| Mn-56 | $^{55}Mn(n,\gamma)^{56}56$ | 2.6 h | $\beta$-$\gamma$ |
| Cu-64 | $^{63}Cu(n,\gamma)^{64}Cu$ | 12.8 h | $\beta$-$\gamma$ |
| Zn-65 | $^{64}Zn(n,\gamma)^{65}Zn$ | 245 d | $\beta$-$\gamma$ |
| Ni-65 | $^{64}Ni(n,\gamma)^{65}Ni$ | 2.6 h | $\beta$-$\gamma$ |
| Zr-95 | $^{94}Zr(n,\gamma)^{95}Zr$ | 65.0 d | $\beta$-$\gamma$ |
| Zr-97 | $^{96}Zr(n,\gamma)^{97}Zr$ | 17.0 h | $\beta$-$\gamma$ |

**Table 4.2** Principle coolant activation products

| Radionuclide | Production processes | Half-life | Major radiations |
|---|---|---|---|
| N-16 | $^{16}O(n,p)^{16}N$ | 7.1 s | $\beta$-$\gamma$ |
| N-17 | $^{17}O(n,p)^{17}N$ | 4.0 s | $\beta$-$\gamma$ |
| F-18 | $^{18}O(p,n)^{18}F$ | 1.8 h | $\beta^{+}$ |
| O-19 | $^{18}O(n,\gamma)^{19}O$ | 29.0 s | $\beta$-$\gamma$ |
| N-13 | $^{16}O(p,\alpha)^{13}N$ | 10.0 m | $\beta^{+}$ |
| Na-24 | $^{23}Na(n,\gamma)^{24}Na$ | 15.0 h | $\gamma$ |

produces nitrogen-13. Other isotopes of oxygen (i.e., oxygen-17 and 18) may also undergo activation and result in the production of additional radionuclides. In general, these products have short radioactive half-lives and are of concern only in those plant areas where the coolant transit time from the core is short (<1 min). These products contribute to radiation fields in the containment building and areas of the CVCS letdown line in PWR's. While the drywell, main steam system (including main steam lines, turbine and reheaters), off-gas system and areas of the reactor water cleanup system may be affected in BWR facilities. The principal coolant activation processes are listed in Table 4.2.

Nitrogen-16 is by far the most significant of the coolant activation products. It is produced in abundance and emits an extremely high-energy gamma ray of 6.1 MeV. The presence of this nuclide is of major concern when evaluating shielding requirements for those systems and components that contain RCS or steam (BWRs) whose transit time since leaving the core is such that any N-16 present has not had time to decay. Nitrogen-16 may be the major contributor of radiation fields in those plant areas immediately adjacent to primary system components (e.g., inside loop rooms at PWRs) during power operations.

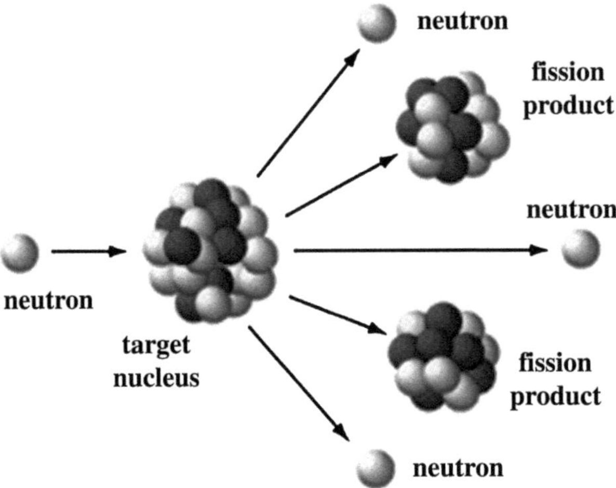


**Fig. 4.1** Fission of U-235 nucleus (www.mtholyoke.edu)

Sodium-23 is present in water as an impurity which when activated produces sodium-24. Another activation product that may be encountered is argon-41. Argon-41 results from the activation of argon-40. The concentration of argon in air is less than 1%. Argon-41 may be produced in measurable quantities in the event of air in-leakage into the primary system. Under normal operational conditions the presence of Ar-41 should not pose any significant radiological concerns.

## 4.3 Fission Products

Fission is the process in which a heavy nucleus is split (by a thermal neutron in the case of U-235) into two or more components with the release of considerable amounts of heat energy. The resulting species are referred to as fission products. Figure 4.1 depicts the typical fission process for a U-235 nucleus. Many of these fission products are radioactive and some give rise to decay chains that produce additional radionuclides. The fission process ultimately produces over 200 radionuclides either directly or indirectly. Most of these fission products are beta-gamma emitters and may be grouped into three basic categories: particulates, gases and halogens. Halogens are chemically active non-metals which belong to group 17 (formerly VII and VIIA) of the periodic table. The isotopes of iodine represent the halogens of major concern at a LWR. Gases encountered are radionuclides of xenon and krypton. These are inert, noble gases, which are chemically inactive. Tritium may also exist as a gas in plant areas.

**Table 4.3** Principle particulate and gaseous fission products

| Radionuclide | Half-life | Radionuclide | Half-life |
|---|---|---|---|
| (a) *Particulate radionuclides* | | | |
| Rb-88 | 17.7 m | I-134 | 52.6 m |
| Sr-89 | 50.0 d | I-135 | 6.6 h |
| Sr-90 | 29 y | Cs-136 | 13.0 d |
| Sr-91 | 9.5 h | Cs-137 | 30.1 y |
| Sr-92 | 2.7 h | Cs-138 | 32.2 m |
| Y-90 | 64.0 h | Ba-134 m | 2.6 m |
| Y-91 | 58.6 d | Ba-139 | 83.3 m |
| Mo-99 | 66.0 h | Ba-140 | 12.8 d |
| I-131 | 8.0 d | La-140 | 40.2 h |
| I-132 | 2.3 d | Ce-144 | 184.4 d |
| I-133 | 20.8 h | | |
| (b) *Gaseous radionuclides* | | | |
| Kr-85 m | 4.5 h | Xe-133 m | 2.2 d |
| Kr-85 | 10.7 y | Xe-133 | 5.3 d |
| Kr-87 | 76.0 m | Xe-135 m | 15.3 m |
| Kr-88 | 2.8 h | Xe-135 | 9.2 h |
| Xe-131 m | 12.0 d | Xe-138 | 14.2 m |

Many fission products do not contribute significantly to shutdown radiation fields due either to their short radioactive half-lives or their low fission yields. The principle fission products of concern at LWRs are listed in Table 4.3.

## 4.4 Tritium Production Processes

Tritium may be produced by either activation or fission. Every so often a ternary fission produces tritium as one of the fission products. Tritium is also produced by activation of boron-10, which comprises about 20% of natural boron. Boron may be utilized in control rods and in the case of PWRs; it is dissolved in the primary coolant via the addition of boric acid for reactivity control purposes. This use of boron in PWRs can result in annual production of tens of terabecquerels (several hundred curies) of H-3 whereas the other production processes produce annual tritium amounts on the order of tens of gigabecquerels (tens of curies). The use of soluble boron in the RCS of PWRs results in tritium posing a greater radiological concern at PWR facilities then that of BWRs from an occupational exposure perspective. Lithium is present as an impurity in some plant metals and may be introduced as lithium hydroxide for pH control in PWR facilities. Neutron activation of lithium-6 (7.5% of natural lithium) also results in the production of tritium. Utilization of lithium hydroxide with low lithium-6 concentrations will minimize the amount of tritium produced in this manner. Tritium is also produced by neutron activation of the small amounts of deuterium present in water. While

| Table 4.4 Tritium production processes | $^6Li(n,\alpha)\ ^3H$ | $^{235}U(n,fission)\ ^3H$ |
| --- | --- | --- |
| | $^{10}B(n,2\alpha)\ ^3H$ | $^2H(n,\gamma)\ ^3H$ |
| | Half-life: 12.3 y, $\beta$-emitter | |

this is a significant source of tritium in BWRs, this process accounts for less than 1% of the tritium produced in PWRs. Tritium production processes are summarized in Table 4.4.

## 4.5 Parameters Effecting Source Terms

The radiological source term associated with a given system or component is predicated on several parameters. These radiological source terms are not necessarily constant and may vary by orders of magnitude under various plant conditions. Factors influencing these radiological source terms are discussed below. It is essential that radiation protection personnel understand the role these factors play when assessing the radiological status of plant components. The proper analysis and evaluation of these parameters allows for the establishment of adequate radiological controls under varying plant conditions.

Factors influencing system radiation levels include component design, system inter-relationships, operating mode, operational history, coolant chemistry, plant age, construction materials, and fuel integrity among others. Obviously radiation levels depend primarily on the amount of failed fuel present in the core and the inventory of activated corrosion products circulating in the primary system at any given time. Each of these parameters has a direct bearing on the radiological conditions associated with various systems and components.

### 4.5.1 System Design and Inter-Relationship

In general, the reactor coolant system will contain the greatest quantities of fission products and activated corrosion products at any given time. Systems directly connected to the RCS will contain somewhat lesser amounts while systems not directly connected to the RCS should be non-contaminated or minimally contaminated. Some systems (e.g., steam generator blowdown, component cooling water, or closed-loop cooling water systems) may become contaminated due to internal leaks in such components as heat exchangers, equipment operational issues or as a result of a plant incident.

Such systems as the spent fuel pool cooling system are obviously of no radiological concern until such time that spent fuel is introduced into the spent fuel pool, assuming that the system does not otherwise become contaminated by some other means (e.g., transfer of contaminated water to the spent fuel pool). Other systems by virtue of their function or location may not pose radiological concerns during plant operation. For instance, reactor coolant pumps, steam generators (for PWR units), equipment located in the drywell (for BWR units), have high dose rates associated with them during periods of reactor operation. However, these components are not considered assessable to personnel, and therefore do not pose radiological issues during periods of plant operation. In addition access to these areas at power may be further restricted due to possible low oxygen concentrations in these areas, the possible need for self-contained breathing apparatus or based on heat stress stay-time limitations. Systems such as the safety injection and containment spray systems, which are in standby mode during normal operations, usually contain water that is relatively free of radioactive contaminants. Additionally, these systems and others, such as the RHR system that is not in operation, may have water inventories that have not been recently mixed with fresh sources of contaminated water. Under these circumstances dose rates in the vicinity of these system components may be relatively low compared to dose rates when in operation. These systems may experience an increase in their associated dose rates when placed into service; for example, when an RHR train is placed into operation during an outage to remove decay heat from the core. In the extreme case, in the unlikely event of a loss-of-coolant accident (LOCA) when make-up water for these systems is highly contaminated, radiation levels will be prohibitive in the vicinity of these systems.

System components should be designed to minimize crud deposition. Components and those pipe sections that influence process stream flow rates (e.g., pipe bends) which could lead to crud deposition should be designed to minimize deposition of material at system or component low points. Locations susceptible to deposition (i.e., crud traps) should be minimized. These areas and other components (e.g., spent resin sluice lines and radioactive waste process piping) should have a mechanism to allow for periodic flushing in order to prevent the long-term buildup of crud and the resulting increase in area dose rates. Pipe plugs may be installed at strategic locations in a resin sluice line, for example. The pipe plugs could be removed to allow a water lance to be inserted into the pipe header to flush deposits that may be present contributing to elevated radiation levels in the area. Materials should be selected to minimize corrosion and purification systems sufficiently sized and designed to provide for efficient removal of impurities. Equipment design and physical layout should be such as to minimize crud deposition.

### *4.5.2 Operational History and Operational Mode*

System radiation levels vary as a function of the operating mode and operational history of the plant. Neutron radiation fields are of concern during periods of reactor operation and usually do not become significant until power levels reach

10% or higher. The numerous short-lived activation and fission products that contribute to plant radiation fields during operation soon decay to insignificant levels after shutdown. Consequently, radiation levels in certain plant locations are drastically reduced after shutdown due to the elimination of neutron radiation and short-lived radionuclides (e.g., N-16). Therefore radiation levels in the vicinity of certain plant systems (e.g., primary system components, CVCS, and RWCS) are typically higher during periods of operation versus shutdown conditions. The presence of short-lived activation products drastically impacts radiation levels in such areas as inside the drywell and the biological shield wall, often preventing access to such locations during periods of power operation. Short-lived activation products will directly impact radiation levels in the vicinity of steam lines in a BWR. Basically if fluid transit times from the core are not long enough to allow sufficient time for short-lived activation products to decay then the presence of these species will contribute to radiation levels in affected plant areas.

Plant radiation levels tend to increase rapidly over the first few fuel cycles as corrosion and activation products are initially produced and buildup to equilibrium values, most notably Co-58 and Co-60, and level off thereafter. Due to the accumulation of crud, the gradual buildup of fission and activation products, degradation of plant components and in general, effects of prolonged operation, plant radiation levels will tend to increase over time. The rate of source term buildup may gradually diminish and may remain constant over a long period of time. Higher radiation levels will result in increased personnel exposure and increased operating expenses and may adversely affect equipment repair and maintenance activities.

If a unit has experienced a high number of scrams while at power or if it has undergone many startup cycles during its lifetime, radiation levels may be higher than those of a comparable unit that has experienced a rather smooth, event free, operational history. The sudden temperature changes and the effects on coolant chemistry associated with plant shutdowns and startups will have a detrimental effect on plant components over the long run, resulting in higher plant radiation fields. The presence of any minor cladding defects may be aggravated by these-type operational excursions that could promote increased leakage of fission products into the RCS. Operational excursions often challenge chemistry controls, resulting in periods whereby optimal chemistry parameters are difficult to maintain. Over the long run these conditions may contribute to higher levels of crud production and resulting increases in radiation source terms.

### 4.5.3 Coolant Chemistry

The maintenance of good plant chemistry will minimize the corrosion rate of plant components. Uncontrolled corrosion leads to excessive crud build-up and increases the inventory of activated corrosion products. As the amount of these species increase the result will be higher radiation levels in affected plant areas.

Therefore, the adherence to good plant chemistry will minimize plant radiation levels over the operational life of a plant.

Corrosion rates are affected by a number of factors. In general as water temperatures increase the corrosion rate is accelerated. System flow rates effect the formation of protective oxide coatings. If system flow rates are sufficiently high, protective oxide layers may be removed, exposing underlying metal surfaces to additional corrosion. Hydrogen gas present in the water forms a protective barrier on metal surfaces. Dissolved oxygen present in the water reacts with hydrogen gas, removing this protective barrier, as well as promoting various types of corrosion processes. The concentrations of hydrogen ions in water affect the corrosion rates of various metals. The pH value provides a direct indication of the concentration of hydrogen ions in water and therefore is a parameter that should be strictly maintained within acceptable limits. Depending upon the reactor type and method of primary chemistry employed the pH of the RCS is usually maintained between neutral and slightly alkaline (e.g., a pH of 6–8).

The conductivity of water is a measure of the ability of water to conduct a current. As this ability increases (i.e., as conductivity increases) the rate of corrosion will typically increase. The amount of dissolved solids in water, primarily determines the conductivity value. If the concentration of dissolved solids is low, conductivity will be low and corrosion rates minimized.

Operating conditions should be controlled to the extent possible in order to minimize corrosion. Operating limits on fluoride and chloride concentrations, pH, dissolved oxygen concentrations, conductivity, among other parameters must be established and maintained. The presence of chloride has a significant detrimental effect on corrosion mechanisms such as stress corrosion cracking. Operating periods should be restricted whenever applicable chemistry specifications are exceeded. This practice will minimize the amount of corrosion during periods when chemistry parameters are not within prescribed values. Plant technical specifications and procedures may specify chemistry limits and sampling requirements associated with the measurement and monitoring of various chemistry parameters. Limiting conditions of operation may also be specified when measurement values are out of specification to minimize adverse impacts. Water chemistry is discussed in more detail in Chap. 8.

Various processes and techniques may be utilized to reduce the long-term build-up of radiation source terms. Chemicals can be added to contaminated systems to induce so-called crud bursts. The added chemicals attack the oxide corrosion layers and crud deposits lining inside surfaces of equipment and components. Various shutdown chemistry regimes may also be utilized to promote source term reduction. Filtration and demineralization may then remove the displaced deposits. Care must be taken to ensure that the chemical reagents used do not promote corrosion of plant metals themselves. If crud burst techniques are to be used, the advantages must be compared to any possible disadvantages (e.g., impact on radioactive waste processing systems, operational considerations and long-term corrosion concerns, associated with the use of a particular reagent).

Chemical techniques and system decontamination processes associated with dose reduction initiatives are discussed in Chap. 8.

## 4.5.4 Construction Materials

Plant components are constructed from various metals and alloys all of which have unique corrosion characteristics. Ideally construction materials will not contain potential activation species, however; this is not always practical. The use of various construction materials will influence plant radiation levels. The corrosion properties of those metals frequently encountered in LWR components are described below.

Carbon steel is probably the most common material used in the fabrication of system components and piping. Carbon steel is primarily iron with a carbon mixture. Small quantities of manganese, silicon and phosphorus may also be present. The corrosion of carbon steel is strongly influenced by pH. Corrosion rates of carbon steel are high at both low and high pH values. Corrosion is minimized in the pH range of 10.5–12. Primary system components constructed of carbon steel are typically lined with stainless steel, preventing contact of the carbon steel with water to minimize corrosion concerns.

Stainless steel is the material of choice for those applications when corrosion is of primary concern. Stainless steel consists of an iron-based alloy that contains chromium and nickel in concentrations ranging as high as 16% and 8% respectively. The presence of chromium and nickel provide increased strength as well as improved corrosion characteristics for stainless steel. Stainless steels are susceptible to chloride stress corrosion cracking. Stress corrosion may occur whenever a physical stress is present in conjunction with sufficient concentrations of chloride and oxygen. In general, if both the oxygen and chloride concentrations are maintained below 0.1 ppm, chloride stress corrosion will be greatly minimized. A large percentage of the cobalt in the primary system originates from corrosion of stellite-hardened valve components. The Electric Power Research Institute (EPRI) on behalf of the nuclear industry has issued several excellent research reports dealing with cobalt reduction and primary water chemistry control techniques to reduce radioactive source terms at LWRs. These topics are discussed in greater detail in Chap. 8.

Inconel is a nickel-based alloy, which contains iron, manganese, copper, and chromium that is less susceptible to chloride stress corrosion cracking. It is however; subject to caustic stress corrosion cracking which usually is not a major problem at LWRs.

Zircaloy cladding is the most common material utilized for fuel element cladding. Zircaloy has a low neutron absorption cross-section, is resistant to corrosion from high temperature water and it develops a tightly adherent oxide film that minimizes corrosion. The presence of fluorides will greatly increase the corrosion rate of Zircaloy and for this reason fluoride concentrations are kept to a minimum in the reactor coolant.


## *4.5.5 Fuel Integrity*

Under routine operating conditions the vast majority of fission products are retained within the fuel cladding and do not contribute significantly to out-of-core radiation fields. Early in the nuclear power industry this may not have been the case when units operated for extended periods of time with high cladding leak rates resulting in fission product release rates that did contribute to higher annual exposures. The vast improvement in fuel designs, coupled with lower cladding leak rates as the industry matured, are primary reasons for the large reduction in fission product source terms over the last 10–20 years. Fuel cladding failure rates are usually quite low, the magnitude of which is dependent upon fuel design, cladding material and plant chemistry and operational history. Additionally, most plants have instituted strict operational controls in the event of excessive cladding leak rates. Such measures as changing control rod configurations to reduce the power level in the core region where the defective fuel element(s) is located and even shutting down prematurely to replace defective elements are practices employed today to help minimize radiation exposures and source terms. If cladding leak rates become excessive plant radiation fields may increase by orders of magnitude in affected plant areas such as waste hold-up tanks, RCS letdown lines and purification loops.

The integrity of fuel element cladding can be verified to ensure that defective elements are replaced as necessary to minimize plant radiation fields resulting from leaking elements. Various techniques are available to identify leaking fuel elements. During refueling operations an individual fuel assembly may be placed in a container (or can) in the fuel pool that may then be isolated from the spent fuel pool water. The assembly is allowed to remain in the canister for a given length of time after which a sample of the water within the isolated canister may be extracted. This sample may then be analyzed and the radioactivity concentration of the water equated to the degree of cladding failure that may be present. Obviously higher activity levels usually signify a greater degree of cladding failure. This process is referred to as wet sipping. Alternatively the can may be drained and the activity of gaseous or volatile fission products determined in a similar manner to evaluate the integrity of the fuel cladding. This process is known as dry sipping.

In recent years ultrasonic inspection (UT) techniques have been increasing in popularity when evaluating fuel cladding integrity. This method is faster than the sipping methods described above (5–10 min per assembly versus 15–45 min). Depending upon the method employed, UT inspections can be performed without the need to transfer assemblies to a special location thus eliminating the need to handle a given assembly multiple times. This method can result in significant timesaving during refueling outages and is particularly beneficial if fuel shuffling is the critical path.

Fuel element cladding failures may cause primary coolant activity to increase while at power. This increase could be rapid and significant depending upon the size and type of cladding damage. Even though a leaking fuel element cannot be specifically identified while at power, using the techniques described above, steps can be taken to identify the general location of a leaking element within the core. Control rod configurations may be changed in an attempt to identify the core region containing the assembly. Control rods can be alternately withdrawn and inserted in different regions of the core while monitoring changes in radioactivity levels of the primary coolant. If activity levels decrease while maintaining a particular control rod configuration the location of the defective fuel assembly may be assumed to be in the core region where the control rods are more fully inserted. Alternatively, if activity levels increase in the primary coolant the defective assembly may be in the core region where the control rods have been further withdrawn. During the next refueling outage the fuel assemblies in the suspect core region may be tested to identify the defective fuel assembly. Utilizing this technique may result in a significant reduction in the number of fuel assemblies requiring leak testing during a refueling outage.

Ambient radiation levels are significantly affected by the degree of fuel cladding failures. Operating with defective fuel assemblies will increase out-of-core radiation levels, resulting in higher radiation exposures to personnel while operating and during outages and maintenance periods. The costs associated with an unplanned plant shutdown to replace a leaking assembly must be weighed against the costs associated with higher radiation source terms and additional long term radiological control measures that may be necessary as a result of higher plant radiation fields. Additionally, depending on the magnitude and number of fuel failures, radioactive species (e.g., transuranic alpha-emitters) that are not normally present in concentrations that influence radiological control measures may become the limiting factor during maintenance periods. For instance if alpha contamination is present in sufficient quantities respiratory protection equipment may be required for certain evolutions not normally requiring the use of such equipment. If cladding defects are sufficient to cause the release of radioactive iodine and xenon in large enough quantities, then operational concerns may result. System leaks that normally do not pose any operational concerns could now result in plant areas being posted as airborne radioactivity areas, workers becoming "contaminated" with volatile airborne species, complicating ingress and egress from the RCA and additional radiation exposure to plant personnel.

Over the years improved fuel assembly designs have greatly improved the operating performance of fuel elements reducing the number of fuel failures experienced by the LWR industry. Plant technical specifications and operational limits and RCS chemistry controls reduce the probability of fuel cladding failures and once identified, strict operational controls have been established to limit the amount of time a plant may continue in operation with cladding defects exceeding specified limits. Long-term radiation exposures will be reduced considerably by identifying and removing defective fuel assemblies from service in a timely manner. If necessary fuel assemblies may be reconstituted to repair or remove damaged or leaking fuel rods and the assembly placed back into service to maximize fuel burn-up. By establishing an operating philosophy of operating with "zero" fuel defects a LWR facility can significantly reduce personnel exposures over the life of the plant.

## 4.6  Ambient Radiation Levels

As discussed previously in this chapter the fission process is responsible for producing the radionuclides encountered at LWRs, whether they are fission or activation products. If these products were confined solely to the reactor vessel personnel exposures could be easily controlled and reduced. However, the transport of these products is extensive and various mechanisms and processes affect numerous plant systems and areas. Activated corrosion products circulate in the primary system and deposit within the reactor vessel and components of plant auxiliary systems as well as throughout the entire primary system. Additionally, many factors influence the radiological conditions associated with plant systems and components at any given time. The component itself, its location, operating history, plant operating mode and other factors previously described all influence plant radiological conditions. It is important that radiation protection personnel understand these parameters in order to adequately assess and anticipate radiological conditions to afford effective radiological safety protection of plant workers.

The numerous parameters that influence plant radiation source terms make it difficult to estimate plant radiation levels over the long-term. When considering the differences in plant designs and unit operating histories, in addition to the factors previously discussed above, the ability to accurately predict radiation levels at a given location using data from other facilities may yield rough approximations at best. A site-specific database should be established and maintained for trend analysis purposes. Various industry studies and technical reports are available that summarize radiation levels encountered at LWR facilities.

Numerous Electric Power Research Institute (EPRI) reports have been published that provide data on radiation levels at various locations and for specific components (e.g., steam generators). One such report, EPRI NP-3432, PWR Radiation Fields Through 1982, summarizes shutdown radiation field measurements obtained at various locations at Westinghouse designed PWRs. The objective of the report was to identify operating techniques and other factors that have an effect on radiation fields. This program was known as the Standard Radiation Monitoring Program (SRMP) and provided the summary of field measurements taken at standardized locations in contact with primary system components and piping. The program was terminated for various reasons in the 1990s.

## 4.7 Sources of Airborne Contamination

Under normal operating conditions the fission products and activation and corrosion products that are produced will remain and accumulate in quantity within the primary system and be removed by filtration systems such as demineralizers. Consequently any breach of the primary system has the potential to release airborne contamination.

Obviously fission products must reach the coolant in order to eventually pose an airborne contamination hazard. There are two general processes by which fission products reach the coolant. During fuel fabrication small amounts of fuel (e.g., ceramic powder) may be present on the outer surfaces of fuel assemblies. Though great care is taken during the pellet loading and fuel rod assembly process trace amounts of uranium powder may be transferred to the surface of fuel rods. These deposits are sometimes referred to as "tramp uranium." Any uranium-235 present in this tramp uranium that fissions may cause recoil fission products to directly enter the coolant. The average path length of recoil fission products is extremely short in the range of 7-11 $\mu$m. If the fission occurs close to the fuel cladding surface there is a finite probability that some of the fission products may reach the coolant. Due to the strict quality control and inspections performed on new fuel assemblies by fuel fabricators, prior to shipment to customers, this is not typically a significant source of fission product inventory in the coolant.

The primary mechanism that contributes to the vast majority of the fission product inventory in the primary coolant is associated with fuel cladding failures or defects. The amount of fission products escaping from fuel assemblies depends on the nature of the cladding defect, the type of fuel, the operating temperature, the reactor power level and the severity and extent of any defects. As noted above there are two general classes of fission products, non-volatile and volatile. Non-volatile products are largely retained within the fuel pellets themselves with the fuel cladding serving as another barrier against leakage into the RCS. Species that do manage to leak through the fuel cladding remain circulating in various systems until removed by purification and cleanup processes or until systems are opened during maintenance and inspection activities. Obviously these species may also pose airborne contamination problems in the event of component or system leakage.

Volatile fission products escape more easily from otherwise non-damaged fuel rods and may pose airborne contamination problems in the event of small system leaks. By their very nature krypton, xenon, and iodine species more readily migrate or diffuse through fuel cladding, especially if small microscopic leaks are present. Valve packing, gasket, or pump seal leaks, that may not represent a serious operability issue for the particular system involved, may give rise to the presence of short-lived decay products of xenon and krypton in addition to the presence of the noble gases themselves. During periods of power operation these type leaks may cause airborne contamination concerns due to the presence of the short-lived daughter products. Even without any observable or detectable fuel cladding defects fission products may escape from fuel rods during power operation. However, by maintaining strict operating controls and maintaining water chemistry within prescribed specifications fuel cladding defects can be minimized.

Several factors affect the amount of contaminant that may eventually become airborne. These factors include fission yield, diffusion characteristics, half-life and solubility in the primary coolant. In addition such factors as ventilation removal rate, gravitational settling characteristics and diffusion deposition on surfaces will influence actual airborne radioactivity concentration levels. Among the fission products with the highest yield are the bromine-krypton- rubidium and the iodine-xenon-cesium chains. The noble gases krypton and xenon are most likely to become airborne. Since these gases decay to form particulate matter, rubidium and cesium radionuclides (e.g., Rb-88 and Cs-138) would also be present as airborne contamination. All these nuclides are beta-gamma emitters. Thus any leaks or maintenance activities involving the breach of systems, in which no significant decay time has elapsed, may present the opportunity for these radionuclides to become airborne. These decay chains are depicted below.

$$^{88}Kr(2.8h) \rightarrow \,^{88}RB(17.7m) \rightarrow \,^{88}Sr(stable)$$

$$^{138}Xe(14.2m) \rightarrow \,^{138}Cs(32.2m) \rightarrow \,^{138}Ba(stable)$$

The airborne potential of corrosion products and the remainder of the fission products depends more on the airborne generation process itself (e.g., welding, grinding, or cutting). Consequently corrosion products usually pose airborne contamination concerns during refueling and maintenance outages. In fact activated corrosion products (e.g., Co-58 and Co-60) typically represent the major contributor to airborne contamination during maintenance activities.

## 4.8 Summary

The corrosion process ultimately leads to the formation of activated corrosion products. Various parameters and mechanisms result in the activation of these corrosion products and the degree to which they will contribute to ex-core radiation fields. The deposition of activated corrosion products in plant systems and components contributes to contamination and airborne radioactivity concerns during the performance of maintenance activities. Programs that aggressively target cobalt reduction initiatives, the maintenance of strict chemistry controls, and the optimization of purification and filtration system performance will result in lower station exposures over the long term.

# Bibliography


1. Electric Power Research Institute, Report NP-3432, PWR Radiation Fields Through 1982
2. Nuclides and Isotopes, Fourteenth Edition, General Electric Company, 1989
3. Radiological Health Handbook, U.S. Department of Health, Education, and Welfare, 1970
4. Shleien, B., and Terpilak, M., The Health Physics and Radiological Health Handbook, Nucleon Lectern Associates, 1984

# Chapter 5
# Demarcation of Radiological Zones

## 5.1 Overview

Survey results are of little value if the data is not effectively communicated and understood by radiation workers. The radiological status of plant areas must be communicated to individuals by means of a zoning and posting convention. To be effective a standardized nomenclature for the various radiological zones should be established and utilized to demarcate radiological areas. Licensing authorities throughout the world have established standardized posting terms and designations for use in their areas of jurisdiction. Various conventions may use a color code or number designation to classify the magnitude of radiation levels or contamination levels present within a given area. The radiological sign posting convention promulgated by the US Nuclear Regulatory Commission in Part 10, Chapter 20 of the Code of Federal Regulations (10CFR20) is presented here.

Numerous areas of a LWR facility may require specific radiological sign postings at any given time. During outage periods or whenever a large amount of maintenance activity is ongoing within the radiologically controlled area (RCA) the number of radiological sign postings can increase significantly. These postings are competing for the attention of employees against a backdrop of a host of other signs and labels. Industrial safety signs, warning signs, general information signs, equipment and component tags and labels in addition to other miscellaneous postings may be prevalent. Additionally, when one considers signs designating the location of firefighting equipment, fire doors, safety and medical supplies, evacuation routes, exit signs and the like, than one can begin to visualize the cacophony of signs present. These signs and postings serve an important function with regards to protecting the safety of employees and many are required by codes or regulation. To ensure that individuals remain cognizant of the importance of various safety-related signs it is important to maintain a standardized program using signs that are easily recognizable and understood. A balance must be maintained between over use of signs and the need to inform workers that a hazard exists. The needless use of signs could lead to an environment that desensitizes individuals to

R. Prince, *Radiation Protection at Light Water Reactors*,
DOI: 10.1007/978-3-642-28388-8_5, © Springer-Verlag Berlin Heidelberg 2012

**Fig. 5.1** Three pocket
radiation caution sign with
examples of inserts (Courtesy
of G/O Corporation,
www.gocorp.com)

the point that safety-related signs may tend to blend into the background. Under
these circumstances an individual's failure to recognize the presence of a radiological safety posting (or any safety-related posting) could lead to serious safety
consequences.

Considering the above, radiological postings should be easily recognizable,
concise, and clearly specify the radiological hazard present within the posted area.
To convey the importance of radiological safety, associated signs should be neat in
appearance with the use of handwritten clarifying information minimized. Standardized, pre-printed, high quality radiological signs are available from various
suppliers. Many suppliers offer a wide range of inserts that can be utilized, as
radiological conditions require. Figure 5.1 displays a radiological posting utilizing
inserts to specify specific controls associated with entry into the posted area.
Certain standards should be established concerning the use of radiological signs
and posting of radiological areas to underscore the importance that these postings
play in ensuring the radiological safety of employees. Posting guidelines could
address such items as those noted below.

1. Radiological areas should be signposted as soon as possible upon evaluation of
   survey results. Any subsequent changes in the classification of a radiological
   area should be accomplished in a timely manner.
2. Radiological areas should not surround or enclose lower classified areas. For
   instance, a high radiation area should not enclose a radiation area and a contaminated area should not surround a clean area. Apart from the obvious
   impracticalities posed by such postings, entry and egress to such areas would be
   administratively complicated.

3. Radiological postings should be located at each entry point leading to a given radiological area.

4. The radiological area should be clearly demarcated, utilizing physical barriers and existing structures when possible. This is especially important when encountering high radiation areas that may be required to be maintained locked. If contamination is present in an enclosed room and the source is unknown or widespread, such that posting will require extensive use of ropes and related materials to adequately demarcate the contaminated area, then it may be preferable to post the entire room. If the room is infrequently accessed by personnel this may not adversely affect work activities or other tasks.

5. The use of stanchions, radiation rope, radiation tape and other radiological posting materials should be used in a prescribed manner. Posting practices should ensure that posted areas are clearly demarcated to preclude inadvertent access to an area. Attention should be given to such items as the placement of stanchions flush to adjacent walls, leaving no gaps in the posting, and securing of radiological rope on stanchions with no excess draped onto the floor. The use of radiological tape to clearly demarcate the extent of localized areas of contamination may be useful. These may appear to be minor items, but when taken collectively, convey a message to radiation workers, whether intended or unintended, concerning the importance of radiological sign postings.

6. When postings include specific radiological conditions (e.g., actual dose rates in the area) that are subject to change, it may be prudent to provide the survey date that the results were obtained and the posting updated on a routine basis. The administrative aspects of updating postings simply based on a time period should be considered. However, caution must be exercised to ensure that posted information is maintained current to ensure that a culture does not develop whereby workers become accustomed to disregarding posted survey data based on the belief that information is not current based on previous experience.

Radiological caution signs are characterized by the standard radiation symbol commonly referred to as the "tri-blade". The caution sign consists of the three-bladed radiation symbol that is purple (or magenta) or black in color on a yellow background. Purple is commonly used in the USA while black is commonly used in most other countries. The details concerning the design of the radiation symbol may be found in the Code of Federal Regulations, Chapter 10, Part 20 (10CFR20). The basic radiation warning symbol along with the standard dimensions of the tri-blade are displayed in Fig. 5.2.

## 5.2 Restricted and Controlled Areas

A "restricted area" is typically established at LWR facilities to essentially serve as a buffer area to radiological areas of the plant. This area may be envisioned as a buffer zone, access to which is controlled and limited to authorized personnel.

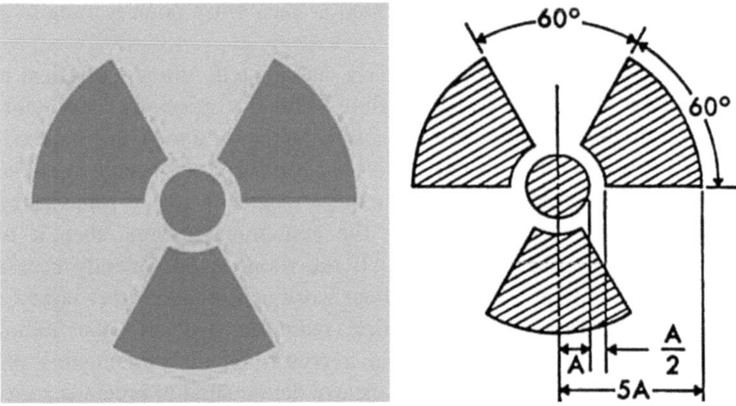


**Fig. 5.2** Basic radiation warning symbol. The symbol may be either *black* or *purple* in color on a *yellow background*

The Nuclear Regulatory Commission (10CFR20) defines a restricted area as an area that a licensee restricts access to for the purpose of protecting individuals against undue risks from exposure to radiation and radioactive materials. The International Atomic Energy Agency defines a "controlled area" in various documents such as the IAEA Safety Standards series. A controlled area is defined as an area where specific measures or safety provisions may be required to control or limit personnel exposures and to prevent the spread of contamination during working conditions. Essentially a controlled area and restricted area serve the same function. Entrances to restricted areas (or controlled areas) should be posted with a sign stating "Restricted Area" with the radiation-warning symbol displayed on the posting. Other appropriate information that may be posted at entrances to the restricted area may include such items as the need for personnel dosimetry, no eating, drinking or use of tobacco products and perhaps the category of the zone.

## 5.3  Radiological Controlled Area

The radiological control area, for purposes of this text, is that area of the plant that encompasses the various zone classifications described below. Areas within the controlled area posted as radiation areas, high and very high radiation areas, airborne radioactivity areas or contamination areas would be contained within an RCA. The primary RCA includes such areas as the containment building, auxiliary building, fuel building and perhaps radioactive waste handling and processing areas that are contiguous to plant buildings at a PWR. The primary RCA for A BWR will encompass the turbine building and perhaps the control building in addition to those plant areas noted for a PWR. Depending upon the plant and site layout including the number of units at a given site there may be more than one

RCA. Satellite RCA areas could include instrument calibration laboratories, radioactive waste handling and storage areas, and radiochemistry laboratories for example.

There are obvious advantages in minimizing the number of entry and exit points into the RCA. This may not be possible at those multi-unit sites where individual units are not physically connected. The need to establish radiological monitoring stations along with the associated staffing requirements should be taken into consideration when multiple entry and exit locations are necessary. A balance between providing ease of access to the RCA to support plant operations and the ability to maintain established radiological monitoring standards at each location must be achieve.

## 5.4  Radiation Areas

Again using the definitions provided in 10CFR20, a radiation area is defined as an area in which radiation levels could result in an individual receiving a dose equivalent in excess of 50.0 µSv (5 mrem) in 1 h at 30 cm from the radiation source. Radiation areas are required to be conspicuously posted with a sign or signs bearing the radiation-warning symbol and the words "Caution Radiation Area". Similarly a high radiation area is defined as an area in which radiation levels from sources external to the body could result in an individual receiving a dose equivalent in excess of 1 mSv (100 mrem) in 1 h at 30 cm from the radiation source. High radiation areas are required to be conspicuously posted with a sign or signs bearing the radiation-warming symbol and the words "Caution High Radiation Area" or "Danger High Radiation Area". Radiation area and high radiation area postings are displayed in Fig. 5.3.

A very high radiation area is defined as an area in which radiation sources external to the body could result in an individual receiving an absorbed dose in excess of 5 grays (500 rads) in 1 h at 1 meter from a radiation source. Very high radiation areas are required to be conspicuously posted with a sign or signs bearing the radiation-warning symbol and the words "Grave Danger, Very High Radiation Area". These areas must also include measures to ensure that an individual is not able to gain unauthorized or inadvertent access. Common measures, in practice, include entombing very high radiation areas behind shield walls of some sort, typically a fixed structure that requires elaborate measures for removal. Metal frames, bolted in-place, that secure concrete or lead shield blocks from removal would serve this function. Figure 5.4 depicts an example of a very high radiation area sign. Note the different shape often employed with very high radiation area signs to draw the attention of workers.

Another category of radiation area that is typically established at LWR facilities is referred to as a "locked high radiation area". Though not specifically defined in 10CFR20 a locked high radiation area is defined in plant technical specifications. A locked high radiation area is an area in which radiation levels from sources

**Fig. 5.3** Examples of
radiation area and high
radiation area warning signs.
The radiation symbol may be
either *black* or *purple* in color
on a *yellow background*

**Fig. 5.4** Example of a very
high radiation area sign
(Courtesy of Frham Safety
Products, Inc.,
www.frhamsafety.com)

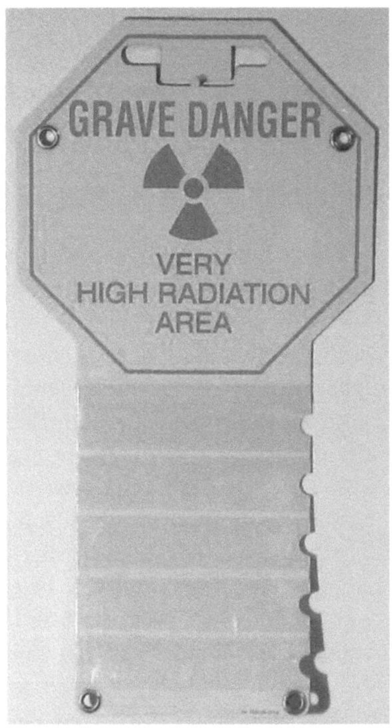


external to the body could result in an individual receiving a dose equivalent in
excess of 10 mSv (1,000 mrem) in 1 h at 30 cm from the radiation source. These
areas are posted with a sign or signs bearing the radiation-warning symbol and the
words "Caution (or Danger) Locked High Radiation Area". Sometimes the term
"Technical Specification" is substituted for "locked", specifically designating the
area as a locked high radiation area as defined by plant technical specifications. As
the name implies, in addition to the posting requirements, these areas are required

to be maintained locked. If it is not feasible or practical to lock an area exceeding 10 mSv/h (1,000 mrem/h) at 30 cm, then a flashing light may be provided at each entrance to the area or positive administrative controls established (e.g., RP coverage or remote video coverage) during entry to the area.

Various other regulatory controls are established for high radiation areas in addition to the posting requirements noted above. High radiation areas (i.e., 1 mSv/h at 30 cm) must be equipped with a control device that actuates a visible or audible alarm upon entry or the area may be maintained locked if the configuration allows. Doors that are maintained locked for high radiation area control purposes should be self-closing and shall have the capability to be opened from the inside so that no individual will be prevented from exiting the area. Additionally as noted above in the USA standard technical specifications for operating LWRs specifies various administrative controls for entry into high and locked-high radiation areas. These standard technical specifications require for high radiation areas, in addition to the 10CFR20 posting requirements, that each entryway be barricaded. Entries into high radiation areas must be governed by a radiation work permit or equivalent work control document. The RWP or appropriate work document should include or reference the anticipated radiation dose rates in the area and the necessary radiological protection equipment and measures required for entry.

The barricade requirement for high radiation areas may be satisfied by a physical posting. However, to meet the intent that each entryway be barricaded in accordance with plant technical specifications the barricade must be present at all times other than when workers are entering or exiting the area. To prevent the barricade from inadvertently being left down, after workers past through the posting, the use of swing gates are commonly used. The swing gates are designed to automatically swing back in the closed position after opening. Figure 5.5 depicts two examples of swing gates in common use. Both designs are mounted on a wheel frame to facilitate placement in the field. These type swing gates are often employed during outages when transient high radiation areas may be present in plant areas that are not surrounded by physical walls or capable of being enclosed. High radiation areas, due to the presence of components that have been dismantled for maintenance purposes that are located in open plant areas may be simply posted with radiation rope and stanchions with entryways consisting of a SOP or a length of radiation warning rope strung between adjacent stanchions. Under these circumstances a swing gate could be positioned at each entryway into the zoned area. Note that the swing gate designs also incorporate holders to place the high radiation area postings. Thus, when the swing gate returns to the closed position it satisfies the requirement that each entryway into the area be signposted.

A convenient way to control access to high radiation areas, assuming that the access configuration to the area is amenable, is by means of an access turnstile. The use of such a turnstile provides another administrative barrier to prevent unauthorized access to these areas. The turnstile shown in Fig. 5.6 is equipped with a dosimeter reader, a card scanner and a computer interface. Individuals allowed access to a specific high radiation area would be authorized entry under a governing document. The turnstile logic would be established to allow access to

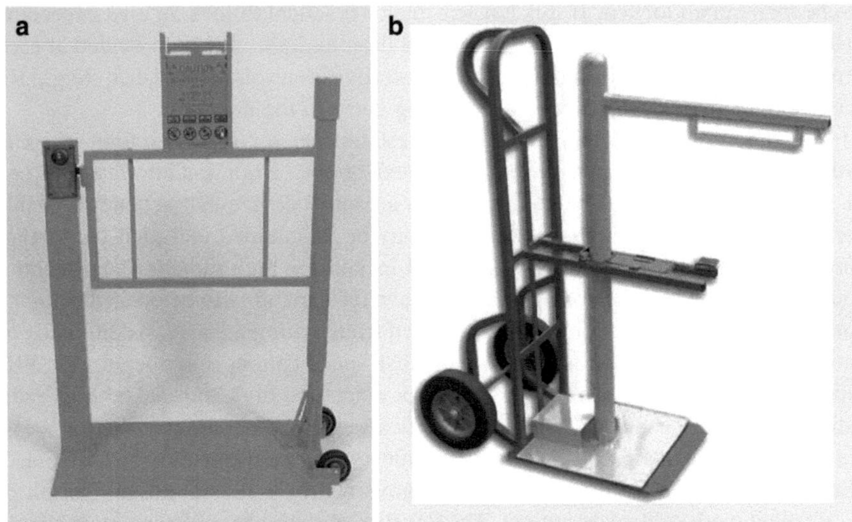

**Fig. 5.5** Two examples of swing gates used as barriers to high radiation area entrances. **a** Courtesy of G/O Corporation, www.gocorp.com. **b** Courtesy of Frham Safety Products, Inc., www.frhamsafety.com

**Fig. 5.6** The Mirion Technologies turnstile with advanced features that include a dosimeter reader, card scanner and computer interface (Courtesy of Mirion Technologies, www.mirion.com)

pre-authorized individuals entering under a recognized work document (e.g., RWP number or work package number). To access the area workers would scan their identification badge in the badge reader and enter the work package number. Upon confirmation of a valid work package number and that the individual is authorized to enter the area the green light on the turnstile would illuminate and the worker could pass through the turnstile. It must be emphasized that this is an administrative barrier to ensure that properly authorized workers are allowed access. A turnstile would not normally constitute a lockable barrier due to its physical arrangement that could allow a person to willfully duck under the turnstile and enter the area.

Radiological protection equipment and measures are also specified in plant technical specifications and require that each individual or group entering a high radiation area:

- Be provided a radiation monitoring device that continuously displays radiation dose rates in the area, or
- A radiation monitoring device that continuously integrates the radiation dose rates in the area and alarms when the dose alarm set point is reached, or
- A radiation monitoring device that continuously transmits dose rate and cumulative dose information to a remote monitoring location monitored by radiation protection personnel responsible for controlling worker radiation exposure within the area, or
- Be provided a self reading dosimeter.

The entry group is required to be under the surveillance of an individual qualified in radiation protection procedures who is equipped with a radiation survey meter. The RP is also responsible for controlling personnel exposures while within the high radiation area. Alternatively the entry group may be monitored remotely by closed circuit television. The individual providing the remote monitoring must be qualified in radiation protection procedures and be in direct communication with individuals in the high radiation area. Entries whereby an RP is not present require a pre-job brief that, at a minimum, should ensure that individuals are knowledgeable of dose rates in the area.

Likewise standard technical specifications require additional administrative controls for entry into locked high radiation areas equal to or greater than 10 mSv/h (1 rem/h). In addition to the plant technical specification controls noted above for entry into high radiation areas, locked high radiation areas must be maintained locked. If the area cannot be maintained locked then the area should be continuously guarded to prevent unauthorized entry.

Other control measures such as the use of electronic surveillance techniques (e.g., use of video monitors during periods of entry) or positive controls established during periods of entry could be established. Positive control measures could include stationing attendants outside the entry point or providing direct communication between RP and the entry team via headsets, radios, or other suitable means.

Based on the above it may be advantageous to incorporate an entire room or area into a given high radiation area. This facilitates locking the area and precludes the need to establish direct surveillance, or supply control devices or otherwise maintain administrative controls. This could lead to a situation whereby areas not meeting the criteria for that of a high radiation area are controlled as high radiation areas. However, if the affected area is not routinely entered the benefits of affording more effective entry controls to the high radiation area may outweigh any disadvantages caused by zoning a larger area.

Entry into high radiation areas and very high radiation areas could result in significant personnel exposures if activities are not properly planned and controlled. This topic is discussed in Chap. 7. To help prevent inadvertent entry into high and locked-high radiation areas or at a minimum to draw attention to these areas prior to entry, various human performance factors could be utilized with regards to sign postings. Brightly colored inserts may be used for the high radiation area wording for areas with multiple zone classification. The use of the universal "stop-sign" symbol for locked-high radiation areas is a common practice. These and other practices serve to distinguish high radiation area postings calling attention to the radiological conditions within the posted area.

## 5.5  Surface Contamination Areas

Unlike the other radiological zone classifications discussed in this chapter, no criteria are specified in 10CFR20 for purposes of designating surface contaminated areas. Consequently, a range of values used for designating contaminated areas may be encountered. The value of 17 Bq/100 cm$^2$ for beta-gamma emitters has gained wide spread acceptance at LWR facilities.[1] Keep in mind if disc smears have been used to evaluate surface contamination levels, actual contamination values may be a factor of ten higher than the established posting criteria (assuming a smear pick-up factor of 10%). The IAEA Safety Series publication 50-SG-DP, Design Aspects of Radiation Protection for Nuclear Power Plants, references values used in England that define contamination zones. The Central Electricity Generating Board (CEGB)[2] had established values based on the energy of the beta emitters present and provided upper and lower values to designate different contamination zones. These values are summarized in Table 5.1.

The obvious benefit of these values is that they are specified in terms of an activity level that provides consistency. Additionally, the establishment of more than one control zone allows for the designation of plant areas that are more highly contaminated. This provides a convenient mechanism to signify to workers that additional radiological controls may be required for entry into such areas.

In lieu of the alpha contamination values noted above in Table 5.1, common practice in the USA has been to use values in the range of 1–2 Bq/100 cm$^2$ range for designating contamination zones based on alpha contamination. Experience

---

[1] This value is based on the previous units of dpm per unit area and is approximately equivalent to 1,000 dpm/100 cm$^2$ or $4.5 \times 10^{-6}$ μCi/cm$^2$. Based on practical considerations this value may be rounded up to 20 Bq/100 cm$^2$.

[2] The CEGB is no longer in existence.

| **Table 5.1** Contamination Zone Limits | Radiation | Control zone 1 Bq/cm$^2$ | Control zone 2 Bq/cm$^2$ |
|---|---|---|---|
| | Alpha (high toxicity) | 0.37 | 3.7 |
| | Alpha (all others) | 3.7 | 37 |
| | Beta (max. energy $\geq$ 0.2 MeV) | 3.7 | 37 |
| | Beta (max. energy $\leq$ 0.2 MeV) | 37 | 370 |

has shown these values to be acceptable for posting of contaminated areas at LWR facilities. The limits are low enough to prevent the spread of significant amounts of surface contamination, provided that individuals follow good radiological work practices and monitoring procedures when exiting contaminated areas. These limits are not overly restrictive in the sense that posting areas at these values should not be overly burdensome. Depending upon the contamination history of a given facility, the number of personnel contamination events experienced, or to maintain more stringent contamination control measures, it may be prudent to demarcate contaminated areas at values lower than those noted above.

Each surface contaminated area should be conspicuously posted with a sign bearing the radiation caution symbol and the words "Caution Surface Contaminated Area" or simply "Contaminated Area". Since high levels of surface contamination may pose more significant radiological concerns, and entry into such areas may require the use of multiple sets of protective clothing, the use of respiratory protection devices or other contamination control measures, it may be beneficial to designate "high" contamination areas. It may be convenient to establish a graded approach when posting contaminated areas. Basically the purpose of such an approach is to allow designation of more highly contaminated areas from those of lower contamination. This practice should not be based on arbitrary values but rather should be utilized to clearly demarcate contaminated areas that require additional radiological safety controls based on the level of contamination present. A value of 1,000–2,000 Bq/100 cm$^2$ is commonly used to designate high contamination areas. The values listed in Table 5.1 for Control Zone 2 areas could also serve this function. Alternatively, surface contaminated area postings could specify the actual contamination levels, or range of levels, present in posted areas. If this approach is taken then mechanisms need to be established to ensure that posting information is maintained current. Figure 5.7 shows an example of a contaminated area posting.

The entrance to a surface contaminated area is usually demarcated by the use of a step-off pad (SOP) or similar method. The SOP is typically inscribed with directions that instruct individuals to remove protective clothing prior to exiting the area and before stepping onto the SOP. Figure 5.8 displays a standard step-off

**Fig. 5.7** Example of contamination area warning sign. Note the use of the term "danger". The term "caution" may be used in lieu of "danger"

**Fig. 5.8** A standard step-off pad. Note the requirement to remove protective clothing prior to stepping of the pad (Courtesy of G/O Corporation, www.gocorp.com)

pad. To minimize the spread of contamination to clean areas of the RCA, the preferred practice is to treat the SOP itself as being on the clean side of the contaminated area or otherwise maintain the SOP free of contamination. Since SOPs may often be located directly in front of doors leading from contaminated rooms, the SOP may protrude into general walkways and hallways where they are subject to foot traffic. If an individual were to inadvertently step onto a contaminated SOP then contamination could be tracked to clean areas of the plant.

To minimize the spread of contamination from highly contaminated areas to areas of lower contamination, two SOP's (or a double SOP) arrangement may be utilized. The worker would be required to remove an outer set of protective clothing, which may only include an outer set of gloves and shoe covers, prior to exiting the inner SOP area, before proceeding to the outer SOP. Normal contaminated area exit procedures would be used at the outer SOP area.

There are many conventions that may be used that are suitable for demarcating entrances to contaminated areas. In lieu of the SOP method described above, equally effective methods may employ the use of a physical barrier, such as a bench, located at the entrance to a surface contaminated area. The use of a bench may also facilitate the actual removal of shoe covers, providing a place whereby individuals can sit, making it easier to remove shoe covers versus the "balancing act" that may be required when using a simple SOP. Individuals' may sit on the bench keeping their feet on the contaminated side of the boundary. Once the shoe covers are removed the respective leg would be moved over the barrier, setting the foot down on the clean side of the bench. Procedures would have to be established to ensure that the seating area(s) are routinely monitored and maintained free of contamination.

## 5.6  Airborne Radioactivity Areas

Airborne radioactivity areas are posted to minimize potential internal exposures to workers. Annual limits on intake (ALI) are established for each specific radionuclide. The International Commission on Radiological Protection (ICRP) periodically issues radiological protection recommendations. These recommendations specify ALI values for the various radionuclides. Values for ALI's are specified as an amount (i.e., Bq) of activity. These intake limits, based on the current basis incorporated in 10CFR20, are established to ensure that individuals will not exceed either an annual committed effective dose equivalent of 50 mSv (5 rem) stochastic ALI; or the nonstochastic ALI to an organ or tissue that would result in a committed dose equivalent of 0.5 Sv (50 rem). Thus if an intake of radioactive material equal to one ALI were to occur the individual would receive either a whole-body annual dose equivalent of 50 mSv or an organ dose equivalent of 500 mSv. The ALI is the more limiting amount of a radionuclide that if taken into the body would result in either a whole body effective dose of 50 mSv (5 rem) or an organ dose equivalent of 500 mSv (50 rem) in a year.

The ALI is directly related to the established annual dose limit. To determine compliance with the dose limit a parameter that can be directly measured in the work environment would be beneficial. Obviously if one knew the airborne radioactivity concentration and the time of exposure then one could calculate the potential intake of a given radionuclide. Applying conservative assumptions with regards to annual occupational exposure times and breathing rates of workers then an "average" airborne activity concentration value could be determined. This "average" value would represent the allowable airborne radioactivity concentration that a worker could be continuously exposed to in an occupational year without exceeding the ALI. As long as the underlying assumptions apply then measurement of airborne radioactivity concentrations could be utilized to ensure that exposures do not exceed the applicable ALI.

Values for a reference person have been specified by regulatory agencies and the ICRP. It is assumed that a reference person works 2,000 h in a year (40-hour week for 50 weeks per year) and inhales air at a rate of 20 l/min, or 2,400 m$^3$ during the work year. These reference figures allow determination of the average airborne radioactivity concentration that could exist during a working year that would result in one ALI of exposure. This average airborne concentration is referred to as a derived air concentration (DAC) and measured in units of Bq/m3. The DAC values are derived by dividing the annual limit on intake (ALI) by the volume of air inhaled by a Reference Person in a working year.

$$DAC = \frac{ALI\,(Bq/yr)}{2,400\ m^3/yr}$$

The DAC value is then calculated for each radionuclide based on the ALI and the annual occupational exposure period. The DAC serves as a practical value since airborne radioactivity concentrations are measured in units of activity per unit volume (e.g., $Bq/m^3$). To obtain the DAC in μCi/ml the following relationship applies:

$$DAC = \frac{ALI\,(\mu Ci/yr)}{2.4\ \times\ 10^9\ cm^3}$$

The DAC values are derived limits intended to control chronic occupational exposures due to airborne radioactivity concentrations in the work environment. DAC values relate to either internal committed dose equivalent resulting from inhalation of radioactive materials or in the case of the noble gases (e.g., isotopes of Kr and Xe) external exposure due to immersion in a semi-infinite cloud of these gases. Again it must be emphasized that DAC and ALI values are based on the annual exposure limit of 20 mSv/yr other than for the case of the USA where 10CFR20 values are predicated on a 50 mSv/yr exposure limit.

An airborne radioactivity area is any area in which airborne radioactive material exists in concentrations in excess of the derived air concentrations specified in Appendix B to 10CFR20. Radionuclide-specific DAC values (based on ICRP 60) are listed in Table 1, Column 3 of Appendix B. Each airborne radioactivity area is posted with a sign bearing the radiation caution symbol and the words "Caution Airborne Radioactivity Area". Furthermore, if airborne concentration levels are below 1 DAC, but based on occupancy times in the area, an individual could exceed an exposure of 12 DAC-hours in a week, then the area is required to be posted as an Airborne Radioactivity Area. When evaluating the need to post such areas credit may not be taken for the use of respirators for posting purposes. These airborne posting guidelines are based on 10CFR20 requirements applicable in the USA (Fig. 5.9).

Airborne radioactivity concentrations are usually calculated for gross activity values. Measuring the gross beta-gamma activity on a given air sample filter for example does not necessarily provide sufficient information to determine if an area requires posting as an airborne radioactivity area. Specific radionuclide concentration values are required in order to determine if DAC values are exceeded. Various techniques and specialized analysis equipment (e.g., germanium detectors coupled with a multi-channel analyzer) are required to perform gamma spectroscopy analysis of air sample media. However during outages and other heavy workload periods it is often more practical to determine gross activity

**Fig. 5.9** Example of
airborne radioactivity area
warning sign

concentrations due to the large number of air samples that may require analysis. Gross activity determinations offer a convenient and relatively quick method to determine airborne concentration levels in lieu of extensive laboratory analyzes or the use of gamma spectroscopy systems. Under these circumstances it may be beneficial to determine an "effective" beta-gamma screening value that could be compared to gross radioactivity airborne concentration values. Gross airborne radioactivity levels below a screening value, based upon appropriate conservatisms, would be indicative that airborne radioactivity concentrations do not meet posting criteria.

The most restrictive ALI's, and hence DAC values, are associated with alpha-emitting radionuclides such as the long-lived transuranics. The most limiting beta-gamma emitter that could reasonably be expected to be present in airborne contamination at LWRs is Sr-90. Since Sr-90 is a fission product in order to pose a significant airborne concern fuel cladding defects would have to be present. The same situation obviously applies to any transuranics. Procedures could be implemented to eliminate consideration of these more restrictive radionuclides. If it could be shown that these radionuclides are not present, it would prove useful in reducing the use of respiratory protective equipment and the unnecessary establishment of airborne radioactivity areas, without jeopardizing the radiological safety of workers.

The potential presence of transuranics and Sr-90 may be estimated by primary system coolant radiochemistry analyzes obtained during the current operating cycle leading up to an outage. If these samples do not indicate the presence of these species or if primary system radiochemistry concentrations are below a specified value, that would not pose significant airborne concentrations when primary systems are opened for maintenance, it may be possible to preclude consideration of these radionuclides in airborne radioactivity. Alternatively, or perhaps in conjunction with primary system radiochemistry data, representative contamination smear samples could be obtained from contaminated areas and analyzed for the presence of Sr-90 and transuranics. These smears could be obtained from plant areas that based on history or experience have been shown to be leading indicators for the presence of fission product contamination. Such smear samples could be obtained before an outage or during the initial stages of an outage.

Appendix A summarizes the DAC values for selected radionuclides most likely to be encountered in airborne radioactivity at LWRs. Under certain circumstances other radionuclides may also be encountered in significant amounts and should be considered when necessary. It is often convenient to perform a preliminary screening analysis of air samples in the field. This may be necessary to support radiological posting of plant areas, maintenance activities, or to initiate interim radiological control measures prior to the availability of laboratory analysis results. The effectiveness of field screening of air samples may be greatly simplified if the likely constituents of airborne contamination are known or can be reasonably estimated based on source terms. The primary focus could be directed towards determining that transuranics (i.e., long-lived alpha emitters) and the more restrictive beta-gamma emitters are not present.

A review of the 10CFR20 DAC values indicates that the Sr-90 DAC value of $2.0E-9$ μCi/cm$^3$ is one of the more limiting DACs. If no significant fuel cladding failures have occurred and the presence of airborne transuranic radionulcides and Sr-90 can be neglected, then a screening value for beta-gamma emitters based on Co-60 could be utilized. The DAC value of $1E-8$ μCi/cm$^3$ for Co-60 is one of the more limiting values after Sr-90. Additionally, Co-60 may represent one of the major constituents of airborne radioactivity in many circumstances. A screening value of $1.0E-9$ μCi/cm$^3$ (approximately 40 Bq/m$^3$) may be appropriate. This activity concentration is 10% of the Co-60 DAC of $1E-8$ μCi/cm$^3$ and under many circumstances could represent a conservative value utilized for field screening of air samples or to establish airborne radioactivity postings. The use of a gross beta-gamma screening value may be particularly beneficial when evaluating the need for respiratory protection.

The other major radionulcides of interest may include Co-58, Cs-137, Mn-54, and Mn-56 among others. The DAC values for these radionulcides are greater than the screening value proposed here. Air samples exceeding this value could be subjected to laboratory isotopic analysis. The screening value of $1.0E-9$ μCi/cm$^3$ (40 Bq/m$^3$) is lower than the DAC values for all other primary beta-gamma emitters that could reasonably be expected to be present in airborne radioactivity under the conditions assumed here. Based on practical situations a radionuclide may be considered not present if the airborne concentration of the radionuclide is less than 10% of its DAC value.

The approach detailed above may assist in the decision to post an area as an airborne radioactivity area without the need to perform time consuming isotopic analyzes of air samples. It must be emphasized that the establishment and use of an effective beta-gamma screening value to facilitate field analysis of air samples or posting of airborne radioactivity areas must be based on appropriate consideration of the anticipated radionuclide mixture and operational parameters specific to the given situation. This approach is just one of many that could be considered. The proposed screening value is based on 10CFR20 DAC values and hence predicated on an annual exposure limit of 50 mSv. The establishment of a screening value based on the ICRP annual dose limit of 20 mSv/h, utilizing appropriate ALI values would result in a different screening value than that proposed here.

A similar approach may be taken with regard to long-lived alpha airborne concentrations. Assuming that a basis can be supported that ensures that airborne concentrations are less than 10% of the applicable DAC value then a screening value based on U-235 and U-238 DAC values may be considered. A screening value of $2.0E-12$ $\mu Ci/cm^3$ (approximately $0.07$ $Bq/m^3$) may serve the same purpose as that described above for beta-gamma emitters.

Engineering controls may be necessary to ensure that airborne radioactivity does not spread to other plant areas. Ventilation in affected rooms should be operable and a negative pressure differential maintained, relative to areas in the immediate vicinity of the airborne area. Tenting, glove boxes, or some type of enclosure may be erected to confine the source of the airborne contamination. Additionally portable ventilation units could be used to filter the local air and to control the spread of any airborne radioactivity that may be generated during the course of a task. Precautions must be taken to ensure that the presence of airborne radioactivity stays within the confines of the demarcated area. Though not necessarily a requirement it is good practice to maintain airborne radioactivity areas locked if the physical arrangement allows.


## 5.7 Radioactive Material Storage Areas

Areas in which radioactive material is stored or used may also need to be posted even if the presence of these materials does not result in radiological conditions necessitating posting under one of the criteria noted above. Wherever license radioactive material is used or stored in amounts exceeding 10 times the quantity of such material specified in Appendix C to 10CFR20, then the area shall be posted with a sign bearing the radiation caution symbol and the words "Caution Radioactive Materials".

Oftentimes in LWR environments the handling of radioactive material collected from job locations may result in the need to designate temporary radioactive material storage and handling areas. These temporary storage locations should be evaluated with regards to posting requirements associated with the presence of radioactive material. If specific areas are not designated as such then labeling and tagging of individual bags or containers of radioactive material may be necessary. Containers that are not maintained within a designated storage area that may be left unattended for periods of time, as a minimum, should be labeled or otherwise identified as radioactive material, even if the contents of such containers do not require posting based on radiation levels.

## 5.8 Hot Spots

Often times it may be advantageous to worn individuals of localized areas where radiation levels are significantly higher than ambient or general area radiation levels. If such localized areas are easily accessible or present in areas frequently accessed by personnel that may be prone to loitering, the identification of such areas serves to minimize personnel exposures. Such localized areas of high radiation levels are commonly referred to as "hot spots". There are no standard guidelines pertaining to the posting of hot spots. However, to be effective hot spots must be selectively posted, balancing the need to warn individuals of a radiological hazard, while not over-posting hot spots to an extent that workers become desensitized to the presence of these postings. For instance, posting a "hot spot" of 30 mSv/h (3 rem/h), as measured by a teletector-type instrument, located at a distance of 2 m into a pipe chase too small for an individual to enter, when general area radiation levels are 10 mSv/h (1 rem/h), may be of limited value. A rule of thumb for posting hot spots that balances radiological risks while minimizing over posting concerns would be to identify localized areas with contact dose rates exceeding general area dose rates by a factor of five when general area dose rates exceed 1 mSv/h (100 mrem/h). Note that hot spot measurements are usually obtained at contact to system components or the source of the radiation hot spot. A hot spot sticker is affixed to the area of concern or posted in the immediate vicinity of the hot spot in a location that is readily visible to personnel with perhaps a notation that describes the exact location of the hot spot (e.g., hot spot of 50 mSv/h one meter in the overhead on the letdown line). Hot spot stickers or postings usually have provisions for recording of the dose rate. The above rule-of thumb at least provides a minimum guideline with regards to posting of hot spots. If no guideline is established then RP personnel will have to decide on the need for hot spot postings on a case-by-case basis. This could result in a multitude of hot spot stickers present in various areas of the plant, some of which may provide limited radiological safety benefit.

## 5.9 General Posting Guidelines

Obviously in a LWR environment and due to the varying radiological conditions that may be encountered based on operating mode and during outage periods, more than one radiological zone classification may be associated with a given location. For instance a contaminated area may be located in an area containing components or in which radioactive materials are stored that result in radiation levels sufficient to require posting as a radiation area or a high radiation area. Perhaps the area also contains a contaminated system that has a valve leak, resulting in the need to post the area as an airborne radioactivity area. In fact many zoning combinations are possible, depending upon the radiological conditions present in a given area.

Various types and designs of posting materials are available from vendors. The important feature is to standardize on the posting design to be employed at a particular facility. Standard postings should be selected or established for the major radiological posting classifications described above and these postings displayed at each entrance to the respective zone. Additional Radiation Protection related information is often posted in conjunction with the major radiological posting classifications. This information could include such items as the need to obtain RP coverage prior to entry or to contact RP prior to entry. Surface contamination levels, general area dose rates, location of any leaking components, the location and magnitude of any localized hot spots and similar information may also be provided. To maintain these type postings in an orderly configuration and to facilitate the posting of such information a convenient posting method should be established. The use of clip-on signs displaying the radiation caution symbol, equipped with several insert pockets, has found widespread use in the LWR industry. Standard size, pre-printed inserts are available that fit into the pockets provided. These pre-printed inserts could be inscribed with such notations as "RWP required for entry", "Survey required prior to entry", the specific major radiological posting classifications or customized inserts to meet specific needs of a given facility. The use of pre-printed inserts and signs conveys a professional appearance of radiological postings and enhances the readability of postings on the part of employees.

The word "danger" may be substituted for "caution" when posting radiological areas. In addition it must be emphasized that individuals should not be able to access a posted area without encountering a posting depicting the radiological conditions in the area. Each entrance leading into a room or area meeting one of the area posting criteria should be posted. This is fairly straightforward for enclosed rooms with established doorways at entrance and exit points. However, for open areas or locations without physical walls the number of postings may be extensive. Under these conditions postings should be provided along each length of rope between two adjacent stanchions or as minimum, postings should be visible from each direction of approach allowing access to the area.

Radiological postings should include clear nomenclature and be neat in appearance and should not interfere with any other plant postings or signs. This is especially important when utilizing radiation rope and stanchions to demarcate a given area, which can easily become disorderly or moved out of position. Disorderly postings can have a detrimental effect on worker attitudes towards radiological safety. Radiation Protection personnel must demonstrate by their actions that the establishment and proper maintenance of radiological zoning and posting play a vital role in ensuring the radiological safety of employees.

## 5.10  Summary

Radiological sign postings serve a vital function to inform and warn workers of radiological conditions in plants areas. Radiological areas should be properly demarcated with postings that accurately reflect radiological conditions in the posted areas. Consistent format and terms should be used to convey the meaning of the various radiological postings that may be encountered by workers. Radiation worker training programs should ensure that the meaning and purpose of radiological sign postings and the definitions of radiological areas are addressed in these training programs. Radiation protection personnel should diligently inspect the condition and accuracy of radiological postings when conducting routine tours of the RCA. Radiological postings should be professional in appearance to signify the important safety role they serve.

## Bibliography


1. International Atomic Energy Agency, Safety Guide No. NS-G-1.13. Radiation Protection Aspects of Design for Nuclear Power Plants, Vienna, 2005
2. International Atomic Energy Agency, Safety Guide No. NS-G-2.7, Radiation Protection and Radioactive Waste Management in the Operation of Nuclear Power Plants, Vienna, 2002
3. United States Nuclear Regulatory Commission, Consolidated Guidance: 10CFR20 – Standards for Protection Against Radiation, NUREG-1736, Washington, D.C., 2001
4. United States Nuclear Regulatory Commission, Control of Access to High and Very High Radiation Areas in Nuclear Power Plants, Regulatory Guide 8.38, Washington, D.C., 1993

# Chapter 6
# Operational Radiation Protection

## 6.1 Overview

This chapter addresses the operational radiological surveillance aspects of a light water reactor radiation protection (RP) program. Operational aspects include those functions associated with daily surveillance and maintenance activities in support of plant operations. Those activities typically associated with operational radiation protection functions at the working level are presented. These activities include performance of radiological surveys, radiological signposting, demarcation of radiological areas, access control, preparation of radiation work permits and job coverage activities.

The primary objective of the RP program is to ensure the radiological safety of plant workers and the environment through effective implementation of the radiological safety program. Operational RP program activities should strive for excellence and continuous improvement, staying abreast of new innovations and techniques to promote improved radiological safety performance. The overall effectiveness of a RP program is measured by the successful implementation of sound radiological safety practices. The success of the radiological safety program is directly related to the quality and effectiveness of those activities performed on a daily basis that support operational program elements. There are many excellent operational radiation protection programs both in the USA and abroad. Health physicists should draw on this vast wealth of experience and knowledge when implementing industry standards of excellence. Routine benchmarking of recognized industry performance leaders is an important element in supporting continued improvements in radiological safety programs within the nuclear industry. Utilization of this experience will aid in the establishment and maintenance of efficient and effective operational RP programs.

A standardized, consistent approach, towards implementing various elements of the operational RP program will increase the effectiveness of radiological safety measures. This is particularly important for those utilities operating multi-unit stations or those who own and operate units at multiple site locations. Consider the difficulties faced by station workers when they encounter changing radiation protection requirements for a given task. Plant workers expect the RP section

R. Prince, *Radiation Protection at Light Water Reactors*,
DOI: 10.1007/978-3-642-28388-8_6, © Springer-Verlag Berlin Heidelberg 2012

(and rightly so) to protect them from the radiological hazards associated with the performance of their duties. Nothing will undermine the confidence that employees have in the station's RP program more quickly than the perceived inability on the part of RP personnel to implement consistent and effective radiological safety measures. As many station health physicists can testify, this attitude once established, is difficult to overcome. If protective measures are changed on the apparent whim of RP personnel, for no apparent reason, this confidence can be quickly eroded. Under these circumstances workers are more apt to consider radiological safety measures specified for a given activity to be merely guidelines or worse, may ignore requirements altogether. Consequently, changes in radiological safety requirements driven by individual preferences should be minimized and when necessary, the reasons clearly communicated to the job supervisor and workers. The establishment of standardized radiological safety measures for various activities may minimize these concerns. However, it is recognized that when experience or knowledge dictates the need, RP technicians must be afforded the opportunity to exercise their professional judgment, and given the authority to deviate from established guidelines to enhance the radiological safety of workers during the performance of radiological work activities. Clearly a balance has to be achieved in providing consistent radiological safety measures while at the same time allowing RP personnel to utilize their skills.

## 6.2  Radiological Surveillance

Primary functions of the RP group are to conduct radiological surveillance activities, to evaluate plant radiological conditions and to utilize this information to prescribe appropriate controls. These activities include the evaluation of area radiation levels, surface contamination levels and airborne radioactivity concentrations. The objective of the surveillance program is to adequately assess the radiological status of plant areas for both specific tasks and routine situations. The establishment of effective radiological protection measures is predicated on the availability of complete and accurate survey data and the correct interpretation of this data. Personnel must be trained in radiological survey techniques and the proper operation of associated monitoring instrumentation prior to conducting surveys independently. Radiological surveillance activities must be diligently performed to ensure the establishment of effective radiological protection measures. The principles and techniques associated with the performance of various types of radiological surveys are described below.

### 6.2.1  Radiation Surveys

Radiation surveys are performed to evaluate radiation levels in plant areas, to ensure that no unanticipated personnel exposures occur and to maintain worker exposures within established guidelines and limits. Surveys may also be performed to determine personnel dosimetry requirements. The primary objectives are to verify that plant areas are properly posted, to inform workers of radiation levels in a given area and to confirm radiation levels in work areas both prior to and during the performance of work activities. Radiation monitoring may include the use of both portable and fixed area radiation detection equipment. Radiation monitoring equipment is discussed in detail in Chap. 11.

Radiation surveys at LWRs are conducted on both a routine and non-routine basis. From a practical perspective the radiations of concern include gamma, beta and neutron. Obviously under various circumstances (e.g., fuel failures and breach of primary systems) the presence of alpha contamination can be the controlling factor pertaining to the establishment of radiological safety measures. However, since alpha contamination does not represent an external radiation exposure concern, the measurement and control of alpha radiation is discussed in the next section dealing with contamination surveys.

Survey frequencies are predicated on various parameters, the primary ones being the potential fluctuation of radiation levels in a given area based on plant conditions, location of the survey, and how often a given area is routinely accessed by plant personnel. The nature of the task such as inspection activities, routine tours and whether or not actual maintenance work is to be performed are other factors that influence the frequency and extent of routine surveys. Areas with stable radiation levels (based on experience or plant location) may be surveyed less frequently than areas where rapid fluctuations in radiation levels could be encountered due to plant operating conditions or work activities. When establishing routine survey frequencies, consideration should be given to exposures received by technicians to ensure ALARA principles are maintained. For instance, it may be appropriate to simply lock high radiation areas, infrequently accessed, and survey these areas only when necessary, prior to entry.

Before performance of a radiation survey, technicians should ensure that all necessary equipment and supplies are available and functioning properly. This includes appropriate protective clothing, dosimetry devices, and other protective equipment that may be required. Survey instruments should be confirmed to be in current calibration. A battery response check should be performed as well as an overall operability check. Survey meters should be source response checked to verify detector operability. This response check prior to use may simply be a quick check, not necessarily recorded, which verifies that the survey meter responds to radiation. The purpose for the survey (e.g., routine verification survey or specific pre-job survey) should be clearly understood to afford an opportunity for the RP technician to properly preplan activities to ensure that all necessary survey information is obtained and recorded. Thorough preparation will allow surveys to be performed efficiently, thus minimizing exposures attributable to survey activities. Personnel should be familiar with the plant location where the intended survey is to be performed and the impact that nearby equipment and components may have on the local radiation environment. Technicians should also be aware of the radiological status, actual and anticipated, normally associated with the area to be surveyed. This knowledge should be utilized to assist in the recognition of unanticipated or off-normal conditions in order to implement timely corrective actions to preclude unnecessary personnel radiation exposures.

After all prerequisite activities have been satisfied the technician may proceed to the survey location. Upon entering the area to be surveyed, monitoring equipment should be in the operating mode, preferably on the scale (or next highest) that corresponds to the range of anticipated exposure rates. This practice minimizes personnel exposures during the time required to adjust the instrument controls to obtain an on-scale reading. In the majority of cases, this will not be a significant factor; however, for those occasions when exposure rates are high, the time saved will contribute to dose savings over the long term. This practice also minimizes unnecessary exposures by preventing individuals from unknowingly lingering in areas with elevated radiation levels. Survey meters equipped with digital displays and auto-ranging features alleviate this problem. Figures 6.1 and 6.2 depict RP technicians performing radiation surveys in plant areas.

Routine surveys ensure that established radiological controls are appropriate for the area(s) surveyed. Routine surveys usually involve a complete check of the surveillance area to verify that area dose rates have not changed significantly, necessitating a change in sign posting or area classification (e.g., from a radiation area to high radiation area) since the previous survey. General area readings are obtained as necessary to adequately assess the radiation profile of the area or room. General area readings are obtained at those locations that would normally be occupied or transverse by an individual entering the area. General area readings should be obtained in the immediate vicinity of the work area. Normally these readings should not include survey points taken within 30 cm or less from components unless the physical restriction of the area necessitates personnel to routinely approach within these distances during the course of routine entries. General area readings are not necessarily the highest dose rate levels measured in an area, especially if they are localized and not indicative of actual radiation levels that may be encountered during routine egress. Contact readings are obtained on tanks, heat exchangers, piping and valves, and other components of concern. The location of any "hot spots" (see Chap. 5) should be identified and posted as appropriate.

Measurements should be made at those locations most likely to be occupied by personnel (e.g., control panels or gauge and meter locations). Since survey results are valid as long as conditions remain unchanged, notice should be taken of those factors that may influence radiation levels. Factors such as the operating status of nearby components, fluid level in storage or process tanks, the presence of radioactive material, the use of temporary shielding and similar type items may all influence survey results. Good survey practices should provide mechanisms to identify and notate these factors on radiological survey forms. While performing surveillance activities RP personnel should assess other conditions in the area such as lighting, general housekeeping conditions, material condition of components, presence of leaks and any other condition that may affect the health and safety of employees.

**Fig. 6.1** A radiation protection technician performing a radiation survey (Courtesy of Eastern Technologies Inc, www.orex.com)

If unusual or non-expectant high radiation levels are encountered during the performance of a routine survey then an in-depth investigation should be performed to ascertain the source or reason for the abnormal levels. Under these circumstances other individuals in the area should be informed of the unexpected radiation levels and if necessary instruct individuals to move to a low dose rate area or vacate the area as appropriate. The reasons for the existing conditions should be notated on survey documents and communicated to RP supervision. Numerous industry incidents, resulting in unnecessary personnel exposures, have been attributable to lack of follow-up or complacency towards investigating the cause of unanticipated radiation levels.

Established record keeping systems should allow workers to readily retrieve current and past survey results for review as necessary prior to entering work areas. Computerized radiological survey databases greatly facilitate the dissemination of this information. The ability for workers to retrieve information from personnel computers or centralized computer workstations to review and discuss radiological conditions prior to entering radiological areas is a convenient and effective means of communicating radiological survey data to plant personnel in a timely manner.

Pre-job radiation surveys are conducted in much the same manner as a routine survey with an added major element being to ensure that the correct work area is surveyed. Attention must be directed at ensuring that survey results accurately reflect the exposure rates in the work area. This is especially important if the job will result in significant personnel exposures and the pre-job survey data will be

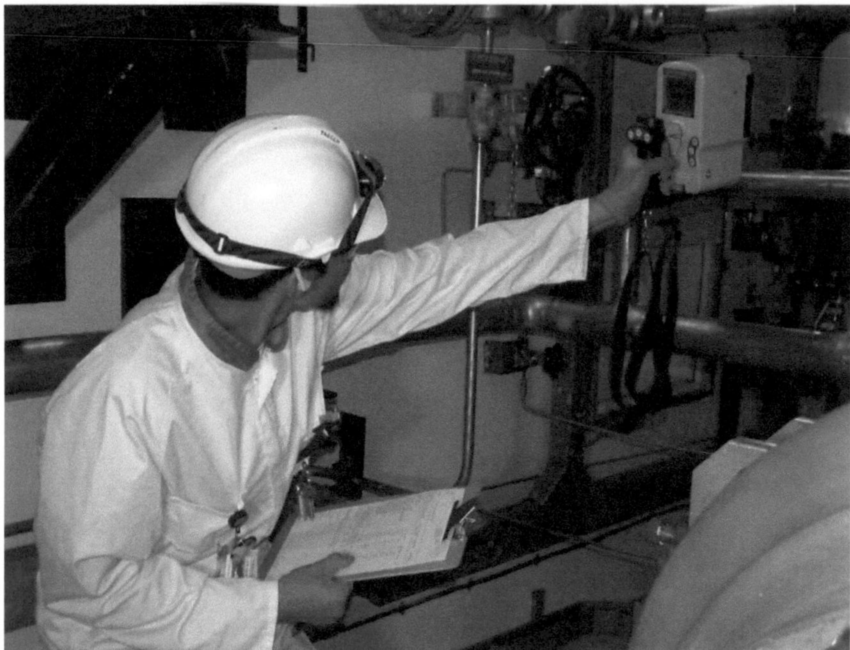

**Fig. 6.2** A radiation protection technician performing a radiation survey in the field (Courtesy of ESKOM and Koeberg Radiation Protection Department)

used to develop person-sievert exposure estimates prior to the job. If a graded ALARA planning approach is utilized, and the extent of pre-planning activities is determined by pre-job dose estimates, then it is essential to ensure that pre-job survey data is sufficient for these purposes. An incomplete understanding concerning the location and nature of the task could lead to surveys being performed on the wrong component or in the wrong location, or even the wrong unit at multiple unit facilities. Numerous industry events have been reported concerning unnecessary or unanticipated exposures to workers as a result of inaccurate pre-job surveys (e.g., incorrect component or wrong room surveyed or lack of understanding of the work to be performed). The RP technician must have a proper understanding of the purpose and reasons for the pre-job survey and be aware of the exact locations to be surveyed.

Surveys conducted for specific job coverage purposes are performed to verify that radiation levels are within acceptable limits or are as anticipated for the particular activity. The identification of localized areas with elevated radiation levels and detailed knowledge of the work area radiation environment provides valuable data in implementing ALARA measures and for use during the planning stages of a task. Radiation protection technicians must be cognizant of those activities to be performed during the course of a job that may influence radiation levels. Such items as the alteration of temporary shielding, the breach of systems, changes in storage tank water levels, or other eventualities which may arise during the course of a job, should be identified and provisions established to confirm radiation levels during critical stages of a given task.

Unscheduled surveys are usually more exacting when compared to requirements associated with routine surveys. Non-routine or special surveys may result from plant incidents, equipment breakdowns and malfunctions, spillages of radioactive material among others. Under these circumstances the primary objective is to evaluate radiological conditions in affected areas and to demarcate radiological boundaries as soon as possible, to prevent unnecessary personnel exposure and the inadvertent spread of contamination. Obviously since these surveys may result from an unscheduled event, the exact nature of the radiological conditions may not be fully known prior to surveying the area. Consequently, the initial survey team must take appropriate precautions to minimize exposures, the spread of contamination, and implement appropriate radiological safety measures related to entering areas of unknown or changing radiological conditions. Initial response actions must not only evaluate the extent of radiological conditions in the immediate area, but also in other areas, possibly remote, that may have been impacted by the incident. For example, a spill involving a large volume of high activity water, may require consideration of dose rates emanating from impacted floor drains and associated drain lines and waste collection tanks, floor sumps, or at other locations where spillage may collect.

Even though the above discussion pertained primarily to gamma radiation surveys there will be situations requiring a determination of dose rates due to the presence of beta radiation. Obviously routine area radiation surveys are primarily measuring the presence of gamma radiation emanating from closed system piping and components. Beta radiation dose rates should be obtained whenever workers are exposed to an open radioactive system or any time exposure to significant beta dose rates is suspected. Beta radiation dose rate surveys for work involving hands-on activities or that require workers to be in close proximity to open components should be performed to ensure that prescribed protective clothing, such as the type of work glove, provide the necessary protection.

Radiation survey results should be documented on a radiological survey form and may be maintained electronically to facilitate dissemination of survey data. Ideally a computerized data base depicting plant components and equipment and the physical layout of the area should be maintained. Survey data should be capable of downloading to a centralized data base to eliminate hand written survey data and to facilitate review of the data by workers. A standardized nomenclature should be developed to allow easy interpretation of survey data. The capability to distinguish and notate such parameters as the type of radiation recorded on the survey form, the distance at which measurements were made, and the units of measurement should be taken into account when recording survey data. An example of a survey form is depicted in Fig. 6.3. Note the explanatory footnotes included on the survey form. A standard convention facilitates the recording and interpretation of survey data. Workers should be trained in the meaning and purpose of any specialized notations and the proper interpretation of radiological information provided on survey forms.

| Survey No.: | RWP No.: | RX. Power:          % |
|---|---|---|
| Instrument | Instrument ID# | Instrument Cal. Date |
| Surveyed By | Survey Date | Survey Time |

**POSTING ABBREVIATIONS:** RA Radiation Area;   HRA High Radiation Area;  LAS Large Area Smear; OH Overhead; LHRA Locked High Radiation Area;  CA Contamination Area

**Remarks:**

| Smear # | dpm/100 cm$^2$ |
|---|---|
|  |  |
|  |  |
|  |  |
|  |  |
|  |  |
|  |  |
|  |  |
|  |  |
|  |  |
|  |  |
|  |  |
|  |  |
|  |  |
|  |  |
|  |  |

Circled numbers on map designate location of smears. Contamination levels are for beta-gamma. Numbers on map not circled indicate dose rate readings in μSv/h unless otherwise specified.

Survey Abbreviations:  CT = on contact; WB = represents whole-body dose rates; n = neutron dose rates; LAS = large area swipe; β = beta dose rates

Reviewed by:_______________________        Date:____________

**Fig. 6.3** Radiological survey form

The ability to download plant-specific photographs or area floor plans allows the development of radiological survey maps tailored to a specific plant location. The open area on the radiological survey form, where survey details are recorded, could be used to depict components in a specific room or plant area. Depending on the specific photo documentation capabilities available actual photographs could be down loaded to a survey from a master data base. Survey maps for RHR pump rooms, the spent fuel pool heat exchanger room, charging pumps, or a floor plan for a given elevation of the auxiliary building could be interposed within a designated location of the survey form. These capabilities provide an easier to interpret survey form, especially for members of the general work force.

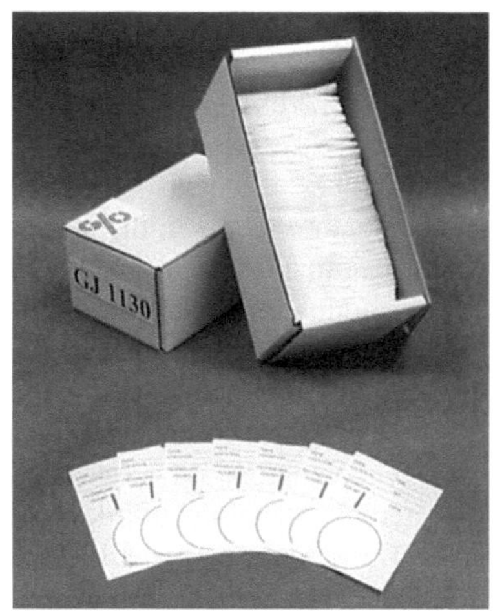

**Fig. 6.4** Individual disc smear envelope design offered by the G/O Corporation (Courtesy of www.gocorp.com)

## 6.2.2  Contamination Surveys

Contamination surveys performed in radiological areas are predominantly directed towards evaluating transferable contamination levels. The purpose of these surveys are to ensure that radioactive contamination is not present in unwanted areas, to confirm surface contamination levels are not excessive in designated plant areas and to verify that contamination has not spread to clean areas. Fixed contamination that is not readily transferable usually does not pose radiological safety concerns. Work control processes should ensure that adequate pre-job planning processes evaluate those work activities that could loosen or remove fixed contamination from floor or component surfaces. The potential for grinding, welding, sanding, abrasion type work, use of needle guns and similar activities to loosen fixed contamination, should be considered during the job planning and preparation phase. Contamination surveys include smear (or disc or wipe) tests to determine the amount of loose (or transferable) activity over a given surface area (e.g., 100 square centimeters). Direct measurements are obtained to evaluate the presence of fixed contamination. Thin window G-M and scintillation type instruments are often used for these purposes. Survey instrumentation types and models are discussed in further detail in Chap. 11.

Pre-survey preparations and precautions noted for radiation surveys such as ensuring equipment readiness and operability, availability of supplies, reason for and location of the survey and other preparation activities are equally applicable to contamination surveys. In addition, precautions may be necessary to guard against cross contamination of smears and equipment. Smears may be individually packaged to minimize the chance of cross-contamination and to facilitate handling of smears in the field. Figure 6.4 displays a common variety of disc smear used at LWRs. The smears come prepackaged in individual envelopes to minimize cross contamination concerns. Each smear envelope allows for the recording of the smear number, location, or other identifying information. Prior to entering the survey area, disc smears should be numbered or otherwise suitably notated, if practical, so survey results can be referenced to a given survey location. A sufficient number of smears should be obtained to ensure an adequate assessment of contamination levels in the area.

Contamination surveys are performed in those plant areas where the possibility of contamination exists or in areas with known contamination to verify the magnitude and extent of the contamination. When surveying for transferable contamination, clean disc smears are wiped over the surface of interest. Floors, walls, piping, plant components and equipment are surveyed when performing a routine contamination survey in a given area. It should be emphasized that transferable contamination levels obtained via smear techniques are qualitative at best. These type surveys allow an estimate to be made of the magnitude of loose contamination present. For many situations contamination surveys performed at LWR facilities, an estimate of the actual contamination level may be sufficient, in the majority of cases. If the purpose of the survey is to determine if an area should be posted as a contaminated area, or to confirm the results of decontamination efforts (i.e., post decontamination surveys), qualitative results may be sufficient. No significance should be attached to whether or not the contamination levels are 2,000 dpm/100 cm$^2$ or actually 2,500 dpm/100 cm$^2$ (approximately 40 Bq/ 100 cm$^2$). More importantly, RP technicians should evaluate survey results in their proper context and take appropriate action. For instance, if the value of 1,000 dpm/ 100 cm$^2$ (transferable) for beta-gamma contamination is used as the level for designating plant areas as contaminated, then survey results indicating 900 dpm/ 100 cm$^2$ should indicate to a technician that the area may indeed be contaminated, and a more comprehensive survey of the area should be performed or simply post the area as a contaminated area.[1]

On the other hand, when performing surveys at RCA boundary areas or when performing unconditional release surveys on items leaving the RCA, then more exacting survey techniques should be utilized. Oftentimes this may require the use of more than one type of survey instrument (e.g., beta-sensitive and gamma-sensitive meters capable of detecting low levels of contamination). Under these circumstances contamination survey results should be quantifiable and results subject to rigorous review and appropriate calibration factors utilized to ensure adequate quantification of results based on the importance of these type surveys. A laboratory counting system with a known background level may be required when analyzing smears obtained for these purposes.

---

[1] It is not uncommon to find contamination survey results reported to three or four significant figures. Such results are of dubious value and health physicists should ensure undue significance is not applied to data unnecessarily. Based upon counting system efficiencies, smear survey area and associated parameters, smear results may be "technically "accurate. However, it should be realized that the actual survey area may have transferrable contamination levels that vary significantly from the reported results.

Semi-quantitative survey results may be obtained if appropriate procedures and precautions are followed. A standard survey area may be established over which a smear is to be taken. This survey area should be directly (or at least easily) related to the applicable contamination control values utilized at the facility to simplify data manipulations. A common type smear material should be utilized which can be directly analyzed by available counting equipment without modifications. For example, if the detector chamber is sized for 5 cm diameter smears, this should be the size of the smears used for surveys. Any other size will introduce geometry errors, especially if the smear has to be reduced in size prior to analysis. Smears made of different materials may have different "pick-up" factors depending upon the type of surface (e.g., wet, dry, rough, painted, etc.) surveyed. Detectors should be standardized and calibrated for the energy of the radiations comprising the contamination. The extent that these and other parameters are addressed will determine the final accuracy of transferable contamination survey data.

When one considers the conditions under which contamination surveys are oftentimes performed (e.g., use of extensive protective clothing, respiratory protection, or working from platforms or ladders and perhaps dealing with heat stress issues) the difficulty of consistently obtaining smears over a pre-defined surface area becomes apparent. Individuals will exert varying degrees of pressure when obtaining a smear and the pressure used by any single person will differ from one smear to the next. These variables will influence the pick-up factor when performing smear surveys. If a high degree of accuracy is necessary or to ensure conservatism in survey results it may be appropriate to utilize a smear pick-up factor. A pick-up factor takes into consideration that a disc smear collects a certain amount of the transferable contamination actually present. The collection efficiency when performing loose contamination surveys, regardless of the filter media utilized, should be understood to be a value less than 100%. A pick-up factor of 10% is commonly used in lieu of specific data. In other words assume that measured smear results represent 10% of the true or actual transferrable contamination present. Obviously if data exists or tests have been performed to evaluate specific pick-up factors for smears, then an appropriate pick-up factor may be utilized.

A convenient and efficient method to monitor for the presence of loose surface contamination, especially in normally clean areas of a plant, is by means of a treated oil cloth which fits onto specially designed mop heads (e.g., masslinn surveys). This technique for large area contamination surveys has gained widespread acceptance within the industry. These treated cloths have good pick-up characteristics and allow for large surface areas to be surveyed in a minimum amount of time. Typically a large area such as a general walkway is surveyed with the treated cloth. The cloth is then monitored for contamination by means of a count rate survey instrument (e.g., a hand-held frisker). If contamination detected is greater than administrative limits (e.g., 100 net cpm above background) then a comprehensive smear survey may be performed to locate and to decontaminate localized areas of contamination. The use of these type cloths reduces the number of smears and the time required to handle and analyze individual disc smears,

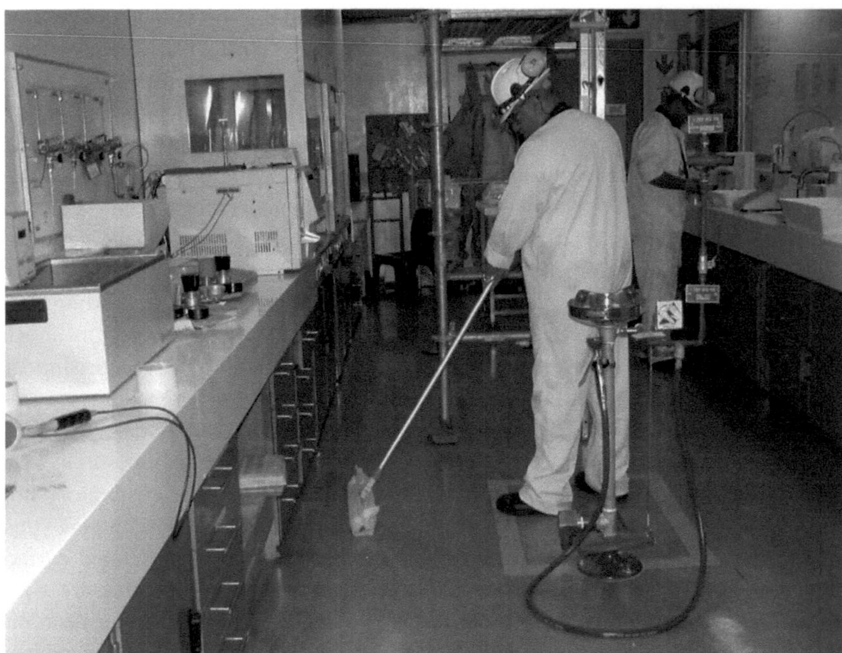

**Fig. 6.5** An RP technician performing a contamination survey using a masslinn mop to obtain a large area swipe (Courtesy of ESKOM and Koeberg Radiation Protection Department)

resulting in dose savings. Figure 6.5 depicts an RP technician utilizing a masslinn mop while performing a contamination survey.

When performing contamination surveys in clean areas, smears should be taken at those locations most likely suspect of being contaminated. This could include exit points from contaminated areas, in the vicinity of components or equipment containing contaminated fluids, floor drains and other suspect areas. Smears should be taken in general walkway areas and other locations routinely frequented by personnel. Radiation protection personnel should use their professional judgment and experience in selecting key or suspect locations when performing contamination surveys. When performing routine contamination surveys in established contamination zones, smears should be obtained immediately outside and adjacent to the zone to verify that contamination has not migrated to clean areas. Smears should be taken at the primary entrance/exit location (e.g., on the step-off-pad itself or the barrier or demarcation line separating the contaminated area from the clean area of the plant) to confirm that contamination is not present at these locations. Smears should be taken in sufficient numbers and locations to accurately ascertain the extent and magnitude of contamination levels in the survey area. If routine entries are required to be made in a given contaminated area, then smears should be obtained in the vicinity of the areas workers are likely to frequent. For instance, if operators enter an area to obtain meter or gauge readings or to manipulate equipment, then smears should be obtained on local control panels, valve handles, and floor areas in vicinity of accessible components to ensure that contamination levels are acceptable. Obviously, RP personnel should ensure that equipment is not inadvertently altered while performing contamination surveys, or any radiological survey. Under no circumstances should a survey be performed if the actual act of performing the survey could move a valve handle or alter readings on plant equipment or gauges. Attention should also be given to surveying door handles, plant telephones, handrails and other areas that may go unnoticed, but are subject to becoming cross-contaminated.

One of the major objectives of contamination surveys is to identify the source of the contamination. If the zone was established due to the presence of contaminated system leaks, then the contamination survey should demonstrate that the source has not increased in extent or magnitude. If leaks or standing water are encountered, then they should be surveyed to verify their contamination status. If wet smears are obtained then appropriate procedures should be followed to ensure that smear results are properly analyzed. If a smear is saturated with moisture it should be realized that analysis results might indicate a somewhat lower contamination level than what may actually be present, unless appropriate self-absorption or efficiency correction factors are applied.[2] A portable or external count rate meter should be utilized to analyze wet smears to preclude the possibility of cross-contaminating laboratory counting equipment, especially when equipment design may make decontamination difficult. Alternatively, wet smears may be dried prior to analysis to minimize self-absorption effects.

The presence of oil or hydraulic fluids is of special concern due to the tendency of these type fluids to retain particles that may be radioactively contaminated. Petroleum-based products (e.g., lubricants) have the tendency to be easily transferred to other locations if individuals inadvertently step onto an oil spot for example, and subsequently move about plant or room areas. When oil or hydraulic system leaks are encountered they should be surveyed, contained and cleaned as soon as possible, even if the leak is within a designated contaminated area. This practice helps to minimize the spread of contamination and will facilitate subsequent decontamination efforts, since the presence of these-type fluids typically have a detrimental impact on decontamination efforts. If significant amounts of oil or lubricant fluids enter liquid waste streams (e.g., building sumps or floor drains), processing time and costs associated with the handling and disposal of such wastes may increase significantly. Radioactively contaminated oil may also pose unique radioactive waste disposal concerns.

---

[2] Even though contamination survey results are often reported as gross beta-gamma activity, analysis equipment often used employs a GM or proportional type detector that has had its efficiency determined by use of a beta emitter calibration source. Under these circumstances and due to the low gamma sensitivity of these type detectors, the reported activity may be primarily due to the beta component of the contamination. Any moisture present will result in self-absorption of the beta particles to some degree. Consequently, if this affect is not accounted for, the measured activity may be lower than that actually present.

If ambient radiation levels allow, contamination surveys may be performed directly by means of a portable count rate meter. This method should be used whenever actual or expected contamination levels are high. This may prevent the need to handle highly contaminated smears, which may also pose significant cross-contamination concerns, as well as the potential to contaminate laboratories and analysis equipment. In fact, when considering the nature and purpose of most in-plant surveys, there is no significant advantage gained by analyzing highly contaminated smears in laboratory counting equipment. For practical reasons it is useful to establish an upper contamination limit on smears that are allowed to be analyzed by laboratory equipment. An upper limit of 170–800 Bq (10,000–50,000 cpm) as measured by a pancake type, thin window, G-M probe may prove sufficient for most purposes. Administrative controls or procedures should be established to ensure that smears are routinely monitored prior to being allowed into the radiological analysis laboratory to preclude the possibility of contaminating equipment or laboratory work areas. When using hand-held probes to measure contamination levels precautions should be taken to prevent contaminating the survey probe or puncturing the detector window.

Fixed contamination is of interest when work involves grinding, welding, machining or similar operations that may cause fix contamination to become dislodged. Under these circumstances fixed contamination could pose contamination control as well as airborne exposure concerns. Fixed contamination may be difficult to assess in plant locations where ambient general area radiation fields are present. Radiation levels of a few $\mu$Sv/h (or a few mrem/h) may be sufficient to render fixed contamination surveys meaningless. This is especially difficult when attempting to measure fixed contamination levels in the range of a couple hundred Bq/100 cm$^2$ (several thousand dpm/100 cm$^2$) in a radiation area for example. If fixed contamination levels are to be evaluated then the component or item may need to be relocated to a low background area or measures implemented in order to determine the amount of fixed contamination present. The need for accurate assessment of fixed contamination levels is usually limited to those situations that may produce loose contamination during the course of work activities (e.g., grinding and cutting). Samples of the material such as floor coatings or paint from the surface of a component may have to be collected and surveyed at a remote location. Occasionally fixed contamination levels may be encountered that produce measureable dose rates. Under these circumstances the amount of fixed contamination may pose significant airborne and contamination concerns depending upon the nature of the task. Appropriate radiological control measures would be necessary to minimize worker exposures and the spread of contamination.

Fixed beta–gamma contamination surveys are usually performed with a thin window G-M type survey instrument or scintillation detector. Alpha contamination surveys typically utilize scintillation detectors. In either case the detector should be held as near the surface being surveyed as possible (e.g., 10–15 mm for beta-gamma surveys and a few mm for alpha surveys) without coming in contact with the surface of the item. The probe should be moved at a rate slow enough to allow adequate response time of the survey meter. In the event that contamination is detected then survey results should define the area and boundary of the contamination if possible.

The most critical contamination surveys are typically encountered during the course of work activities at those stages which may influence contamination levels. It is essential that RP personnel review work packages to identify those steps critical to contamination control. Breach of contaminated systems with resulting spillage of residual water, grinding and machining operations, vigorous maintenance activities and handling containers of radioactive material or waste, all pose potential concerns pertaining to the spread of radioactive contamination. Key elements associated with planning and coordination of radiological work activities are discussed in detail in Chap. 7.

Though the above discussion was primarily directed towards the measurement of beta-gamma contamination under certain circumstances alpha contamination could be encountered in significant quantities. In order for alpha contamination to pose a concern fuel cladding damage would have to be experienced. Cladding damage would have to be relatively severe and the plant operated for an extended period of time with the damaged fuel. Normal operational controls employed in the LWR industry typically limits the amount of time a unit is allowed to operate in the event that significant fuel cladding damage is detected. Strict operational controls in conjunction with improved fuel fabrication and performance, together with established chemistry controls, greatly reduces the probability of experiencing fuel cladding damage during operating cycles. Notwithstanding, in the event that alpha contamination is present, the primary radiological safety concern would most likely be related to control of alpha airborne radioactivity concentrations. Due to the low DAC values for transuranic alpha-emitters, respiratory protective measures could be dictated by the presence of alpha contamination.

Contamination survey results are recorded in units of activity per area. A standard sized survey area should be utilized in order to compare survey results to established action levels. A convenient measurement value is the amount of contamination per 100 $cm^2$. The following equation may be used to determine contamination levels per unit area:

$$dpm/100 \ cm^2 = net \ cpm/[(EF)(AF)]$$

where:
EF = is the detection or counting efficiency of the survey meter or laboratory counting system expressed as a decimal (e.g., 30% efficiency would be 0.30), and;
AF = is the size of the area surveyed expressed in increments of 100 $cm^2$ (e.g., AF would be 1.0 for a survey area of 100 $cm^2$ and 3.0 for a 300 $cm^2$ area).

The efficiency factor (EF) for the counting instrument or survey probe for the specific type of contamination measured is utilized together with the survey area. If the survey area is something other than 100 $cm^2$ then appropriate area correction factors should also be applied. As noted earlier if quantitative results are required, a pick-up factor for the smear material may also be applied when evaluating the amount of transferable contamination present.

Contamination survey results should be documented on a radiological survey form and as previously noted maintained electronically to facilitate dissemination of survey data. Again a standardized nomenclature should be developed to allow easy interpretation of survey data. The location at which smears were obtained should be noted using an established format. This format could be a sequential numbering system depicting the location of each smear. If dose rate values are also recorded on the same survey form the smear locations may be enclosed in a circle or box to distinguish smear locations from dose rates. Contamination values should denote the types of contamination present; beta-gamma, beta or alpha. This could be accomplished by affixing the appropriate radiation symbol adjacent to the smear location on the summary table. Contamination survey results may be summarized in a table on a standard radiological summary form as depicted in Fig. 6.3.

### 6.2.3 Equipment and Unconditional Release Surveys

Materials and items removed from the RCA or radiologically restricted areas that will have no radiological controls imposed on their release require special attention. These unconditional release surveys or so-called "free-release" surveys require exacting survey techniques. This is the last line of defense in preventing contaminated items from being released from the RCA. Unconditional release surveys should be performed in a quantitative manner. The presence of fixed contamination must also be assessed when unconditionally releasing items from the RCA or controlled area. Plant procedures should specify guidelines associated with the performance of free-release surveys. These guidelines could include such aspects as those noted below.

- Release surveys performed with a standard frisker should be equipped with a probe with a known efficiency. Ideally the efficiency should be determined for the frisker-probe unit. Background limitations should be specified to ensure that an acceptable lower limit of detection may be achieved. For instance guidelines could require that background radiation levels be stable or no greater than a few Bq (or 100–300 cpm) above background.
- Items released from the RCA that were not in a contaminated area could be released based on direct surveys, only if the geometry or physical characteristics of the item do not preclude representative measurements based on frisking. Under these circumstances it is vital to confirm that the item was not potentially exposed to contamination. Additionally strict release limits (e.g., no activity detectable above background or a minimal level such as <1 or 2 Bq (or 100 cpm) above background should be applied.
- Items released from the RCA that were in a contaminated area or otherwise potentially exposed to contamination should be surveyed for both fixed and removable contamination prior to release.
- Materials or items with probable internal contamination (e.g., valves, pumps, process gauges and other items whose internal surfaces come in contact with contaminated fluids or gases) should be disassembled and internal areas surveyed prior to unconditional release. Items for which disassembly may prove impractical or may otherwise damage the item will require special attention. If station

procedures or governing regulations establish limits below which items may be unconditionally released then various survey methods may be used. Highly sensitive dose rate and contamination survey meters capable of measuring radiation and contamination levels to demonstrate compliance with applicable release limits may be acceptable. Items that can be placed inside radiation monitoring equipment, such as a small article monitor, with acceptable detection capability could be used to monitor such items for free release. Caution should be taken when applying these type methods to ensure that monitoring procedures are adequate to verify that acceptable detection limits are achieved.

- Liquid, porous-type materials, or materials capable of absorbing contamination should not be unconditionally released from the RCA until verified by survey or isotopic analysis to be free of contamination or below established release limits.

Methods associated with the monitoring of hand-carried personal items released from the RCA should be established. This area could be fraught with difficulties if radiation workers are not properly indoctrinated concerning the procedural requirements associated with this matter. Procedures should clearly define what constitutes a "personal" item and what controls are associated with the use of these items that are allowed to be surveyed for release from the RCA by the general workforce. It may be perfectly acceptable to allow workers to survey such items as pens, notebooks, work documents, and other personal hand-carried items that have not entered contaminated areas. This allowance may be predicated on the type of survey equipment available at the RCA exit. Automatic monitors with pre-established alarm thresholds that do not require individuals to interpret survey results are ideally suited for this purpose. On the other hand the use of portable survey equipment whose detection capabilities are dependent upon prescribed survey methods (e.g., rate of movement of survey meter and the distance from the object) greatly increase the risk of inadvertently releasing items that exceed unconditional release limits.

Notwithstanding, the degree to which workers are allowed to monitor items when exiting the RCA should be clearly defined and strict monitoring guidelines established. Worker responsibilities for monitoring items should be clearly understood and individuals trained in the proper use of monitoring procedures and equipment. These guidelines should address such actions as the need for qualified RP personnel to perform expanded surveys of personal items in the event that an individual alarms a PCM upon exit.

The use of small article monitors eliminates many of the problems associated with hand frisking items being released from the RCA. These user-friendly units employ high sensitivity detector designs capable of detecting low levels of contamination. Users simply initiate the measuring sequence after placing an item within the monitor. There are no other settings or adjustments required on the part of the user to activate these type monitors. After the measuring period an indicator informing the worker that the item is clean either via a message to that affect or a "green" status light is displayed. This informs the individual that the item may be retrieved from the unit and is acceptable for release.

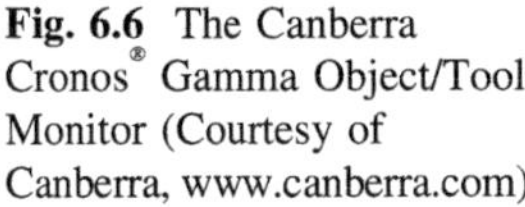

**Fig. 6.6** The Canberra Cronos® Gamma Object/Tool Monitor (Courtesy of Canberra, www.canberra.com)

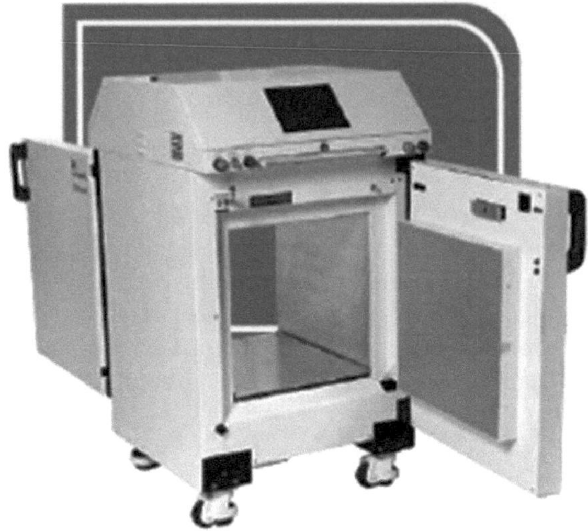

Units are commonly equipped with large plastic gamma-sensitive scintillation detectors. Models available to the industry are provided with various thicknesses of lead shielding to provide the necessary detection sensitivity. Counting conditions, including count duration and allowable background count rate, may be specified to ensure that the established alarm setting can be detected based on ambient background radiation levels. Units are commonly equipped with a "high background" alarm indication that warns when the alarm set point cannot be achieved due to current background conditions.

Two models for monitoring items leaving the RCA are depicted in Figs. 6.6 and 6.7. These units should be positioned in such a manner at the RCA exit location to prevent removal of an object by a potentially contaminated individual. The RCA exit flow path should be such that workers place items into the monitors then proceed to an exit PCM for monitoring. Once the individual clears the PCM they exit the RCA and proceed to the "clean" side of the tool monitor. Assuming that the items were cleared by the monitor then the individual may retrieve the item from the back door of the unit (i.e., the door on the non-RCA side). Essentially units equipped with a front and back door are positioned to straddle the RCA demarcation line.

Tool monitors equipped with only one door are typically positioned before the exit PCMs. Individuals exiting the RCA place items into the monitors and wait for the counting sequence to end and for the items to be cleared. The cleared items are then placed on a table or other suitable location. Individuals proceed to the PCMs for monitoring and then exit the RCA. The items previously cleared by the tool monitors may then be retrieved. The physical arrangement should be such that once individuals exit the PCMs they do not have to physically enter back into the RCA to retrieve the previously cleared items. Whatever method is used whether a drop-off location on a table or counter or if an RP technician is responsible for transferring the cleared items out of the RCA, administrative controls must be in place to ensure that

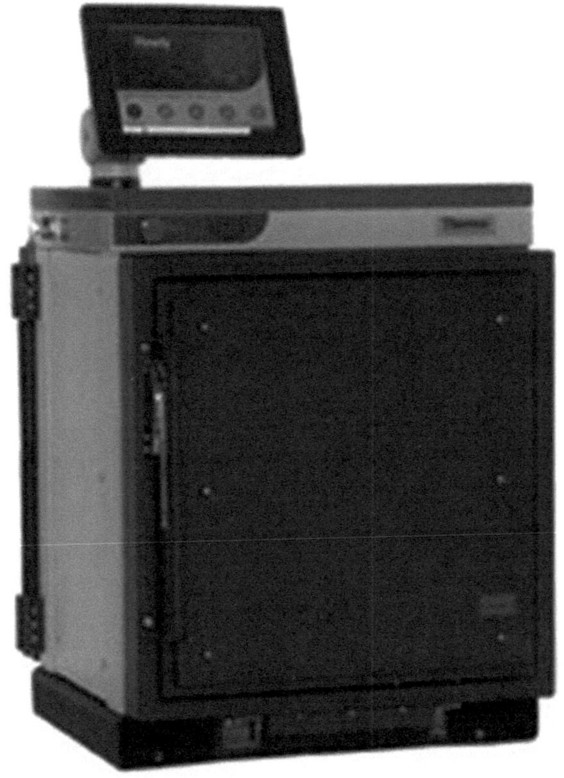

**Fig. 6.7** The Thermo Fisher Scientific Small Articles Monitor SAM-12 (Courtesy of Thermo Fisher Scientific, www.thermofisher.com)

inadvertent cross-contamination of the monitored items does not occur. Contamination detected on an individual by a PCM, should be reason to retrieve any item carried from the RCA by that individual for survey, even if previously cleared by a tool monitor. Since these items are removed from the tool monitor by workers who have not been monitored by a PCM there is a chance that a contaminated individual could transfer contamination to a clean item after it has been monitored.

### 6.2.4 Airborne Radioactivity Surveys

Control of airborne radioactivity is one of the more demanding tasks associated with operational RP programs. The effective control of airborne radioactivity is essential in minimizing impact on worker efficiency. Airborne radioactivity may result for various reasons and under certain circumstances can present exposure problems, the degree and magnitude of which depends upon several factors. Ineffective control of airborne radioactivity may result in increased personnel exposures and radioactive waste volumes, general costs associated with decontamination efforts, longer time required to affect repairs or perform tasks, and associated administrative burdens. The radionuclides of interest and their production modes were presented in Chap. 4. This section presents those elements associated with an effective airborne radioactivity-monitoring program.

Airborne radioactivity concentrations may consist of noble gases, and volatile radionuclides as well as particulate species. Consequently the determination of airborne radioactivity concentrations may require a combination of measurement techniques. Noble gases may be collected in a vial or small container for analysis. Radioactive isotopes of iodine require a specialized filter medium (e.g., impregnated charcoal cartridges) to obtain reasonable collection efficiency; while particulates may be collected on various types of filter media. These air sampling methods are discussed further in Chap. 11.

The presence of airborne contamination represents the primary pathway for internal personnel exposures at LWR facilities. The purpose of a comprehensive air sampling program is to evaluate airborne radioactivity concentrations that workers may be exposed to under varying conditions. The main objectives of an airborne radioactivity monitoring program include the assessment of the amount of radioactive material inhaled by workers during routine conditions and while performing work activities, to provide an indication of when airborne concentration levels exceed established administrative limits, and to evaluate concentration levels for specific tasks so that proper respiratory protective equipment may be prescribed.

Air sampling at LWRs may be performed to evaluate particulate, gaseous, iodine or tritium airborne concentration levels and usually consists of a combination of continuous air sampling and grab sampling. A grab sample is a short-duration air sample usually obtained at a critical juncture of a task. The sample collection time may be 5 minutes or even less under most circumstances. For this reason grab samples are usually obtained using a high volume air sampler to ensure the collection of a sufficient volume of air in order to meet minimum level of detection criteria. Grab sample results represent conditions at a specific time and location. Grab samples are obtained for such activities as initial breach of a contaminated system, upon commencement of welding or grinding, response to a spill of radioactive material, or as a result of a radiological incident. Continuous air samples on the other hand provide results over an extended period of time and essentially provide an integrated sample. Assuming that a portable pump with a fixed filter head is used to collect the continuous sample the results do not provide any information concerning changes in airborne radioactivity concentrations over the sampling period. The ability to provide continuous air monitoring with local read-out and alarm functions should also be available.

Airborne particulates primarily consist of radionuclides that become airborne as a result of some process that supplies the motive force that disperses contamination, allowing it to become airborne. Airborne particulates and suspended matter comprise a wide range of sizes and chemical states. They may be produced by mechanical processes such as grinding, cutting, welding, drilling and machining or by the movement of air currents over contaminated surfaces. Maintenance activities involving the breach of contaminated systems and the handling of contaminated components and ventilation flow patterns may all contribute to airborne radioactivity concentrations under varying conditions. Activation and fission product species typically comprise the majority of airborne particulates of concern at LWRs.

The radioactive gases most commonly encountered are radionuclides of xenon and krypton. These are inert gases that are chemically inactive. These gases are highly volatile and easily escape from enclosed systems if provided an opportunity, such as via leaking components or when opening, breaching or venting contaminated systems. The half-lives for these radionuclides range from seconds to days with the exception of krypton-85, which has a ten-year half-life (see Chap. 4). Tritium may also be present as a gas, which is readily converted to tritiated water vapor. Halogens are chemically active nonmetals that belong to Group 17 (formerly Group VII A) of the periodic table. The radionulcides of iodine and bromine fall into this category with the isotopes of iodine of primary concern.

In BWRs the vast majority of fission gases are transported from the primary system and carried over with the steam and subsequently processed and removed on a continuous basis. While in PWRs, these gases remain and accumulate to greater amounts within the primary system. The CVCS system does provide limited removal and clean up of potential airborne species; however, the inventory of potential airborne radionuclides in the primary system of a PWR at power in general exceeds that of a BWR.

Several factors determine the available inventory of radionuclides that could eventually become airborne. These factors include fission yield, dispersion and diffusion properties, half-life and solubility in the primary coolant. Other factors may include ventilation removal rate, gravitational settling properties and diffusion deposition on surfaces that could influence airborne concentration levels. Among the fission products of highest yield are the bromine-krypton-rubidium and the iodine-xenon-cesium chains. The noble gases krypton and xenon are most likely to become airborne. Since these gases decay to form particulate matter, rubidium and cesium radionuclides (e.g., Rb-88 and Cs-138) may also be airborne in the event of a system gas leak. Thus any leaks or maintenance activities involving the breach of systems affording no significant elapsed decay time, may present the opportunity for these particulates to become airborne. The airborne potential of corrosion products and the reminder of the fission products depend primarily on the airborne generation mechanism itself (e.g., welding, cutting and grinding). The sources of these airborne radionuclides are usually of greater concern during refueling and maintenance outages.

To obtain airborne radioactivity concentrations all that are required is a known sample volume and the measured activity associated with the sample volume. To obtain the sample volume you need to know the collection time and average sample flow rate over the collection period. A standard equation to calculate airborne activity concentrations is presented below.

$$\frac{(\text{Sample net cps})(1 \text{ Bq}/\text{cps})}{(\text{Detector efficiency})(\text{sample volume in m}^3)} = \text{Bq}/\text{m}^3$$

**Fig. 6.8** Particulate glass fiber filter paper used for sampling of airborne particulates (Courtesy of HI-Q Environmental Products Company www.hi-q.net)

The appropriate conversion factors may also be used to calculate airborne concentrations in units of $\mu Ci/cm^3$ as necessary.

The various air sample collection methods for the different airborne species encountered at a LWR are presented below.

### 6.2.4.1  Particulate Air Sampling

Sampling for radioactive particulates is commonly performed by filtration. This sampling method is relatively simple and consists of drawing air through a filter medium. Air is drawn through a filter at a known flow rate over a measured time period to obtain the total volume sampled. A wide selection of filter media is available with essentially 100% collection efficiencies. Glass fiber filter media is commonly used for particulate air sampling. Glass fiber filters can be used at high flow rates and are ideally suited for the collection of micron and sub-micron particulate matter. Filters should have known collection efficiencies for the particle sizes they are designed to collect. Figure 6.8 displays a typical glass fiber filter used for particulate air sampling.

**Fig. 6.9** Charcoal cartridge filter holder with a companion particulate filter holder (Courtesy of F&J Specialty Products, Inc.; www.fjspecialty.com)

### 6.2.4.2 Radio-Iodine Air Sampling

The presence of radioactive iodine, primarily I-131, requires specialized collection media. Iodine exists as a vapor and consequently standard filter media is not efficient for collection of airborne iodine species. The collection efficiency for airborne iodine can be greatly increased by utilizing an absorption medium specifically designed to collect iodine. The absorption medium facilitates the collection of the contaminant of interest by chemical bonding. To improve the overall collection efficiency of the absorption medium the physical design of the filter should have a large surface area. Activated charcoal is a common absorption medium incorporated into iodine collection filter cartridges. The activated charcoal's physical form is granulated to increase the surface area available for absorption for a given sized filter cartridge. The most common filter medium used for radio-iodine air sampling consists of activated carbon impregnated with triethylene di-amine (TEDA). The use of TEDA with activated carbon improves the overall retention of radio-iodine within the filter.

Significant airborne concentrations of radioiodine, if encountered, may only be present during the early stages of an outage. Under these conditions airborne concentrations of noble gases may also be present. A disadvantage of activated charcoal is that it also traps noble gases such as xenon and krypton. Under these conditions silver zeolite cartridges are used for radio-iodine air sampling to minimize the interference of xenon when analyzing the cartridge for radio-iodine concentrations. Silver zeolite minimizes the retention of xenon within the carbon collection medium. Figure 6.9 depicts a charcoal cartridge filter holder. Note that a particulate filter is positioned upstream of the charcoal holder. When iodine airborne radioactivity concentrations are a significant fraction of DAC values for the iodine isotopes or if the use of respiratory protection equipment is based on sample results; the particulate filter should be analyzed in conjunction with the charcoal cartridge. Even though the collection efficiency for radio-iodine of a glass fiber particulate filter is low some iodine will be collected on the filter. The vast majority of cases probably does not warrant the gamma spectroscopy analysis of

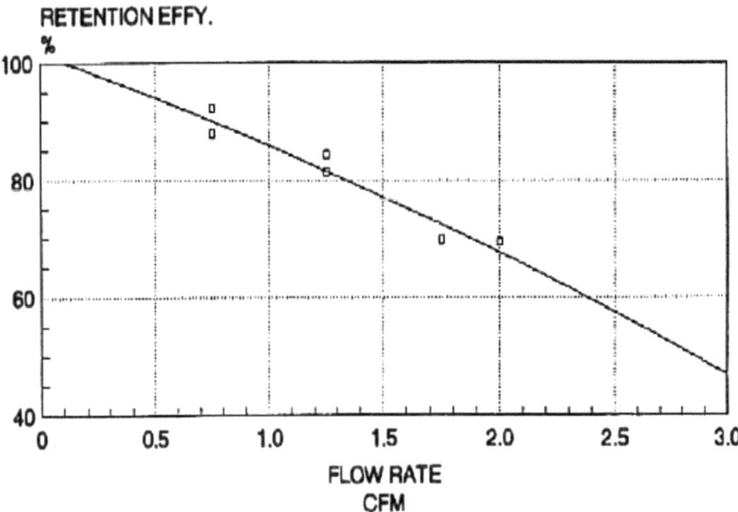



**Fig. 6.10** Collection efficiency versus flow rate values for a typical TEDA impregnated cartridge (Courtesy of F&J Specialty Products, Inc.; www.fjspecialty.com)

the particulate filter; however RP personnel should be aware that under certain circumstances the contribution on the particulate filter may have to be considered.

The collection efficiency for radio-iodine is a function of flow rate. The collection efficiency decreases as the sampling flow rate increases. Figure 6.10 displays the collection efficiency versus flow rate values for a typical TEDA impregnated cartridge. The changes in collection efficiency for flow rates below 0.5 cfm or approximately 14 l/min, may be neglected in most cases. However for flow rates >20–25 l/min corrections for retention efficiency are significant and should be taken into consideration. Since iodine air samples are usually analyzed by gamma spectroscopy, the charcoal cartridge collection efficiency values may be incorporated into the software algorithm used to calculate the activity concentration. Input data to calculate the iodine airborne concentrations requires the sample volume or the collection time and sample flow rate. Once the sample flow rate is entered as an input viable the data base could be utilized to automatically correct for the collection efficiency. The various suppliers of charcoal filters provide test data for each batch of filters. The test data, as a minimum, includes retention figures for iodine and the collection efficiency versus flow rate.

### 6.2.4.3 Noble Gas Air Sampling

Since noble gases are chemically inert, normal filtering mechanisms are not effective for the collection and analysis of noble gas airborne radioactivity concentrations. However, field sampling methods are relatively straight forward and oftentimes

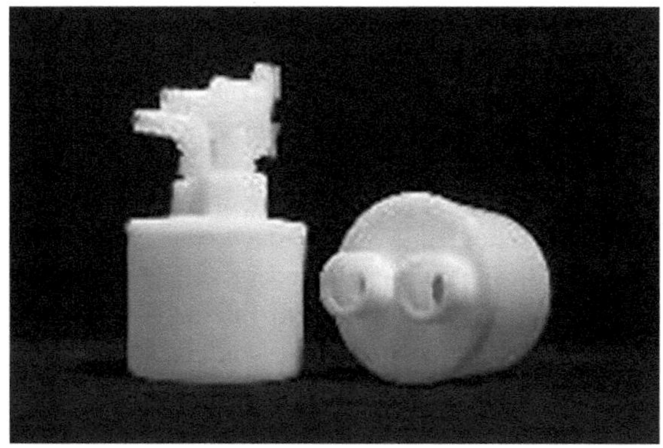

**Fig. 6.11** Typical container for collection of noble gas air samples (Courtesy of F&J Specialty Products, Inc.; www.fjspecialty.com)

do not require specialized sampling equipment. One simple method is to fill the sample container with demineralized water, proceed to the plant location to be sampled and simply empty the demineralized water from the container. The ambient air back-fills the sample container as the water is emptied. Another method involves the use of a low volume air pump to evacuate a sample container (see Fig. 6.11) to allow ambient air to flow through the container. A lapel air sampler is ideally suited for this function. Sampling time should be sufficient to allow the exchange of several air volumes equivalent to the volume of the sample container. Sample containers with volumes of 20–100 $cm^3$ are sufficient for most purposes.

### 6.2.4.4  Tritium Sampling

Tritium is the radioactive isotope of hydrogen. The radioactive decay of tritium results in the emission of a beta particle and a neutrino and the production of a helium atom. The decay process is depicted below.

$$^3H \rightarrow {}^3He + \beta^- + \nu$$

where $\nu$ represents the neutrino.

The sampling and evaluation of airborne tritium concentrations offers unique challenges. Chemical and nuclear characteristics of tritium contribute to these challenges. Tritium is a weak beta-emitter with maximum beta energy of 18 keV and average beta energy of less than 6 keV. The presence of any significant airborne tritium concentrations, under most circumstances, will be accompanied by other airborne radionuclides that emit higher energy beta particles and gamma rays.

Tritium primarily exists as a gas (HT) or as water vapor (HTO), often referred to as tritiated water and usually comprises the majority of tritium airborne activity. Tritiated water vapor poses the more significant exposure hazard. Exposure pathways include skin absorption of tritium water vapor and inhalation. The relatively high DAC value of 2E−5 µCi/ml for tritium makes tritium one of the more innocuous radionulcides. However, under certain circumstances airborne tritium concentrations could approach levels requiring monitoring for dose assessment purposes. The uptake of HTO vapor is essentially 100% for inhalation and ingestion exposure pathways. The presence of HT gas does not represent a significant exposure concern due to the low deposition in the lungs. Tritiated gas is readily exhaled by the lungs. Exposures to HT may be neglected in most circumstances due to the small amount of HT taken into the lungs coupled with the short residency time within the lungs. Tritiated water uptake into the body is assumed to be completely and instantaneously absorbed and rapidly mixed with body water. The biological half-life of $^3$H is about 10 days. Due to these biological characteristics a couple hours after uptake HTO will be evenly distributed throughout the body's' fluids.

Tritium is one of the few radionulcides whose biological half-life or retention time within the body can be changed relatively easily. The excretion rate of tritium can be increased by increasing the fluid intake into the body. The biological half-life can be decreased to 2–3 days by increasing liquid intake by 3–4 l/day (NCRP-report 65, 1980). The concentration of $^3$H in urine samples obtained shortly after exposure will be essentially the same as that in body water. These retention and elimination properties of $^3$H facilitate dose assessment resulting from tritium uptakes. Consequently, if the need arises to assess exposures due to airborne tritium, urine analysis should be initiated.

The presence of ambient background radiation fields in plant areas where airborne tritium concentrations are to be measured may severely impact the detection capability of portable tritium-in-air monitors. Additionally the presence of radon and noble gases may also influence the sensitivity of these units. To provide reasonable sensitivity portable tritium-in-air monitor designs include both gamma and radon compensation features. Gamma compensation is achieved by using multiple ionization detectors in a side-by-side or cruciform configuration requiring the use of two or four ionization chambers, respectively. This allows for reasonable detection capability in radiation fields of 100–200 µSv/h (10–20 mR/h). Radon compensation is typically achieved electronically by differentiating the radon signal from that of the tritium signal. Portable units are capable of measuring tritium airborne concentrations of less than 1% of the DAC. State-of-the-art handheld models provide a useful means for detecting elevated tritium airborne concentrations.

Various tritium sampling methods may be employed when an extended sampling period or continuous monitoring is required. Several techniques may be utilized to obtain samples for airborne tritium analysis. Sampling methods for the collection of tritiated water vapor include the use of "bubblers", a desiccant, or by condensation or freezing. These methods all involve the collection of a sample from a given location and require the samples to be analyzed by laboratory equipment. Therefore a period of time will be required that could be up to several hours long, depending upon the sampling and analysis method, before results are obtained. Collection of tritiated water vapor may be obtained by pumping the air to be sampled through a water filled container. The air flow will bubble up through the column of water in the container removing water vapor from the air introduced

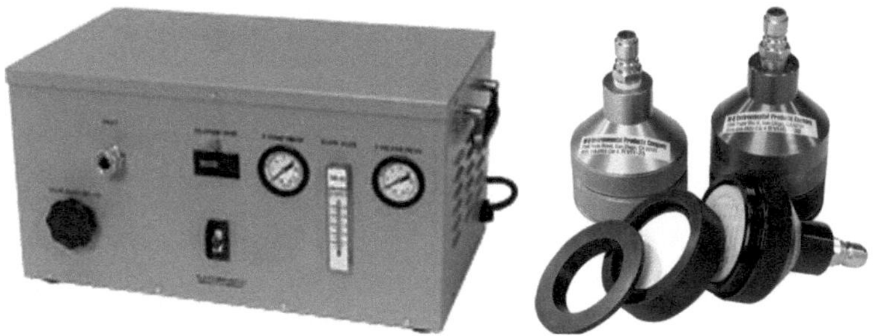

**Fig. 6.12** The HI-Q self contained continuous duty air sampling unit and filter holders (Courtesy of HI-Q Environmental Products Company www.hi-q.net)

into the sampling device. Samples collected for airborne tritium analysis utilizing these methods are analyzed for tritium content by means of liquid scintillation counting.

### 6.2.4.5 Air Sampling Guidelines

Plant areas generally accessible to personnel should be continuously monitored at strategic locations. This monitoring data serves to confirm actual airborne radioactivity levels present in various plant locations routinely frequented by workers. Additionally, this data could be used for trend analysis purposes and to establish baseline airborne concentration levels associated with specific plant locations. Component leaks or an unanticipated release of radioactive material as a result of work activities, or an operational transient, could be identified and corrective action taken in a timely manner to minimize personnel exposures or impact on plant operations. Areas that may require routine air sampling to verify airborne concentration levels include general work areas in the auxiliary building where operators and other personnel routinely tour or frequent on a regular basis. Plant areas that have a greater potential for the presence of airborne radioactivity may include pump rooms for those systems directly connected to the RCS, the fuel building or in the vicinity of the spent fuel pool, the hot sample room and radiochemistry laboratory, radioactive waste handling and processing, areas or rooms containing waste holdup or storage tanks, and decontamination facilities among others. A sufficient number of airborne monitoring locations should be established to provide an adequate profile of the airborne concentration levels within selected areas of the RCA. Figure 6.12 depicts a continuous duty air sampling unit offered by HI-Q Products that may be used at LWR facilities. These models are available with either a digital or analog display features. These units have sampling flow rates up to approximately 170 l/min depending upon the model. Various filter holders used in conjunction with particulate filters and charcoal cartridges are also depicted.

The potential for significant internal exposure resulting from the inhalation of airborne contaminants occurs during those maintenance activities conducted on primary system components or other highly contaminated systems that result in the generation of airborne contamination. In order to accurately assess internal exposures via inhalation, it is necessary to ensure that representative air samples are obtained for analysis. Consideration should be given to sample location, stage during the task at which the air sample is acquired, particle size and chemical properties of the airborne contaminant(s) and duration of sampling. Air sample results are of little value if the sample collected does not adequately represent the concentrations present in the work area.

A conservative approach is usually taken at LWR installations with regards to the minimization of internal exposures, resulting from the inhalation of airborne contamination. The establishment of strict access controls for entry into airborne radioactivity areas, the use of engineering controls, and prescription of respiratory protection equipment all serve to minimize internal exposures. Measurable internal uptakes resulting from exposure to airborne contamination should be a rare occurrence. Under normal or routine conditions airborne exposures should be minimal and the need to perform elaborate exposure assessments should be the exception. Consequently, the need to perform routine chemical analyses or the determination of particle size distributions of airborne contaminants is normally not warranted. In the event of a significant internal uptake (e.g., >10% of an ALI) or radiological event, special analyses may be necessary and appropriate to provide an adequate dose assessment. Under these circumstances particle size sampling, determination of solubility characteristics for the radionuclides involved, as well as other analyses may be necessary.

Airborne monitoring at LWRs for the most part is concerned with evaluating the amount of radioactive material that may have been inhaled by personnel while performing work activities. When WBC results indicate internal deposition, some attempt is usually made to review airborne sampling data, coupled with worker stay-times, to quantify the magnitude of any possible internal exposure. Airborne sampling can provide useful information if there is some assurance that survey results are meaningful and can be directly related to exposure periods. So-called personal air samplers are best suited for this purpose. These type air samplers are often referred to as "lapel" air samplers since they attach to the lapel area of the individuals' protective clothing. Lapel samplers are capable of being placed in close proximity to the breathing zone of the individual. Consequently, these type air samplers provide a more representative sample of the air breathed by individuals during the course of an activity. A primary drawback with lapel air samplers is the low flow rate achievable from these lightweight units. Flow rates are in the range of 1–5 l/min, and depending on the nature and length of the task, the volume of air sampled may not be sufficient to obtain the required sensitivity of detection for the radionuclides present in the air.

When work activities result in relatively long-duration exposure to airborne contamination, especially if alpha-emitting transuranics may be involved, breathing zone sampling may afford more accurate exposure estimates. The need for representative breathing zone air samples for workers under these conditions may be necessary not only to obtain more accurate exposure estimates but also to eliminate the need for overly prescriptive protective measures (such as the elimination of air hoods or SCBA units). When low DAC-value alpha emitters are present, oftentimes conservative control measures may be utilized based on general work area air samples or "worst case" scenarios. Under these circumstances additional radiological control measures may be employed based on the difficulty of measuring these type radionuclides or when live-time monitoring equipment does not provide the necessary detection capability in a timely manner. Alternate methods for determining worker exposures, after the fact, may not be cost effective in lieu of the establishment of additional field radiological control measures. For instance, the costs associated with the administrative aspects of issuing a large number of lapel air samplers to workers and analyzing the survey data, should be weighed against the costs to perform bioassays or more sensitive analyses of air samples. This comparison may be necessary when evaluating the most cost-effective approach when dealing with airborne alpha contamination while maintaining acceptable standards of radiological safety.

Many work activities at LWR facilities that pose airborne contamination concerns may involve strenuous tasks, the handling of heavy equipment, or may be performed in highly confined work areas. Under these conditions the use of personal air samplers must be closely monitored to guard against cross-contamination of air filters from sources other than airborne radioactivity. If lapel air samplers come into contact with contaminated components in the work area or are improperly handled by workers, monitoring results may not be indicative of actual breathing zone airborne concentrations. Additionally, if the activity involves the use of extensive protective clothing, teledosimetry, communication devices and perhaps industrial safety equipment, the ability to obtain accurate breathing zone samples may be impacted. Under these circumstances adherence with strict contamination control measures when wearing and removing personal air samplers is crucial. If the air sampler is inadvertently allowed to come into contact with a contaminated surface, resulting in the transfer of contamination onto the filter media, then exposure estimates may be unduly high, resulting in unwarranted corrective actions or the assignment of inaccurate exposures. Bear in mind that upon the completion of many tasks, the worker may be sweating profusely and physically tired, with the worker's primary focus being the removal of protective clothing and equipment, and to exit the work area as quickly as possible. Diligent RP coverage must be provided and proper contamination control techniques utilized by individuals during these stages to ensure the proper handling of lapel air samplers. For these and other reasons various precautions should be implemented to ensure that valid data is obtained when personal air samplers are utilized.

Based on the discussion above it becomes apparent that maintaining airborne radioactivity levels below concentrations requiring the need for respiratory protective equipment (RPE) or implementation of measures to maintain airborne contamination levels below those values requiring RPE is beneficial. The approach taken by the LWR industry, pertaining to the use of RPE has gone from a position of over-subscription to one of last resort. Improved radiological safety measures, diligent control and minimization of contamination and overall source reduction efforts play a crucial role in controlling airborne radioactivity levels and reducing the use of RPE.

Portable air samplers are oftentimes more practical and convenient than lapel air samplers when assessing airborne radioactivity concentrations in support of work activities. Major benefit of portable air samplers is the higher flow rates that may be obtained, and the resultant shorter collection period to obtain sufficient sample volumes for analysis. Elevated airborne radioactivity concentrations encountered in conjunction with maintenance activities are typically experienced for short-duration periods on the order of minutes and seldom exceeding 30 or more minutes. These conditions negate the need for the use of lapel air samplers for the majority of LWR applications. Portable, higher flow rate, air samplers are ideally suited for air sampling under these conditions.

Portable air samplers may be placed in close proximity to the breathing zone or in a location were the highest or most representative airborne concentrations are expected, based upon a given work activity. Portable air samplers should be kept off the floor and away from other secondary sources of contamination, to ensure that the filter media is not contaminated due to the collection of loose surface contamination, or other sources not representative of actual airborne concentrations. When evaluating results obtained from general work area air samples it should be realized that results might be lower than those in the actual breathing zone. When air sampling is performed to confirm the magnitude of airborne concentrations or the need for, or the type of respiratory protection device is not predicated on air sample results, then the use of portable air samplers are often the preferred method of air sampling. Figure 6.13 displays two air sample models that provide a convenient means of obtaining breathing zone air samples. These designs facilitate the placement of the air sample head in close proximity to workers. The units may be rolled into place during critical stages of a maintenance activity such as during the initial breach of a contaminated system.

Various correction factors should be taken into consideration if the choice of respiratory protection equipment is based upon the results of air sample data. Air sample results should reflect the peak airborne concentrations that workers are exposed to during work activities, in order to verify that proper protection was afforded to workers. It is essential that air samples be obtained during the critical stages of the task to ensure that peak airborne concentrations are measured. This is especially important when peak airborne concentrations are present for short periods. Consequently, the sampling duration should be chosen to cover those periods when peak concentrations are most likely to occur. Precautions must be taken to ensure that air samples obtained for a specific task are not "diluted" as a result of an unnecessarily long run period that may cover pre or post-job activities during which workers were not present. Continuous air monitors (CAM) may be utilized for measuring airborne radioactivity concentrations during the course of a given activity. Various model types are available that integrate and provide live-time indication of airborne concentrations. Use of these type monitors simplifies

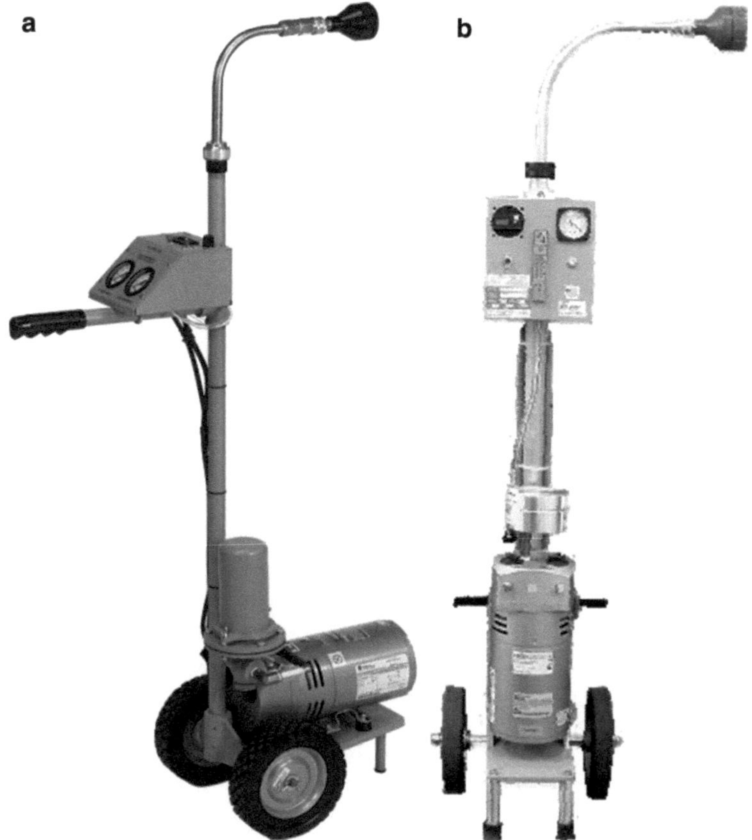


**Fig. 6.13** Portable air samplers with the "goose neck" design. **a** Courtesy of the HI-Q company www.hi-q.net. **b** Courtesy of F&J Specialty Products, Inc.; www.fjspecialty.com)

air sample data collection and analysis. Portable CAM units are discussed in Chap. 11.

As mentioned above air samples should be collected during those periods when the potential for generating peak airborne concentrations is most likely to occur. Such activities as breaching of contaminated systems, machining on contaminated components, decontamination tasks, work in highly contaminated or poorly ventilated areas should be sampled during those stages of the task that are most likely to generate airborne contamination. If the magnitude and duration over which the generation of airborne contamination is variable or unknown, it may be beneficial to obtain a parallel sample to cover the entire period that workers were in the area. If live-time monitoring is utilized this may not be necessary. This practice provides added insurance that individuals were not inadvertently exposed to unknown airborne concentrations. The availability of "negative" results could prove useful for dose assessment purposes.

## 6.3 Frequency of Surveys

Radiological surveys are conducted on both a routine and non-routine basis. In general, survey frequencies will depend upon the nature of the work activity or task, the magnitude and extent of the radiological hazards involved at the survey location, the existence of radiological control measures (e.g., enclosures, contamination barriers, or portable filtration equipment), procedural controls, and the use of protective clothing or equipment. Survey frequencies should be predicated on the establishment of effective radiological controls in order to minimize both personnel exposures and the spread of radioactive contamination. Survey frequencies should be selected with due consideration given to exposures received by monitoring personnel while performing these surveys. Frequencies should be based on the premise that the benefit gained by performing the survey outweighs the exposure received. For instance, high radiation areas not frequently accessed by personnel (e.g., waste hold-up tank rooms, various pump and heat exchanger rooms or the volume control tank room) may not need to be surveyed routinely. These type areas could be posted, with the requirement for a radiological survey to be performed prior to entry or obtain RP coverage for entry. Additionally, these type areas are ideally suited for the placement of teledosimeters or area radiation monitors equipped with remote read-out capability that could further reduce the frequency of entries into such areas. The need to routinely perform surveys in high radiation areas should be evaluated by RP supervision. If access is required into these areas on a regular basis, then the reasons should be reviewed and if possible procedures or processes revised to reduce the need for such entries. Ideally, steps should be taken to eliminate or reduce the physical size of the high radiation area. If entries are made to obtain readings from gauges or other monitoring type devices, the use of camera's, mirrors or remote video monitors or even the relocation of a gauge to a low dose rate area may be cost justified.

A routine survey program should be established to provide sufficient radiological data to support daily activities. The frequency of routine surveys may be based on historical data as well as anticipated and known radiological conditions. The program should be incorporated into procedures or other suitable means to ensure that required surveys are performed in a given area at the desired frequency. The program should be flexible enough to allow revisions to be made concerning the frequency and location of routine surveys based on plant conditions. Various factors should be taken into consideration when determining the frequency of routine surveys. Consideration should be given to such factors as those described below.

1. Plant areas where changing radiological conditions may exist due to plant operations or as a result of work activities should be surveyed on a frequency commensurate with the potential radiological hazard. The frequency for these type surveys may range from once per shift to daily or weekly.
2. Exit points from the radiological control area (RCA), which encompasses the major work locations (i.e., auxiliary building, fuel building, reactor building,

radwaste building and containment building), should be surveyed for contamination daily or perhaps even more frequently. During outages or high workload period's consideration should be given to performing these surveys on a more frequent basis, perhaps to coincide with break periods when large numbers of individuals are exiting the RCA. These survey frequencies could be altered depending on past experience. Obviously, these surveys serve an important purpose since they indirectly verify the effectiveness of in-plant contamination control measures. If contamination is detected at exit points, then corrective actions may be warranted, which could include increased survey frequencies within the RCA. The importance of ensuring that contamination is not tracked beyond RCA boundaries cannot be overstated. If contamination is inadvertently tracked to worker residences or other offsite locations the resulting problems associated with response measures, negative publicity and possible legal concerns could be time consuming and expensive.

3. Eating areas used by personnel who routinely work in the RCA should be surveyed routinely and perhaps more frequently during outages. Since the presence of contamination in eating areas is of special concern, due to the potential internal exposure pathway, these areas must be diligently surveyed.
4. General areas within the RCA that workers frequent on a regular basis should be surveyed routinely (e.g., multiple times per week). These surveys serve to confirm that radiological conditions have not deviated from the expected conditions and that personnel are not inadvertently exposed to unknown or changing radiological conditions.
5. Plant areas, entry to which is controlled by a standing or generic RWP, should be surveyed at least weekly to verify that the protective requirements specified by the RWP remain applicable considering the radiological conditions that exist. Survey frequencies in support of standing RWPs may also be predicated on past experience and changed accordingly.
6. Radioactive material storage areas should be surveyed on a weekly basis. These areas could include radioactive material and waste storage areas, contaminated component or equipment storage areas and radioactive source storage areas. If these areas are located in enclosed areas or access controls established for entry then a longer routine survey frequency may be more appropriate. On the other hand, during outages or other periods when large volumes of radioactive waste are generated, survey frequencies may need to be increased; since the opportunity for changes in area radiological conditions may have increased.

The above items should be considered guidelines, with the routine survey program established for a given facility, directed towards meeting the specific needs and requirements of the facility to support daily plant operations. The frequency of routine surveys may be determined primarily on past experience or based on specific radiological conditions stemming from plant design and operating history. Additionally, the effectiveness of radiological controls may be an important parameter in determining routine survey frequencies. For instance if PCMs, with sufficiently low detection capabilities, are available at strategic locations within the RCA, and rigorous contamination control measures established to control contamination at the source, then the chance of tracking contamination beyond RCA exit points will be reduced. Under these circumstances routine contamination surveys at RCA exit points and other locations (e.g., outside RCA areas) may be extended based on these considerations.

The routine survey program should also detail the process for recording survey data and any reviews required prior to dissemination of survey results. This process should ensure that adverse trends in radiological survey data are identified in a timely manner. Any trend depicting an increase in plant radiation or contamination levels should be evaluated and corrective actions initiated before changes in radiological conditions present a radiological hazard. The point to remember is that a given routine survey is performed for a reason. It may be a plant area that has experienced changing radiological conditions in the past, or an area subject to contamination, or an area routinely occupied by personnel for whom area exposure rates must be verified on a regular basis or for some other reason. If routines are simply performed for procedural reasons, and survey results not evaluated in a timely manner, then an opportunity to proactively identify potential radiological problems could be missed. Results must be reviewed in a timely manner and follow-up actions initiated as necessary. Certain conditions may require involvement of supervisory personnel and established reporting mechanisms should be incorporated into the program. Notification guidelines help to ensure that significant changes in radiological conditions are brought to the attention of plant personnel.

## 6.4 Communication of Survey Results

In order to inform workers of radiological conditions existing in specific areas of the plant survey results need to be communicated. The communication of survey results may take several forms, a combination of which is usually the most effective. Processes should allow for the timely transfer of survey data so that supervisors can preplan activities and that those persons who have a need to review survey data have ample opportunity to do so prior to performing tasks within RCA areas. Computer based radiological survey data management systems are a common and versatile method for widespread dissemination of information in a timely manner. Mechanisms should be established that allow workers to retrieve information from plant computer stations to allow work groups to review radiological survey data prior to starting tasks. The ability to take advantage of computer graphics greatly enhances the quality of survey forms and the ability to communicate results more effectively. Workers are provided data that is displayed in a neat, legible fashion, facilitating the interpretation of survey results versus handwritten data. Other advantages of a computer-based system are that data may be presented in various formats to meet specific needs and survey results viewed and retrieved by work groups, supervisors and others as needed. These systems eliminate much of the paperwork and administrative aspects associated with a paper based survey program.

If cost considerations are not an issue these systems offer numerous benefits. Data can be displayed directly onto large monitors at various locations. Actual plant or area displays may be used to afford more realistic depiction of radiological data. Survey data could be updated more frequently and efficiently in a shorter period of time. If these capabilities exist then the versatility of the system could be used for pre-job briefing sessions. Survey data could be overlaid on photographs of the area or equipment where work is to be performed and areas of radiological interest or concern clearly identified to individuals during the briefing. This would greatly enhance the effectiveness of pre-job briefings and help to prevent radiological incidents resulting from miscommunication of information during pre-job briefings or prior to entering the RCA.

If a computerized radiological data management system is not available then other methods of communicating survey data will have to be employed. Pre-printed survey forms and floor drawings or maps could be utilized to facilitate distribution of survey results. Under these conditions practices may include the posting of survey results at the RCA access control point or some method otherwise established to allow individuals to review data prior to entering the RCA. Multiple notebooks containing copies of current radiological survey data could be made available at the RCA access control point for review by workers prior to entering the RCA.

Other useful means of displaying survey results may include use of status boards. These status boards usually consist of a set of floor plans detailing the radiological classification of various areas of the plant. Floor plans may be displayed in the immediate vicinity of the RCA access control point. Plant radiological conditions are summarized on the appropriate floor plan or elevation plan. Results may be written on erasable coverings that allow the data to be updated as necessary. A color or number coded system may be employed that facilitates recording and depiction of the data on the status boards. For instance, unique colors could be used to represent radiation areas, high radiation areas, contaminated areas and airborne areas. Techniques to allow this information to be displayed visually are more versatile and more easily interpreted by individuals.

No matter what method is used to communicate radiological survey information to workers it is essential that survey results be legible and made available to individuals in a timely and effective fashion. To ensure the usefulness of the data, surveys and status boards must be periodically updated. This requires some vigilance on the part of the RP staff and is well worth the effort. If radiological conditions of plant areas are routinely made available to personnel, and they are appropriately trained to interpret and use this information in the performance of their daily activities, then radiological safety performance should be strengthened. Radiological survey results should be displayed and made available in a professional manner not only to ensure proper interpretation of survey information but to impress upon workers the importance of this information.

## 6.5  Personnel Contamination Surveys

A program to routinely monitor personnel for the presence of radioactive contamination should be established. This program should include provisions for monitoring while performing tasks within RCA areas and upon exit from the RCA. These provisions serve a vital role in evaluating the effectiveness of established contamination control measures and in preventing the spread of contamination to clean areas of the RCA or beyond the RCA. It is essential that workers be monitored for contamination and a process established that prevents contaminated individuals from inadvertently exiting the RCA, or worse leaving the site. These programs usually consist of a combination of friskers, hand-and-foot monitors and whole body contamination monitors. Personnel contamination monitors (PCMs), perhaps in conjunction with portal monitors and hand-and-foot monitors are usually located at RCA exit points to provide the necessary detection sensitivity. Frisker stations may be provided in various locations throughout the RCA to provide a means for monitoring gross-levels of personnel contamination not requiring a high degree of accuracy.

Common industry practice is to provide both beta and gamma sensitive personnel contamination monitors at RCA exit points. These highly sensitive PCMs are ideally suited to detect the radionuclides of interest. Gamma-sensitive PCMs provide a means to detect the presence of hard-to-detect, discrete radioactive particles with a significant gamma component. A typical beta-sensitive PCM is comprised of multiple detectors that essentially monitor the entire body, while gamma-sensitive monitors may have a detector configuration of a portal or doorway with detectors positioned to monitor the head, sides of the body and bottoms of feet. Figure 6.14 depicts two whole-body personnel contamination monitors available to the industry. These models are equipped with numerous detectors to provide a large monitoring surface area. The units may utilize either gas flow proportional detectors or solid scintillation detectors and are capable of monitoring for both alpha and beta contamination simultaneously. Both models count one side of an individuals' body at a time and require the person to turn around to repeat the count sequence for the other side of the body. Computer based electronics provide such features as automatic background subtraction, ability to retrieve contamination measurement data for a given count, self-diagnostic capabilities and various alarm functions, among others. Depending upon the ambient background radiation levels and the allowable value set for the release criteria count times are typically 10–20 s per side. To minimize the chances of contaminating these monitors, particularly the beta-sensitive PCMs, which may be difficult or time consuming to decontaminate, a frisker or hand-and-foot monitor may precede their use. The primary purpose would be to detect the presence of relatively high levels of contamination on either the hands or feet, which are usually the body areas most prone to becoming contaminated. If these screening devices detect contamination, the individual may be decontaminated prior to proceeding to a PCM for further monitoring.

Personnel contamination monitors located at RCA exit points must be sensitive enough to detect the activity level corresponding to established personnel

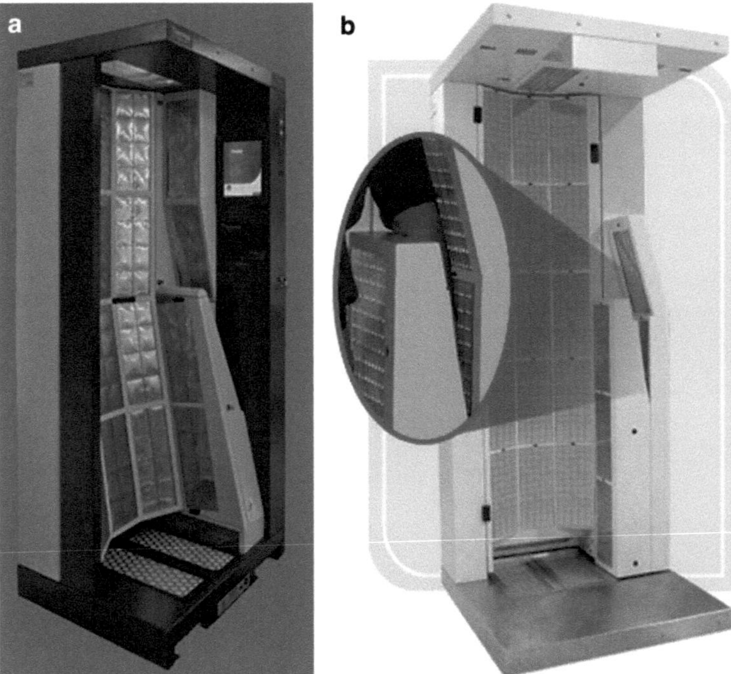

**Fig. 6.14** Examples of whole-body personnel contamination monitors. **a** is the iPCM12 model offered by Thermo Fisher Scientific (Courtesy of Thermo Fisher Scientific, www.thermofisher. com). **b** is the Argos™ model offered by Canberra (Courtesy of Canberra, www.canberra.com)

contamination release criteria. There are no specific regulatory limits pertaining to permissible personnel contamination levels. A common value for the upper limit on personnel contamination used in the USA is 5,000 dpm/100 cm$^2$ for beta-gamma contamination.[3] This value equates to approximately 83 Bq/100 cm$^2$. RCA exit point contamination monitors should have a mechanism that prevents personnel from leaving the RCA upon detection of contamination levels above the release criteria. The alarm signal could be interlocked with a door or turnstile preventing it from opening upon actuation of an alarm signal. Depending upon the configuration of the RCA access control facility, video camera displays of the

---

[3] Throughout this text the use of SI units have been referenced whenever possible. However, when dealing with contamination limits concerning the designation of contaminated areas, high contamination areas, and release values for equipmen and workers exiting the RCA the conventional units have been specified. These limits were based on "rounded" values that were convenient to use in-the-field and had gained widespread acceptance in the USA. Conversion of the various limits to equivalent SI values and using appropriate rounding methods provides some flexibility in determining the final conversion value. Therefore only approximate values are offered for the SI figures so not to imply that these values represent "approved" or "accepted" values by regulatory or standard setting organizations outside the USA.

**Fig. 6.15** Series of PCMs positioned at the primary RCA exit point. The RCA access control office is located immediately adjacent to the PCMs (Courtesy of Progress Energy Incorporated)

PCM locations or an alarm signal may be sufficient if RP personnel are typically stationed in the immediate area of the RCA exit PCMs. Due to the important function served by RCA exit point contamination monitors it is essential that they be maintained in good working condition at all times to ensure that personnel do not transfer contamination beyond the RCA boundary. Figure 6.15 depicts a series of PCMs located at the RCA exit point.

The whole-body contamination monitors described above are typically supplemented with gamma-sensitive portal monitors at the primary RCA exit point. Under certain circumstances personnel contamination may consist primarily of gamma-emitters. The most common situation is usually associated with the presence of discreet radioactive particles. The beta component associated with these very small particles could go undetected by the PCMs. The gamma component associated with these discreet particles though may be sufficient to be detected by a gamma-sensitive portal monitor. These monitors require the individual to step into the "portal" and pause for a few seconds. Portal monitor set points are established at some value above background to provide indication that radioactive material has been detected above background on an individual. A count time of several seconds should be sufficient to allow the detection of any significant gamma contamination. Figure 6.16 depicts two gamma-sensitive portal

**Fig. 6.16** Examples of portal, gamma-sensitive contamination monitors. **a** is the GEM$^{TM}$-5 model offered by Canberra (Courtesy of Canberra www.canberra.com). **b** is the PM-12 model offered by Thermo Fisher Scientific (Courtesy of Thermo Fisher Scientific, www.thermofisher.com)

monitors available to the industry. These models are equipped with large gamma-sensitive plastic scintillation detectors.

Personnel contamination involving discreet radioactive particles (sometimes referred to as "hot" particles) may result in relatively high exposures to small areas of the skin. These particles are typically less than one millimeter in diameter and often microscopic in size. Discreet particles have high specific activity and originate from activated corrosion products or from irradiated fuel. Activity levels associated with these particles range from a few hundred Bq to 30 or 40 KBq or higher (several thousand dpm to the μCi range). Due to their small physical size and high specific activity, skin contamination due to discreet particles can result in high doses to small areas of skin. Consequently, priority is placed on removing these particles as quickly as possible upon detection to prevent skin exposures in excess of regulatory exposure limits. Due to the small area of skin (<1 cm$^2$) over which any such exposure would be received, even exposures in excess of regulatory limits would probably not result in any serious health effects.

Discreet particle decontamination may be achieved by simply lifting the particle from the skin (or clothing) with a small piece of tape. Precautions should be taken to ensure that the particle is captured for analysis. This is especially important when dealing with high activity particles that may result in skin exposures approaching or exceeding regulatory limits. In many instances the particle may not be visible to the naked eye. If removal by tape is unsuccessful than rinsing with water or other techniques discussed in Sect. 6.6 may be utilized. If rinsing or washing techniques are required then arrangements should be made to collect the rinse water to retrieve the particle for analysis. To verify that the particle has been removed the affected area should be surveyed to confirm the particle is no longer present. To minimize the chance of losing the particle the piece of tape, or rinse water, or substance used to remove the particle should be transferred collectively to the sample container that will ultimately be analyzed. The analysis of discreet particles may be performed on a gamma spectroscopy system that has been calibrated for small-sized particle geometry.

If frisking booths or a combination of friskers and hand-and-foot monitors serve as the primary personnel contamination monitoring method at the RCA exit, then efforts must be directed at ensuring that these monitors are used correctly and that individuals are thoroughly surveyed. The human factor aspects of performing "whole-body" contamination surveys with friskers are well known. The need to maintain a slow rate of speed and close proximity of the detector probe to body surfaces being scanned, to ensure that established release limits can be detected, make the use of friskers problematic at best. Compound this with heavy workload periods when perhaps several hundred RCA entries and exits are made per shift, then the use of friskers poses a significant challenge in ensuring proper contamination survey techniques are meticulously followed. Based on these considerations friskers alone should not serve as the final screening device for personnel contamination monitoring purposes.

To perform an adequate whole-body contamination frisk a few key steps are taken into consideration when using a frisker. The frisker probe must be held reasonably close to the body surface (approximately 1 cm or less) and the rate of movement must take into consideration the response time characteristics of the detector. The detector is typically attached to a cable of sufficient length to allow the user to position the probe to monitor all body surfaces. The user must coordinate the movement of the probe while observing the indication on the rate meter or listening for an increase in the audible count rate to ensure that contaminated areas do not go undetected. Frisker units are susceptible to background radiation levels and if a high degree of sensitivity is required (e.g., when used at RCA exit points), a shielded frisking booth may be necessary. To perform a whole-body frisk using a typical 5 cm diameter pancake probe may require 2–4 min. Due to the human factor aspects associated with the use of friskers, when monitoring for the presence of contamination on individuals, good monitoring procedures must be strictly followed. Those facilities that rely solely upon friskers as the primary means for personnel contamination monitoring should ensure that protocols are in place to verify that individuals exiting the RCA follow proper personnel contamination monitoring techniques.

Contamination monitoring stations should be established throughout the RCA to allow individuals to monitor for the potential presence of contamination while performing duties within the RCA. The major aim is to provide a sufficient number of monitoring stations, at strategic locations, to minimize the chances of individuals spreading contamination to clean areas of the RCA. Friskers are ideally suited for this purpose due to their relatively low cost and ruggedness while affording adequate detection sensitivity. The alarm set point for in-plant friskers need not be the same as that established for the RCA exit monitors. The main purpose of in-plant personnel contamination monitoring stations is to provide early detection of gross contamination levels on workers to allow for timely decontamination. Ideally, friskers should be provided at exit points from every contaminated area. In reality this is not always practical due to ambient background radiation levels that may be present and the administrative aspects associated with the maintenance of such a large number of frisker stations. A suitable compromise is to provide friskers in the vicinity of contaminated areas frequently accessed or for the period of time that work is in progress, assuming that background radiation levels afford adequate detection capabilities. These frisker stations may be supplemented with other stations located in convenient, low-background, areas throughout the RCA.

Established practices should require individuals to use plant friskers upon exit from a contaminated area or to proceed to the nearest available frisker immediately upon exiting a contaminated area. Individuals should avail themselves of friskers provided within the RCA to monitor for the presence of contamination whenever contamination may be suspect. This could include such eventualities as when encountering water of unknown origin, retrieving tools or equipment staged adjacent to a posted contaminated area or when an individual transverses large areas of the RCA over an extended period of time. If an individual exits a contaminated area after performing a task and is subsequently found to be contaminated (e.g., bottom of shoes) and in the interim has proceeded to other areas of the RCA, then contamination could be needlessly spread to large areas of the RCA. Individuals should not travel throughout the RCA after exiting a contaminated area and then rely on RCA exit point contamination monitors to detect the presence of contamination. Under these circumstances, contamination may remain on a person longer than necessary, increasing the possibility of accidental ingestion in addition to the inadvertent spread of contamination.

Plant friskers should be set to the lowest scale or range available as background radiation levels allow. Often this corresponds to setting the range selector switch to the "X1" scale or similar position. The alarm setting should be set to actuate at a level two or three times above background. If the ambient background radiation level is a couple of Bq (approximately 100 cpm) then an alarm setting could be established at several Bq (200 or 300 cpm). Under most circumstances an alarm setting in this range should prove adequate for in-plant personnel contamination monitoring purposes.

**Fig. 6.17** Shield booths utilized to provide low-background frisking stations or shielding for a PCM (Courtesy of Nuclear Power Outfitters, www.nuclearpoweroutfitters.com)

Frisker stations should not be located in high background areas of the RCA. Background levels resulting in frisker readings of 10–15 Bq (several hundred counts per minute) or higher may not be suitable for personnel contamination monitoring purposes. If a frisker station must be located in a high background area then shielding of the frisker should be considered, especially if background levels are such that the selector range must be increased to a higher scale. Under these conditions a shielded frisking booth may be beneficial. A frisking booth consists of a shielded frame or enclosure in which the frisker is placed and large enough to accommodate a person. Shielding may consist of lead or metal sheeting or perhaps lead blankets attached to a frame. A three-sided enclosure may be sufficient for most situations. In some cases a simple "shadow shield" wall may be adequate depending upon the spatial source of the background radiation. Figure 6.17 displays two styles of shield booths that could be utilized for in-plant frisking stations. These shield booths provided by Nuclear Power Outfitters may also be sized to enclose a PCM to provide a lower background area.

A standard frisking procedure includes monitoring of the hands and feet and those body areas that are susceptible to contamination (e.g., the knees and elbows) or other body locations that are suspect. For instance, if an article of protective clothing was torn or otherwise noted to be defective during use, then attention should be given to the specific body location that may have been affected. Individuals must be trained in the proper frisking procedure and aware of the need to maintain the probe close to the surface being monitored and to move the probe at a rate sufficiently slow to allow proper detector response. The frisker probe should be accessible while in the resting position to allow an individual's hands to be monitored before picking up the probe. This practice minimizes the chances of inadvertently contaminating the detector probe. Upon detection of personnel contamination, individuals should be instructed to proceed to the nearest phone and contact RP for assistance. A good practice is to maintain a supply of disposable, or single-use, gloves and shoe covers at each frisker station for use in the event of personnel contamination. These items could be placed over the contaminated area (assuming that hands and feet are the most likely body locations that typically become contaminated) before leaving the frisker station, to prevent any further spread of contamination. Alternatively, if a coworker is present the coworker may notify RP while the contaminated individual remains at the monitoring station.

Due to the limitations noted above, regarding the use of friskers for detecting the presence of contamination on workers, it is highly beneficial to place whole-body personnel contamination monitors within the RCA when possible. It is common practice to place PCMs at convenient locations such as near the personnel exit point from the containment building or drywell (e.g., airlock). Often times the PCM location may require extensive shielding to afford acceptable detection sensitivity. However, these efforts may be worthwhile to afford a more reliable method of detecting the presence of personnel contamination in a timely manner. Other high workload areas may also benefit from the presence of PCMs within the RCA proper. Additionally, if a low background radiation area is available within the RCA, which is readily accessible to personnel or is in the vicinity of main traffic corridor(s), the use of PCMs in lieu of frisker stations should be considered.

## 6.6 Personnel Decontamination

When contamination is detected on an individual, whether it is at the RCA exit point or an in plant monitoring station, certain follow-up actions should be implemented. Obviously steps must be taken to decontaminate the person in a timely manner. Additionally, an investigation should be made into the cause and extent of the contamination to prevent recurrence and to ensure that the source of the contamination (if not known) has been identified and contained. Various techniques may be employed for the decontamination of personnel. A brief description of techniques commonly used is provided below. These or similar methods should prove sufficient when performing personnel decontamination procedures.

To ensure effective treatment during the decontamination process, various precautions should be taken. Decontamination should be performed in such a manner as to minimize the spread of contamination to other parts of the body and to prevent accidental ingestion. Ideally decontamination should proceed from the lowest to the highest contaminated areas. An exception to this would be if the contamination were of a magnitude that the limitation of skin exposure is the predominant concern. This could be the case when dealing with highly radioactive discrete particles. This situation may require decontamination efforts to focus on the removal of the discreet particle followed by decontamination of other areas. Decontamination of the eyes or mouth, other than simple flushing with water, should be performed under the direction of medical staff or a suitably qualified individual. Decontamination methods should not cause abrasion of intact skin surfaces. Intact skin often serves as an effective barrier that prevents the absorption of many radionuclides into the body. Consequently, it is vital to maintain the integrity of the skin during the decontamination process. Individuals with cuts or breaks in the skin surface, that could allow absorption of radioactive material into the body in the event of personnel contamination, should not be allowed entry into contaminated areas unless the damaged skin area has been properly bandaged or protected.

Most personnel contaminations encountered consist of localized skin contamination that may often be removed by rinsing with soap and water. Procedures specifying methods for decontaminating skin should start with mild methods unlikely to damage skin areas, progressing to more rigorous methods. Several skin decontamination methods are detailed below and presented in the preferred sequence of administration. As noted earlier, intact skin serves as an effective barrier against absorption of contamination and signs of skin damage should be observed when administering these procedures.

Simple flushing with water is effective for most general skin contaminations encountered at LWR facilities, especially if applied shortly after the contamination is sustained. If simple flushing is not effective then the contaminated area may next be washed with soap and water. The water temperature should be tepid to minimize the opening of skin pores that could decrease the effectiveness of the decontamination process. The wash period should not exceed a couple of minutes, again the aim being not to damage the underlying intact skin surfaces. If necessary, this procedure may be repeated a few times. Anti-bacterial soaps or hand cleansers used at hospitals or medical facilities may be more effective than ordinary hand soaps. Waterless hand cleansers (e.g., mechanics hand cleaner) may also prove effective for decontaminating local skin areas.

If the above steps prove ineffective then the contaminated area could be washed with heavy lather-type soap. A mild scrubbing action could be applied to the lathered area using a soft brush. This technique should be limited to 2 or 3 applications with each application lasting no more than a couple of minutes. If the contamination still persists a mixture of 50% laundry detergent and 50% cornmeal made into a paste and applied with water could be rubbed over the contaminated area. When applying these techniques care must be taken to prevent abrasion or otherwise damaging skin surfaces. It should be noted that several other equally acceptable decontamination methods employing similar types of paste or cream mixtures are available and could be used depending upon the particular circumstances.

An easy and effective technique involves the use of commercially available hand soaps or cleansers specifically made for decontamination purposes. Various complexing agents have been added to these cleansers to facilitate removal of radionuclides from skin. The application and use of these cleansers are performed in accordance with the manufacturers' recommendations and instructions supplied with the product.

The promotion of perspiration often proves effective in decontaminating troublesome spots. Oftentimes if contamination was a result of strenuous work activities or protective clothing became soaked with perspiration the contamination may be deeply imbedded in the pores of the skin. If the area involves the hands or feet the hand or foot could be placed within a rubber or latex glove or shoe cover and taped closed. For other areas (e.g., forearms or calves) a plastic wrap could be applied. When profuse perspiration is observed the area may be uncovered and rinsed with water.

When hair contamination is restricted to a small area, decontamination may involve nothing more than removing the contaminated pieces of hair. If large areas are contaminated then washing with soap and water may be required. Obviously if the contamination involves facial or head hair, then precautions must be taken to prevent the spread of contamination to the face or other parts of the body. After washing allow the area to dry before performing the final contamination check. The face, neck and any other body areas where contamination may have been spread should also be monitored. Washing may be repeated several times if necessary. If these methods prove ineffective, then consideration may be given to shaving the area. If contamination is still present after shaving then the skin decontamination methods described above may be employed.

Decontamination of the eyes, nose or mouth involving anything more than a simple wash or rinsing should be performed under the supervision of medical personnel or other suitably qualified individuals. Strict precautions must be followed to prevent any additional contamination from entering the body. Contamination restricted to the outer areas of the ear may be removed with soap and water utilizing a cotton-tipped swab. Precautions against water entering inner portions of the ear should be taken during the decontamination procedure. If contamination remains, or inner portions of the ear are contaminated, then medical personnel should assist with any further decontamination efforts. It should be stressed that flushing the ear is not a recommended practice since contamination could be easily transported to inner regions of the ear complicating decontamination.

Nose contamination may be effectively removed by having individuals blow their nose. This step may be repeated as long as contamination levels continue to decrease. If contamination remains, a cotton-tipped swab may be used. The swab should be dampened with water and inserted towards the back of the nostril. The swab should not be allowed to touch the sides of the nostril during insertion as a precaution against spreading contamination deeper into the nostril. The swab should be pressed gently against the inside of the nostril and withdrawn in a circular motion, wiping the inside surfaces upon withdrawal. This method may be repeated several times if necessary. Further decontamination measures should only be performed under the direction of medical personnel.

Facilities should be available for decontaminating individuals who may be contaminated over large areas of their bodies. These situations may be encountered in the event of an airborne contamination event or when individuals have worked for extended periods of time in airborne contamination areas. Showering with soap and water typically is adequate for most incidents of this nature. Caution must be taken to prevent contamination from entering body openings or skin breaks. Washing should start at the head and proceed down the body to the feet. Dry and monitor the individual for contamination, paying particular attention to folds in the skin and areas where contamination may be difficult to detect. Body areas that were previously determined to be free of contamination should be surveyed after showering to ensure that contamination has not been spread. Showering with soap and water may be repeated several times if necessary. If localized areas of contamination remain then the appropriate methods detailed above may be implemented.

Any contamination event involving medical issues or injuries should place primary emphasis on the medical aspects of the case. It is highly unlikely that a contamination event would be encountered at a LWR in which the contamination and exposure aspects of the situation pose more serious health concerns when compared to the medical aspects of the event. Minor cuts and abrasions not posing an immediate medical concern may be flushed with water. Medical personnel should perform or direct the decontamination of wounds requiring more rigorous decontamination techniques. If contamination is present adjacent to wounds or cuts after bandaging, then these areas may be decontaminated utilizing the methods described above.

Personnel contamination events may necessitate the need for whole-body counts or other bioassay techniques to either confirm that contamination is no longer present or to assist in evaluating the dose to the individual. Any positive indication of contamination involving the nose or mouth, or otherwise is indicative that ingestion may have occurred, should require a follow-up whole-body count. Obviously, whenever a contaminated wound is sustained it is usually prudent to perform a confirmatory whole body count.

Even if a specific limit has been established for personnel contamination levels at a given facility, every effort should be made to decontaminate individuals to non-detectable levels whenever possible. In the event that decontamination efforts are not successful, it may be necessary to allow a contaminated individual to leave the site. A policy should be established detailing the administrative controls to be employed in such an eventuality.

An administrative process should be established for recording and evaluating personnel contamination events. A reporting mechanism that captures and retains information necessary in performing a dose assessment and summarizing the radiological aspects of the event should be utilized. The summary report should provide a description of the event, the extent, location and magnitude of the contamination levels involved, the length of time that the contamination remained on the person, together with the individuals name and related personal information. These reports should be used for tracking and trending purposes. This data could prove valuable in identifying the location of an unknown source of contamination, weaknesses in contamination control measures, poor radiological work practices or other issues. Corrective actions could be identified and implemented to reduce the number of personnel contamination events or otherwise improve radiological safety measures.

## 6.7  Access Control

Proper control of access to radiological areas of the plant is a key element in ensuring the radiological safety of employees. The main purpose of an access control program is to ensure that only authorized entries into the RCA, by properly trained and qualified people, are allowed. Though this function may be more administrative in nature the importance should not be underestimated. Oftentimes the interface between RP personnel and the worker at the RCA access control office is the last opportunity to ensure that radiological controls have been properly established, that individuals and work crews are prepared for the task, and workers understand the radiological aspects associated with a given activity. Effective control and coordination of entry into the RCA will maintain personnel exposures ALARA and minimize radiological incidents resulting from inadequate coordination and execution of radiological tasks.

Entry into the RCA may be limited to a centralized location, though depending upon the number of units at a given site, the reactor type (e.g., PWR or BWR) and physical layout of support buildings, additional access points may be necessary. Notwithstanding, the number or locations of RCA access control points, there are several distinct functions usually associated with access control activities. A process to assign and track worker dose to a given task should be established, the capability to access and review current radiological survey data, and the opportunity to interface with RP personnel just prior to entry, are typical functions that occur at access control. The RCA access control facility should be designed to facilitate these activities and to ensure efficient control of personnel entry and egress from the RCA. The layout should allow the flow of entry traffic to pass the access control point office while upon exit; traffic flow should be directed to the personnel and equipment contamination monitoring stations. The RP access control point office should be situated to allow visual observation of the RCA access point. Figure 6.18 depicts the location of RCA exit PCM stations located immediately adjacent to the RP RCA access control office.

The requirements to gain access to the RCA may differ somewhat from one utility to the next; however, there are usually several prerequisites that are common to most programs.

1. The individual's current radiation worker status is verified. This usually involves confirming the individual's training date, available dose margins and other details.
2. Confirmation that the individual has a valid reason for entry and is assigned to a proper work package, RWP or other document authorizing the activity.
3. For entries into airborne radioactivity areas or for tasks that require the use of respiratory protection devices, the individual's respirator fit and qualification status is confirmed. If a current whole-body count is a prerequisite for these type tasks then this status should also be confirmed prior to entry.
4. Verification that task-specific radiological control measures and RP coverage requirements are available and have been properly coordinated.

**Fig. 6.18** Workers exiting the RCA are monitored by PCMs positioned at the RCA exit point. The RCA access control office is located to the left of the PCMs in this photograph. A gamma-sensitive portal monitor can be seen in the background behind the PCM shown on the right hand side of this photo. After exiting the PCMs workers then pass through the portal monitors (Courtesy of Luminant)

As the above items indicate a large amount of data must be accessed and confirmed prior to allowing an individual to enter the RCA. The simplest and most efficient method of performing these tasks is by means of a computerized RCA access control system. The computerized RCA access control system should have direct access to relevant employee information that is updated on a live-time or routine basis. The information should be accessible from multiple user terminals and locations. The database should include the names of radiation workers, appropriate personal identification information, current dose data, training status, whole-body count dates, respirator fit data and related information. A computer-based RCA access control facility is usually equipped with computer terminals where individuals may provide an RWP number, work package number or some unique identifier. The entire RCA sign-in sequence is often interactive. The primary purpose of the RCA login process is to assign an individual to a specific task with known entry and exit times and recording of the associated dose received for the duration of the RCA entry. Individuals enter their identification number or dosimetry badge number together with the RWP or work package number. Individuals may also be prompted to acknowledge that they have read and understand

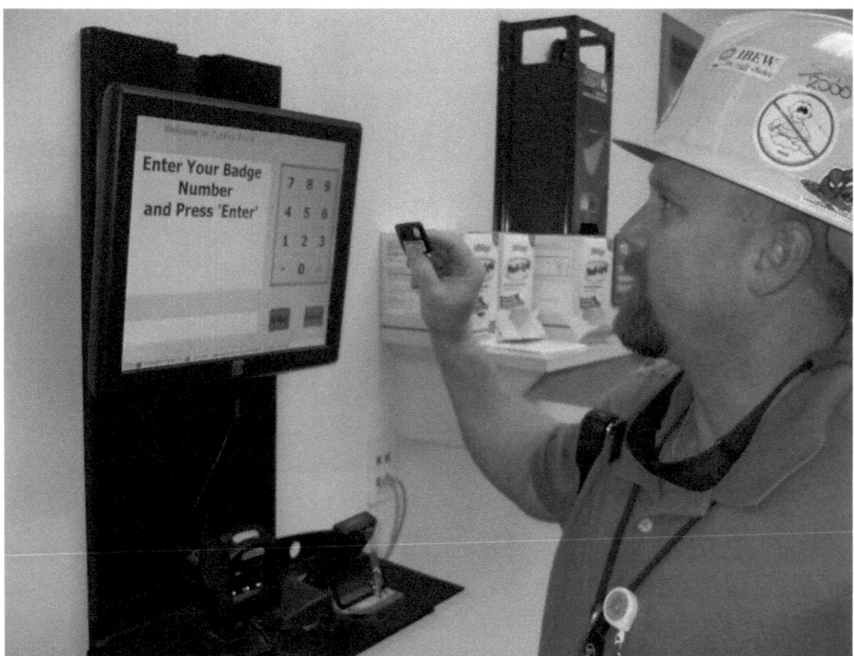



**Fig. 6.19** Worker processing into the RCA utilizing an interactive sign-in process. Note the prompt on the screen for the individual to enter his badge number (Courtesy of Florida Power and Light Company).

the RWP requirements for the task that they are entering under. These steps are usually accomplished by laser scanning of bar codes provided with the work package or on personal identification badges or a combination of both. Utilization of a biometric system, such as a hand or thumb reader, may also be a convenient method for processing entries into the RCA. The major point is that these systems should be automated to the extent possible to support RCA access control functions and the throughput of hundreds of daily RCA entries during high traffic periods that occur during outages. Figure 6.19 depicts a worker entering the RCA utilizing an interactive sign-in process.

The assignment of secondary personnel dosimetry may also occur at access control. The vast majority of nuclear plants utilize electronic alarming dosimeters of various designs. These dosimeters are equipped with visual displays and dose rate and integrated dose alarm features, usually as a minimum. (See Chap. 10 for a more detailed discussion on electronic dosimeters). If electronic dosimeters (EDs) are not assigned to individuals on a permanent basis then an ED is retrieved just prior to entry. After entering the work package and required personal identification information the access system then assigns the ED to the worker. This is accomplished by placing the ED into a reader that communicates to the RCA access control system (Fig. 10.5 depicts an ED reader). During the process the ED

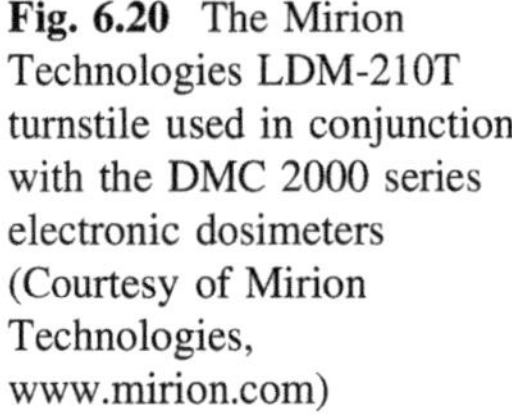

**Fig. 6.20** The Mirion Technologies LDM-210T turnstile used in conjunction with the DMC 2000 series electronic dosimeters (Courtesy of Mirion Technologies, www.mirion.com)

is confirmed to be working properly and the dose rate and integrated dose alarms are established. The alarm settings are predicated on the task to be performed and may be automatically set when the reader activates the ED. Radiation protection personnel or other suitably qualified personnel evaluate the expected radiation conditions in the area where the task is to be performed. Based on the duration and nature of the task dose estimates may be established for a given activity. Dose and dose rate alarm settings are then assigned to the activity for the associated RWP or work package and incorporated into a database. This allows the RCA access control system to access a database of ED alarm settings. During the ED activation process this database is utilized to automatically assign ED alarm thresholds for a given task. If the electronic alarming dosimeter system in use has this feature it affords several advantages. This feature allows RP to establish alarm set points that can provide early indication that radiation conditions are different than what was anticipated or otherwise provide early warning to workers that unforeseen radiation conditions exist. The judicious use of ED alarm settings provides early warning to individuals and in the event of an ED alarm, serves as an effective tool in preventing or minimizing unplanned radiation exposures.

Many RCA access control facilities incorporate a final validation step that may also confirm that the ED is operable. Once this step is successfully completed the individual may enter the RCA. This step often incorporates a positive control device that triggers a door or turnstile to open or rotate. Figure 6.20 depicts the Mirion Technologies turnstile that is used in conjunction with their MGP electronic dosimeters. Once individuals complete the RCA sign-in process they then precede to a turnstile located at the RCA entrance. The worker places the ED in the location that states "place the dosimeter here" (the plaque located on the flat surface of the unit as shown in Fig. 6.20). The turnstile is equipped with a proximity reader and a micro-processor that performs a final validation of the operability of the ED. Upon completion of the validation process the ED will be activated along with a green status light on top of the turnstile and the individual allowed through the turnstile. This entire sequence requires only a few seconds. Once the green status light is illuminated the individual should verify that the ED is operable. The ED display will show a "zero" dose reading. If a fault is detected with the ED unit the turnstile will not unlock and the yellow (or orange) light will illuminate.

A final RCA access validation process as just described utilizing the turnstile arrangement serves a vital function. This is the last barrier to ensure that the ED assigned to a worker is functional. It also requires workers to visually observe the ED to confirm that it is operable. Other arrangements may also be employed other than a turnstile arrangement. The important point is that a final validation step should be incorporated into the RCA access sign-in process. An individual who inadvertently gains access to the RCA with a non-operable ED could receive unnecessary exposure. Workers who have become conditioned to solely relying on an ED alarm versus periodically checking their ED status could be prone to receiving an unplanned radiation exposure.

The RCA exit process should include provisions to record and assign the dose received by the individual for a given RCA entry period. Assuming that an automated RCA access control process is utilized the exit procedure will be similar to those steps performed to gain access to the RCA. Namely, the individual will scan the bar code on their dosimeter badge, or input an identification number at an RCA exit terminal. The individual may be prompted to acknowledge that the correct personnel identifiers have been properly displayed on the screen. The ED is placed into a reader that records the dose received for the individual for the entry. All pertinent data is downloaded to the individual's current dose record. The ED is returned to a storage location and the RCA exit procedure is complete.

## 6.8 Secondary Radiation Protection Control Points

The primary RCA access control point should be capable of handling and managing the activities necessary to prepare and authorize personnel access to the RCA for the majority of the time. However, secondary (or in-plant) RP control points may be established from time-to-time to facilitate support of plant operations and maintenance activities. These control points are typically established during outage periods but may be utilized whenever the need arises. The main purpose of secondary RP control points is to provide a designated location whereby RP and associated work groups may coordinate work activities at a more convenient in-plant location. Secondary RP control points may be established at such locations as the following:

- At the primary personnel entryway to the containment building or drywell.
- On the refueling floor in close proximity to the reactor cavity.
- In the fuel handling building adjacent to the spent fuel pool area.
- Outside the loop rooms to support steam generator inspections for PWR units.

**Fig. 6.21** Access control area established at personnel airlock entrance to a drywell to support outage activities. The RP control point is on the right hand side equipped with video monitors and various radiation monitoring and survey equipment (Courtesy of Progress Energy Inc.)

- On the turbine floor for BWR units.
- Auxiliary building in close proximity to the major work activities.

Oftentimes the primary RCA access control point may not have the facilities or space to coordinate RP support activities for the numerous tasks that take place during an outage. By establishing secondary control points, more conveniently located closer to in-plant work activities, allows for RP support activities to be more closely aligned with the workers who will actually be performing a task. Ideally the proper use of these secondary control points improves the effectiveness and efficiency of RP support activities. If secondary RP control points are established, then it is essential that procedures be followed to ensure the proper coordination and communication of work activities between the primary RCA access control point and the secondary control points. The proper coordination of work activities at control points is a vitally important endeavor. All the relevant issues concerning communication techniques and work coordination activities discussed in Chap. 7 also pertain to secondary control point functions. Figure 6.21 shows a secondary RP control point established at the entrance to a drywell.

To ensure the proper coordination of work activities conducted from a given control point it is essential that RP personnel assigned to these locations understands their role and responsibilities. Understanding of these responsibilities will strengthen the communications between RP and the various work groups. In general, regardless of the location and specific purpose of a particular secondary control point, certain RP responsibilities are typically associated with the functions of these control points. These functions and responsibilities may include such items as the following:

- A lead RP technician responsible for the operation and coordination of the control point should be designated.
- The lead RP technician is responsible for ensuring that work groups check in with RP prior to starting a task and is knowledgeable of the radiological aspects of the work activity.
- Reviewing job-related radiological survey data, RWP or work package requirements with the work group prior to initiating work.
- Any required pre-job briefings should be conducted at this location or alternatively verification that any required pre-job briefing has been conducted.
- The coordination of RP job coverage and surveillance activities, for tasks falling within the domain of a particular secondary control point, should be discharged from that control point.

In general secondary RP control points are responsible for providing the necessary RP support to work groups covered by a particular control point. These responsibilities include ensuring that all radiological aspects associated with work activities are properly addressed.

The ability to properly coordinate work activities and to ensure accurate communication is greatly compounded due to the environment that may be encountered at secondary RP control points during outage periods. High traffic levels, multiple ongoing work activities, work group briefings, the assignment of RP resources to support a task, the preparation and review of radiological survey data, radio and cell phone communications and related activities contribute to less than an ideal communication environment. All these activities contribute to ambient noise levels and in general tend to create confusion and perhaps even chaos if not properly controlled. Since the secondary control points serve as the "hub" for the dispatch of RP personnel, delegation of assignments and control of multiple work tasks, it is important that effective "command and control" be exercised by the RP individual in charge of the control point. Consequently, certain guidelines should be established to ensure that control point activities are effectively managed. These guidelines could include the following:

- A designated lead RP representative should be present and responsible for supervising control point functions while work is in progress
- The control point supervisor should have the authority to control and maintain the level of work activities at such a pace as to ensure that sufficient time is allotted to support work activities.
- If there is any doubt concerning the radiological safety of an activity or the readiness of the work crew to proceed with a task, then the control point

supervisor should have the authority to delay the task until concerns have been addressed.

- The control point supervisor should be responsible for command and control functions at all times. These functions should include the ability to control noise levels and traffic in the vicinity of the control point and to limit access to the immediate area to those individuals with a valid work function.
- In the event that the designated control point supervisor is relieved or otherwise has to vacate the immediate area of the control point, then another individual should be designated as the interim control point supervisor. This designation should be clearly communicated to, and acknowledged by, the individual.

The use of secondary RP control points serves an important function during heavy workload periods. The proper use and management of these control points allows for close and timely coordination of RP support activities during heavy workload periods.


## 6.9  Job Coverage


Radiation protection job coverage simply refers to those activities and services supplied by RP in support of tasks involving radiological safety aspects. The coverage may range from performing a simple review of the activity and confirming the radiological conditions in the work area to an exhaustive pre-job review, requiring specialized training for the work crew, establishment of engineering controls, mock-up training sessions, use of elaborate protective clothing and equipment and extensive radiological surveillance requirements. Radiation protection personnel must be suitably qualified and trained in job coverage techniques with on-the-job training a critical component of the qualification process. Chap. 7 presents the elements of radiological work planning and ALARA controls in more detail.

The essence of a good radiological safety program is embodied in the attention given to the RP coverage provided at the working level and the effectiveness of the coverage provided. Radiation protection job coverage affords an opportunity to observe the quality of radiological work practices and worker attitudes toward the maintenance of an effective radiological safety culture at a given facility. Providing effective job coverage is one of the most important aspects of LWR radiation protection. Technicians providing job coverage must be diligent in their duties and should not hesitate to recommend steps that will minimize worker exposures or improve radiological work practices. A facility may have an excellent RP program with regards to the quality of procedures, use of state-of-the-art equipment, facilities and other program elements, all of which will be of limited value if sound radiological work practices are not implemented in the field. A healthy radiological safety environment is categorized as one in which workers readily assist others in improving performance on a daily basis and where

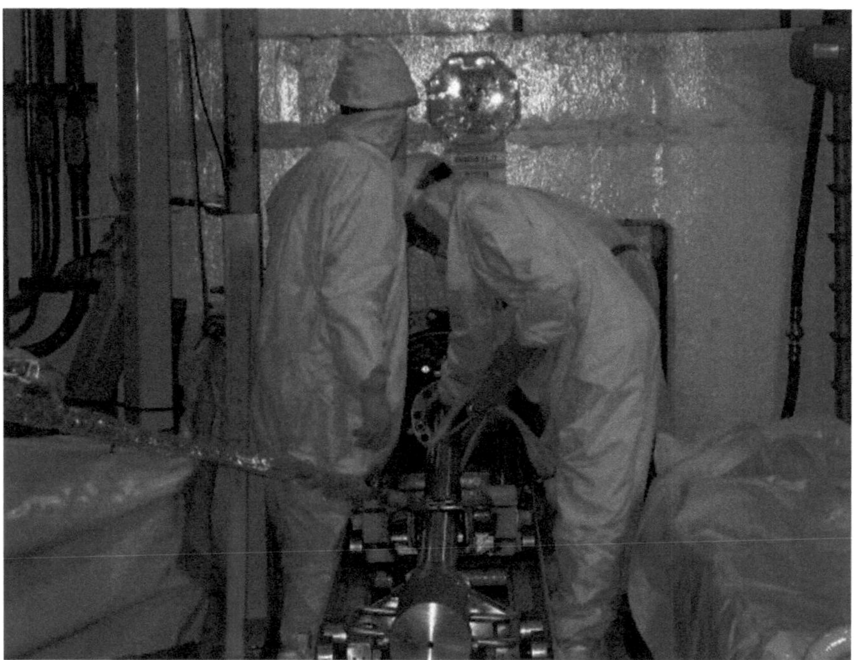

**Fig. 6.22** A control rod drive mechanism being removed from the drywell (Courtesy of Progress Energy Inc.)

individuals understand the importance of identifying and implementing improvements. If these elements are lacking and RP personnel are not diligent in their job coverage responsibilities, annual exposures may be higher than industry averages along with a higher number of contamination events and, in general, weaker radiological performance may be the net result. There is no substitute for ensuring that individuals at the working level are highly motivated and dedicated to the implementation of effective radiological control work practices.

Radiation protection personnel must be diligent when providing job coverage and constantly on the alert to identify activities which are of radiological significance and to take appropriate actions to ensure the radiological safety of workers. The RP department must foster an environment that constantly emphasizes the importance of maintaining an effective job coverage culture and the role it plays in ensuring employee radiological safety. Even though job coverage may be routine and non-eventful at times, the risk of allowing complacency to creep into this element of the radiological safety program could have significant radiological consequences. Handling of BWR control rod drive mechanisms as depicted in Fig. 6.22 involves close coordination with the RP job coverage technician.

Obviously the time during which the actual job is performed is when events arise that could adversely affect the radiological conditions, resulting in unplanned radiation exposures or a radiological incident. The RP technician providing job coverage must be knowledgeable of those activities and eventualities that may be encountered during

**Fig. 6.23** This figure depicts a remote monitoring facility used by radiation protection personnel to monitor multiple ongoing radiological work activities during outages. The facility provides video coverage and is equipped with direct communication to the multiple work locations shown on the monitors (Courtesy of Florida Power and Light Company)

the course of an activity that pose radiological concerns. If circumstances are encountered that can adversely impact radiological conditions or the radiological safety of workers or plant equipment, the RP technician must be prepared to take appropriate corrective actions as necessary. Paramount importance is to ensure that RP personnel are trained to recognize the radiological significance of certain activities and to adapt the RP field coverage to meet changing conditions before radiological problems ensue. The RP technician should be observant of ALARA and exposure management issues or any activity that may pose a radiological safety concern. Lessons-learned should be identified and established mechanisms utilized to capture such items to implement corrective actions prior to performing the activity the next time. The RP technician covering the job must ensure that the established radiological safety measures for a given task are implemented and maintained during the course of the job. Radiological conditions in the work area should be periodically assessed as the job progresses and appropriate surveillance activities performed at specific hold points or during key steps as necessary. The RP technician should periodically verify the dose received by workers to ensure that exposures are as anticipated for the stage of a task or are maintained within specified limits. If radiological conditions deviates to a point where the prescribed radiological control measures are no longer valid or prove ineffective, to adequately protect the workers, than the RP technician must be prepared to stop the job until conditions are such that activities can be safely resumed.

The role of the job coverage technician should be one of a coach versus that of having to ensure that proper radiological safety work practices are followed. The job coverage technician should be observant for momentary lapses in contamination control or dose reduction techniques (e.g., individuals wiping perspiration from facial areas while working in a contaminated area or individuals lingering longer than necessary in higher radiation areas). Even though individuals may have been trained to perform work in radiological areas it must be recognized that good radiological work techniques are not necessarily second nature to individuals not accustomed to performing radiological work activities. If the role requires direct intervention or constant reminders to individuals to follow basic radiological work practices, which were covered in prerequisite radiation worker training, then these issues should be evaluated generically. For example, job coverage observations may indicate the need for additional radiation worker training or improvements in the pre-job planning process.

Job coverage provides a unique opportunity to identify areas to support long-term radiological safety program improvements. All aspects of a given task should be evaluated and monitored during the performance of an activity. Problems encountered with the use of equipment, procedure steps, adequacy of pre-job preparations, knowledge or training issues, lack of adequate contingency planning or any other item pertaining to the safe performance of a task should be identified.

The above discussion pertains to situations whereby the job coverage technician is at the work location in-the-field. Oftentimes remote monitoring techniques may be utilized to provide job coverage. The use of remote monitoring equipment and facilities for certain tasks may be an effective means to provide job coverage. These facilities afford the capability to provide coverage of multiple ongoing radiological work tasks that are typically encountered during plant outages. Remote monitoring also provides an added benefit in reducing radiation exposures to RP and support personnel providing assistance to the work crew when their presence is not required in the immediate work location. Figure 6.23 depicts a radiation protection remote monitoring facility.

## 6.10  Summary

Operational radiation protection is the "heart" of a LWR radiological safety program—it is where the "rubber-meets-the-road. Effective implementation of operational RP program elements dramatically impacts the level of radiological safety afforded to plant personnel. Various support functions (e.g., dosimetry, training, RCA access control measures, respiratory protection, and instrument calibration) are important in providing the tools and processes necessary to implement various facets of the operational RP program. However, it is the proper implementation of the operational program that serves as the final barrier to success or failure. Properly trained radiation workers, aware of the importance of maintaining their radiological safety and that of fellow employees, is a cornerstone of an effective program. These attributes properly embodied in all radiation workers will serve to maintain a constantly improving radiological safety program. There is no substitute for effective job-coverage. These functions must be performed and discharged in a professional and competent manner over the long run to ensure the maintenance of an effective radiological safety program. If these elements are mainstays of a LWR operational RP program, then that program should be one that is characterized by maintaining radiological safety performance indicators at a level consistent with that of recognized industry leaders.

# Bibliography


1. International Atomic Energy Agency, Safety Guide No. RS-G-1.1, Occupational Radiation Protection, Vienna, 1999
2. International Commission on Radiological Protection, General Principles for the Radiation Protection of Workers, ICRP Publication 75, Pergamon Press, Oxford, 1997
3. National Council on Radiation Protection and Measurements, Biological Effects and Exposure limits for "Hot Particles", Bethesda, MD, 1989
4. National Council on Radiation Protection and Measurements, Self Assessment of Radiation-Safety Programs, Bethesda, MD, 2009
5. National Council on Radiation Protection and Measurements, Tritium Measurement Techniques, NCRP Report Number 47, Washington, D.C., 1976
6. U.S. Nuclear Regulatory Commission, Regulatory Guide 8.25, Air Sampling in the Work Place, Revision 1, June 1992

# Chapter 7
# Exposure Management (Minimization and Control of Collective Dose)

## 7.1 Overview

Radiation exposures to personnel must be adequately controlled and maintained to ensure compliance with established administrative and regulatory dose limits. Accordingly, various control measures are established to routinely monitor, track, and trend personnel exposures. The primary objective of a LWR radiation protection program is to maintain the radiological safety of plant employees. A key element of this objective is to prevent any unnecessary exposure and to minimize necessary exposure associated with operation and maintenance activities. This concept is promulgated by international radiation protection societies and the radiation protection community in general. The system of dose limitation established by the International Commission on Radiological Protection (ICRP) in Publication 26 in 1987, and continued in subsequent reports, recommends that any practice involving exposure to radiation produce a positive net benefit and that all exposures be kept as low as reasonably achievable (ALARA) while considering the associated economic and social factors.

The ALARA concept (or principle) has been addressed repeatedly in numerous international reports and publications and has been codified by standard setting organizations in many countries. Consequently, LWR radiation protection programs typically have a formalized process to track and trend personnel exposures along with administrative policies and procedures that outline the requirements for ensuring that personnel exposures are minimized as a matter of routine. These latter requirements may be incorporated into a formalized "ALARA program". Therefore the vast majority of RP activities are either directly or indirectly associated with the measurement, assessment, and control of radiation exposures to station personnel.

As the foregoing discussion indicates a LWR radiation protection program, established on generally accepted radiological safety principles, should strive to minimize collective radiation exposure to personnel. Consequently, the guidelines relating to the minimization and control of personnel exposures should be imbedded in the everyday operation and maintenance activities performed at a nuclear power plant. In principle, these guidelines or operational philosophy

R. Prince, *Radiation Protection at Light Water Reactors*,
DOI: 10.1007/978-3-642-28388-8_7, © Springer-Verlag Berlin Heidelberg 2012

should not have to be embodied in a separate so-called ALARA program. Over the last couple of decades regulatory authorities have emphasized the importance of minimizing collective exposure and strengthened regulatory requirements in this regard (10 CFR Part 20 and ICRP Recommendations for example). The availability of new techniques, equipment and materials has greatly aided exposure reduction initiatives. To aggressively pursue dose reduction initiatives ALARA programs must encompass all departments and should not be deemed as the sole responsibility of the RP organization.

A properly structured and implemented ALARA program has been shown to increase the productivity of workers performing radiological work activities. Unfortunately, this fact may not be recognized or obvious to non-RP personnel, who often regard ALARA related initiatives as another administrative burden to overcome. When dose reduction methods are seen as the responsibility of all workers, then truly great strides in reducing collective radiation exposure at a given plant can be achieved. If ALARA initiatives are seen as overly burdensome then attitudes may be overcome by minimizing the administrative aspects of the program. This could be accomplished by the seamless integration of ALARA program elements into existing work control processes.

This chapter presents those elements commonly associated with an exposure management[1] program. The exposure management program should, at a minimum, specify the requirements for worker training, the processes associated with planning and scheduling of radiological work activities, and the development and review of work packages. Mechanisms to identify improvements and capture lessons-learned to support continued long-term reduction in collective dose totals should be integral to the program.

## 7.2 Historical Industry Exposure Totals

Chapter 1 presented a brief overview of annual exposure totals for the nuclear power industry in the USA. After the TMI-1 accident in 1979 annual industry exposures started to increase significantly. The main factors responsible for the increasing trend were related to post-TMI modifications and the associated extended outages that resulted, increased in-service inspection requirements, and increased staff sizes. Industry exposure totals are tracked and trended based on a 3-year rolling average. The 3-year rolling average provides a more realistic approach in trending actual long-term industry exposures. Many nuclear plants operate on an 18-month refueling cycle while others may be on a 12 or 24 month refueling cycle. Nuclear plants on an 18 or 24 month refueling cycle will have minimal exposures for non-outage years with peaks occurring at intervals of 18 and 24 months. Consequently annual dose totals will have a wide

---

[1] Even though "ALARA" is a commonly used term, "exposure management" will be used in the discussions that follow to signify a broader scope of an "ALARA" program.

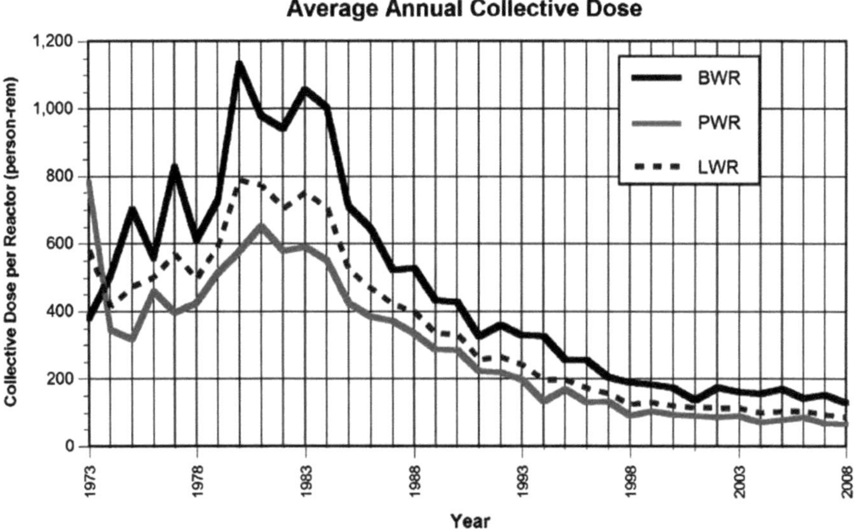



**Fig. 7.1** Average annual collective doses for BWR and PWR reactors from 1973–2006. (Source: US NRC NUREG-0713, Volume 28, www.nrc.gov)

variance from year-to-year. To adequately determine the actual trend in station exposures an evaluation should include outage and non-outage dose totals over repetitive cycles of a given length. The 3-year rolling average is a suitable time period for averaging annual station exposures based on the current length of nuclear power plant operating cycles.

Annual exposure totals for the USA industry for PWRs peaked in 1981 at 6.52 person-sieverts (652 person-rem) and decreased thereafter. Annual exposures for BWRs peaked in 1980 at 11.34 person-sieverts (1,134 person-rem). The 5-year period from 1980 to 1984 represents the highest sustained exposure period for the LWR industry in the USA. The average annual exposures for PWRs and BWRs over this period were 5.90 and 10.23 person-sieverts, respectively. The most recent 3 year averages for radiation exposures through 2008 were 0.7 person-sieverts (70 person-rem) for PWR units and 1.4 person-sieverts (140 person-rem) for BWR units. Figure 7.1 displays the annual collective dose per LWR over the period 1973–2006. Note the drastic reduction in annual exposures since the mid 1980s. Figure 7.2 depicts the annual collective dose per LWR over the period 1994–2009. The highest average annual exposure for PWRs actually occurred in 1973. For purposes of this discussion the 1973 value was neglected for various reasons and has no material impact on current industry dose trends.

Improved outage planning and execution, strict adherence to operating chemistry controls, improved system and plant performance, elimination of repetitive high dose jobs (e.g., replacement of resistance temperature detectors and installation of permanent work platforms in high dose rate areas) and other measures have contributed to the reduction in worker exposures.

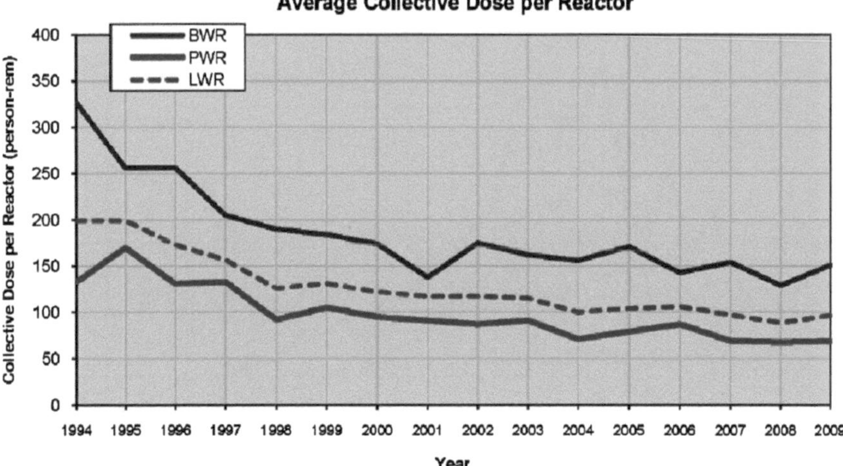



**Fig. 7.2** Average annual collective doses for BWR and PWR reactors from 1994–2006. *Top line* represents exposures for BWR units while the *bottom line* represents exposures for PWR units. The *dotted line* is the average exposure for both PWR and BWR units. (Source: US NRC NUREG-0713, Volume 31, www.nrc.gov)

The Nuclear Energy Agency (NEA) Organization for Economic Co-Operation and Development issues periodic annual reports summarizing world-wide occupational exposures at nuclear power plants. The NEA maintains the Information System on Occupational Exposure (ISOE). The objective of the ISOE is to provide a forum for radiation protection experts from throughout the world to co-operate on matters relating to the radiological protection of nuclear power workers. Dose data is provided on a periodic basis to the ISOE by those participating utilities. Information is collected from various operating nuclear power plants. It should be noted that only those participating utilities or plants from a given country provide dose data to the ISOE. Consequently for a given country dose data may not be provided for all the operating plants since some utilities may not be participating in the ISOE program.

Dose data summarized in the eighteenth annual report issued in 2008 indicated world-wide medium collective radiation exposures for PWR and BWR facilities for 2008 of 0.69 and 1.35 person-Sv, respectively. Similarly the 3-year rolling average dose figures were 0.72 and 1.38 person-Sv for PWR and BWR units, respectively.

## 7.3 Radiological Work Coordination

Probably the most telling aspect of the overall effectiveness of a LWR radiological safety program is embodied in the annual collective dose received at a given facility. To minimize the potential health detriment associated with radiation exposure it is important to establish measures to maintain collective dose ALARA.

This objective can be achieved by close involvement of RP with work groups during the scheduling, planning and execution stages of radiological work activities. To optimize RP job coverage, identified control measures should be arranged sufficiently in advance of a scheduled activity to allow adequate time for planning and preparation activities to proceed. To accomplish this objective, requirements for a radiological safety review should be incorporated into the job planning sequence. If a program exists that requires the completion and issuance of a Work Order, or a Maintenance Request Form, Work Defect Notification, or similar type document, a provision incorporating a review by an RP representative, indicating that the activity has been reviewed for radiological safety considerations may be beneficial. The need to obtain a pre-job survey, an RWP, or specific RP job coverage, for instance, could be noted on the applicable job planning document. This could serve as a means to inform job supervisors or responsible persons that prescribed radiological protective measures will be required. Formalized work control processes that coordinate these activities have proven to be a vital ingredient in controlling and minimizing exposures at LWRs. Work control processes typically coordinate planning, scheduling and performance of preventative and corrective maintenance, inspections, surveillances, and other tasks to allow work to be performed in a safe and efficient manner. Formalized Work Control Centers are now quite common in the nuclear industry as the benefits that may be achieved when work activities are efficiently planned and coordinated have gain widespread recognition. Improved execution of work activities enhances radiological safety aspects as well as industrial safety aspects of work performance. The establishment of coordinated planning and scheduling processes has also had a positive impact on controlling maintenance and operational costs.

Basically the function of work control is to ensure that required resources are properly scheduled to support a given task, systems and components are in a safe configuration to support the task, clearances have been prepared and placed, and that equipment, parts, tools and work package documents are available. The establishment of formal Work Control Centers has been a major factor in improving radiological performance for both at power and outage conditions. Radiation protection organizations should exploit the potential benefits that a formalized work control program offers from a radiological safety perspective.

Integrated Work Control Centers have been instrumental in providing an environment that optimizes the performance of work activities. Work Control Centers typically are comprised of representatives, often dedicated on a permanent basis, from operations, maintenance disciplines (e.g., electrical, mechanical and instrument and controls), planning and scheduling specialists, engineering disciplines and administrative support personnel. Additionally an RP representative should also be assigned to the work control function. As a minimum the RP representative should be intimately familiar with the work control process. This allows RP to be informed of radiological tasks sufficiently in advance to maximize the use of radiological control measures and exposure management techniques. If a formalized planning and scheduling group exists, it is strongly recommended that the necessary RP resources be permanently assigned to this function. This allows

RP personnel to become fully involved with planning and scheduling activities. This affords more effective radiological safety preparation time that is especially crucial when scheduling those tasks requiring extensive radiological safety preparations. Additionally, the assignment of RP representatives to work control provides an opportunity for RP to assist in the scheduling of activities. The RP representative may recognize the importance of taking into account plant operating conditions to recommend steps to minimize personnel exposures or other measures to otherwise improve the radiological safety of work activities. The importance of RP working closely with other departments, whose activities could impact plant radiological conditions, has a major impact on the effectiveness of radiological safety measures. The overall radiological safety performance of a station may be strengthened by fully utilizing the benefits of established work control programs.


## 7.4  Radiological Work Planning and Preparation

Activities associated with radiation exposure management may be broken down into three distinct phases. These phases include those measures considered during the planning stages of a task, radiological field techniques implemented during the performance of the task and those measures to secure from the task upon completion of field work. Many LWR radiation protection programs have implemented or established a pre-job checklist that addresses work in radiological areas and ALARA-related items that are typically considered during the planning and scheduling stages. This topic is discussed in greater detail later in this chapter. Various aspects of these pre-planning efforts may be integrated into the formal work control process. By incorporating the preparation of radiological safety and control measures with pre-planning activities the ability to identify and implement more effective measures are greatly facilitated.

Various controls or techniques may be employed to ensure that pre-planning activities consider measures to minimize exposures and the spread of contamination, or in general, address the radiological aspects of a given task. It should be emphasized that even ordinary or seemingly mundane measures should not be overlooked during pre-job planning stages. Something as basic as ensuring that adequate lighting is available at the work location, or that tools and equipment are operable prior to dressing-out and entering a high radiation area for example, could result in dose savings or the prevention of a radiological incident. Couple this with lost productivity and the potential impact on scheduled activities; the costs (both radiological and economic) associated with inadequate pre-job planning could be significant. The discussion below assumes that RP is an integral component of a Work Control Center.

Various parameters should be considered when evaluating the need and scope of RP job coverage requirements. In order to adequately assess radiological safety needs, the specific details of the job must be known beforehand. No activity within the RCA should be allowed to proceed until a job description, sufficient to establish minimum radiological controls, has been obtained. Once job coverage requirements have been determined, the necessary resources and equipment should be prepared and made available. Ideally the RP technician providing job coverage should be familiar with the activities to be performed and have an understanding of their actual or potential radiological implications.

During the planning and scheduling stages, an RP representative should be assigned to review work packages. The magnitude of the radiological control measures to be employed for a given task will primarily dictate the degree of RP preparation required to support the task. Preparations may involve nothing more than assigning the task to a standing Radiation Work Permit (RWP) for those jobs that do not involve actual maintenance work. Radiation Work Permits are described later in this chapter. These type tasks may be covered by an established set of generic radiological and ALARA-related safety precautions and requirements. Such tasks as inspection type activities, performance of routine plant rounds and surveillance work, or tasks involving the collection of data, not performed in contaminated areas or high radiation areas, could be addressed by an established set of generic radiological safety precautions. Activities that may result in significant personnel exposure or that involve major maintenance work on a radiological system or component (e.g., pump teardowns, valve repairs or breach of highly contaminated systems) or otherwise pose potential contamination control challenges or exposure concerns may require extensive pre-job preparations.

Based upon review of the work package, the RP representative evaluates the extent of pre-planning activities required. This is a key aspect of the work control function. Scheduled activities may be screened several or more weeks in advance of the actual work date. This affords an opportunity for RP to coordinate radiological preparations with the crew assigned to perform the task. If it is a first-time activity or one that poses significant radiological challenges, preparations can be targeted to meet the specific needs of the workers based upon their experience and skill set. Pre-job preparations under these circumstances could include specialized training, the use of mock-ups and practice sessions to improve skills to ensure a successful outcome and minimize time spent in radiological areas. When RP job coverage is required a RP technician can be assigned to work closely with the crew to promote better understanding of the tasks to be performed and the identification of enhanced radiological control measures. Consultation between RP and the work crew has an added benefit in that synergies often result in the identification of alternate tooling, more effective use of equipment and tools, and the enhancement of work procedures that result in efficiency improvements and dose savings. Contrast this to the situation where pre-job preparations start just prior to work. Under these conditions the ability to fully utilize various exposure management measures or to employ techniques to mitigate radiological control issues are limited.

Prior to starting a job, RP should review the activity with due consideration given to the radiological conditions present in the work areas and anticipated radiological conditions that could be encountered when the task is performed. Person-sievert estimates should be developed and if necessary, a detailed review of the task undertaken in accordance with established ALARA planning or exposure management programs. If radiological conditions can be modified to minimize personnel exposures then steps should be taken to do so. Such measures could include the following:

- Possibility of flushing or draining systems or components as appropriate to reduce radiation levels or to minimize the spread of contamination.
- If appropriate, the work area could be decontaminated to reduce contamination levels to minimize protective clothing requirements.
- If the activity involves the breach of a contaminated system or has the potential to spread contamination then the work area should be prepared and properly demarcated to contain contamination to the extent possible.
- Localized hot spots or sources of radiation could be shielded to reduce dose rates in areas to be occupied by workers.
- Measures to reduce airborne contamination levels (either actual or anticipated) may be established to prevent or minimize the use of respiratory protective equipment (e.g., increasing ventilation flow rates in the work area, erecting enclosures or utilizing portable filtration units).
- The task may be delayed or otherwise rescheduled to take advantage of more favorable radiological conditions that may exist at a later time (e.g., different plant power levels or when conditions are such that a component or system may be flushed).

Depending upon the circumstances, various other measures could be considered. Basically any action that could minimize the time in the work area, facilitate the execution of the task, or result in lower personnel exposures should be evaluated.

A convenient means to ensure that pre-job planning activities consider exposure management related aspects and generic radiological safety measures is to provide the work crew with a standardized checklist to evaluate measures that may be appropriate for a given radiological task. This checklist, or other suitable mechanism, affords the work crew or responsible individual an opportunity to review pre-job measures, that may have been overlooked, that may be appropriate for the task to improve job performance. Table 7.1 depicts those items that may be appropriate for inclusion in a pre-job checklist.

The above items should not be considered all-inclusive and any pre-job checklist should be tailored to meet the specific needs of a given facility based upon history, experience, and lessons-learned. A historical data base of previous pre-job checklists may be maintained and cross-referenced to a given task or job. This would facilitate pre-job preparations and minimize the chance of missing a critical pre-job activity. It should be noted that even though the main focus of these efforts may be directed toward the achievement of dose reduction, the elements presented could also apply to contamination control measures and other radiological safety aspects of a task. Measures implemented as part of an exposure management program serve not only to reduce personnel exposures but may also prove to be instrumental in controlling contamination and improved industrial safety performance.

**Table 7.1**  Exposure management pre-job checklist

Availability of tools, equipment and supplies

Training status of workers

Placement of temporary shielding

Placement of contamination control barriers

Confirmation that tools and equipment are operable

Measuring and testing equipment in current calibration

Systems and components verified to be properly configured to minimize radiation levels

Removal of extraneous radioactive materials or other sources of radiation

Placement of portable ventilation and filtration units to control airborne contamination levels

Verification that work area environmental conditions are suitable for the task or have been adequately addressed (e.g., temperature, heat stress considerations, presence of suitable lighting, cooling supplies)

Procedures and work documents are current and available

Confirmation that any required communications equipment, remote monitoring equipment, radios or related items are operable and properly staged

Workers received task-specific training or participated in mock-up training sessions as necessary

Use of special tooling or equipment and an opportunity to train and practice using this equipment

Discipline-specific, dedicated planners or schedulers, who have been trained in the radiological aspects of pre-planning activities, may perform RP-related pre-job planning evaluations. These individuals could consult with department work crews during the planning stages of a task. Alternatively, the RP group may assign individuals to assist work crews with this effort. Radiation protection representatives may be assigned to the work control group, as noted above, or be part of the overall RP organization. Regardless of the process used or who actually performs the pre-job planning evaluation, the earlier this effort is undertaken the more effective the outcome.

Many of the pre-job planning elements discussed above may be intuitively obvious and would typically be addressed for any activity, even those that do not involve radiological concerns. However, the early days of the LWR industry were characterized by events that could have been prevented if these "simple" measures were considered during the planning and scheduling phases. For instance, it was not uncommon for work activities to be scheduled in the vicinity of steam generators when steam generators where drained for inspection or maintenance work. The lack of water in the secondary side of a steam generator (SG) could increase general area radiation levels in the vicinity of the SG by a factor of two or three versus when the secondary side is full of water. It was not uncommon for such activities as scaffold erection to be planned at a time when system status or configuration was not optimal from a dose reduction perspective. Work not otherwise requiring the use of respiratory protection may have been scheduled in parallel with adjacent tasks requiring the use of respirators. The work crew performing the non-respirator task would either be turned-away at the last moment or unforeseen controls applied that may have extended the task or interfered with the performance of work (e.g., communications, use of specific tools or equipment, or the need for additional radiological control measures). Obviously these type situations either resulted in the rescheduling of activities, or the performance of a given task under less than ideal radiological conditions. These aspects coupled with the resultant inefficiencies would often end with the expenditure of additional dose. Inadequate planning and scheduling or the failure to identify adequate radiological control measures prior to the performance of a task, often results in higher or even needless personnel exposures. Negative impacts on scheduled activities (which could pose significant problems during plant outages), the inefficient use of resources (both people and equipment), and additional costs could also ensue. More importantly inadequate pre-job planning may prevent the identification of possible radiological conditions that could be encountered during the performance of a job thus negating an opportunity to plan and prepare accordingly.

## 7.5  Work Coordination and Effective Communications

Proper planning and coordination of radiological work activities are essential elements of maintaining an effective radiological safety program at a LWR. However; the time and effort expended in coordinating work activities could be jeopardized if established controls are not effectively communicated in the field. Numerous incidents have been reported in the LWR industry over the years (e.g., unplanned entries into incore rooms and the seal table area of PWRs, work performed on the wrong component) associated with poorly coordinated work activities or ineffective communication problems leading to radiological events. Therefore sound communication techniques should be followed when coordinating radiological work activities. The intent here is not to discuss all the parameters that affect communication among individuals or groups of workers since that is well beyond the scope of this book. However, it is important to understand those factors that contribute to successful communication, or conversely to unsuccessful communication. The discussion that follows presents an overview of the importance of coordinating job activities and those conditions and factors that may impact effective communications or work coordination activities in a nuclear power plant environment. To minimize (and hopefully prevent) the possibility of events resulting from poor communications the important role that coordinating work activities and effective communication techniques play in ensuring the radiological safety of employees should be understood. It is essential that RP personnel have a unique understanding of this topic due to their position in ensuring that established radiological safety controls are effectively applied at the working level.

### 7.5.1  Complexity of Radiological Work Activities

Many parameters encountered in a nuclear power plant environment potentially affect the ability to properly coordinate work activities or to effectively communicate crucial steps of a job sequence. These parameters may or may not be subject to control. It is important for RP personnel to be aware of these various parameters and

| Table 7.2 Parameters potentially affecting communications | Worker related | Process related | Environmental |
| --- | --- | --- | --- |
| | Individual knowledge | Procedures | Work location |
| | Work experience | RWP | Noise level |
| | Skill level | Work Order | Equipment related |
| | Education | Scheduling | Changing conditions |
| | Communications | Interfaces | |
| | Attitude | Terminology | |

to understand their impact on communication and work coordination. If possible measures should be implemented during the work planning process to address these type issues if the opportunity arises. Again for simplicity and to address the more important elements associated with this subject these parameters may be grouped into three categories as depicted in Table 7.2.

Obviously Table 7.2 does not list all the potential parameters that could impact effective communications. The purpose is to provide some indication of the complexity of dealing with all the elements associated with ensuring clear and concise communication of work controls in the field. Envision the work environment often encountered during refueling outages with ongoing multiple tasks and work crews in adjacent areas and handling complex equipment and this situation clearly increases in complexity.

When considering the above parameters and the complexity of properly communicating and coordinating radiological work activities to diverse work groups it becomes apparent that there are many opportunities to introduce communication-related issues. Additionally, further complexities are added associated with the need to control and communicate radiological conditions during the performance of tasks. For instance RP personnel must communicate among themselves (e.g., job-coverage technician to a control point technician to perhaps an ALARA specialist). In turn radiological parameters that may be subject to change must be communicated to workers, who oftentimes may not fully understand or appreciate the significance of these changes. Important hold points or stages at which radiological conditions need to be confirmed (e.g., survey prior to start of work or survey upon breach of system) must be clearly communicated to the work crew.

Communication and work coordination activities may be further complicated when the use of special protective equipment and clothing is required. Does the worker understand the function and purpose of an electronic dosimeter? Has RP set the electronic dosimeter at the proper alarm threshold? Did the worker hear me through his respirator? These and other questions may come into play while performing a given task. It is essential that these items be properly addressed and recognized during the performance of work activities.

### 7.5.2 Diverse Experience of Personnel

The complexity of maintaining and operating a nuclear power plant, by necessity, requires the skills of a vast array of specialized personnel. There are supervisors and managers, skilled crafts people, engineers, technicians and contractors for example.

Radiation protection personnel interface with these people on a daily basis and many work activities require the concurrent use of personnel from many crafts to perform a given task. Consider how this may influence how RP goes about communicating and coordinating radiological work activities. The needs of the audience must be considered and this audience may be multi-disciplined and changing from one task to the next.

Consider the diverse work group that may be encountered and the attributes that may be associated with the diverse nature of personnel. These attributes may include education level, radiological work experience, past work history (e.g., non-nuclear related), job classification, degree of training and perhaps language and cultural attributes. It is important to understand how this diversity potentially impacts effective communications.

The education level will obviously impact the knowledge and understanding that a particular individual may have concerning radiological terms and requirements. Employee education levels may range from those individuals who did not receive a high school diploma or who did not matriculate to those with advanced college degrees. Many workers may have completed specialized technical training programs while others have not. Consequently some individuals may readily grasp the radiological significance of various terms and units while others may be simply confused. Radiation protection personnel should ensure that required radiological information for a given task has been effectively communicated to workers and the information understood.

The extent and depth of radiation worker training is also variable. Individuals may have only received the minimal general employee training and radiation worker training required for entering and working in radiological areas, while others may have received advanced radiation worker training. Workers employed for several years may have attended various training courses a number of times and possess extensive knowledge of basic radiological terms and concepts. On the other hand, there may be workers who still confuse the meaning of radiation and contamination. Radiation protection personnel must acknowledge this situation and implement measures to ensure radiological work requirements are properly communicated and understood by members of the work crew (e.g., during pre-job briefings).

Obviously the experience level of personnel, both permanent and short-term contractors, varies greatly. The amount of radiological work experience will impact an individuals' ability to understand radiation protection requirements and their ability to work safely within this environment. In all likelihood a wide-range of work experience will be encountered. Some workers may have had no prior work experience, or their experience is limited to their current employment, or perhaps there are individuals with extensive work-related experience obtained at several different companies.

Many plants rely on contractor personnel during those periods of greatly increased maintenance activities which are often the case during refueling outages or extended maintenance periods. Consideration should also be given to both long-term and short-term contractors that may be employed at a given facility. Contractors typically have extensive and perhaps diverse radiological work experience. They may have worked at multiple plants and companies and trained to perform certain activities differently than that of the host utility. Contractors are not usually as familiar with specific plant procedures, policies and programs and may in fact bring different techniques and radiological work practices with them. It is essential that proper training and indoctrination be provided to these individuals to ensure that radiological work activities are planned, coordinated and performed in accordance with established programs. These factors should be taken into account when developing programs to address communication and work coordination efforts for a given group.

## 7.5.3  Variable Work Conditions

Under static conditions effective communications and coordinating work activities is difficult at best. Obviously this is not the situation often encountered at a LWR where radiological conditions, during maintenance activities, are usually not static due to their very nature. Maintenance activities may involve the draining of systems or components, the movement of radioactive material, or breaching of contaminated systems. Consider the potential radiological incidents that could result if the following activities are not properly coordinated.

- Breaching a highly-contaminated system with unknown contamination levels or unknown radionuclide composition
- Sluicing highly-contaminated spent resins from a resin bed to a storage tank involving a spent resin header that runs through multiple rooms and different floor elevations
- Movement of incore probes from regions of the core to retracted storage locations or for replacement
- Draining or transferring the contents of liquid radioactive material storage or hold-up tanks
- Responding to a spill of radioactive material that may have entered floor drains
- Moving and preparing a high-activity waste container for storage or transportation

Obviously many other examples could be encountered that pose equally or more challenging situations whereby poor coordination or a miss-communication could result in unnecessary personnel exposures, the spread of contamination, or other radiological incidents. If a crucial step in a sequence of maintenance activities may result in changing work area radiological conditions it is imperative that the radiological controls associated with the performance of the step be properly communicated and implemented.

Further insight into the importance of ensuring proper coordination of work activities may be gained by reviewing the layers or communication interface chains that may be present when communicating required information at the time it is needed. Radiological information may be obtained from pre-job surveys, work history files, or may be based on anticipated radiological conditions. This data provides important information for job pre-planning purposes and establishment of initial radiological work controls. Therefore it is essential that workers understand the basis of this information and RP personnel properly communicate the relevance of this data and how it applies to the performance of a task. It is possible that "anticipated" conditions may not apply under certain circumstances. Are workers aware of this eventuality and do they know what actions to take in the event that radiological conditions differ significantly from those that the established controls were based? Equally important is to communicate what conditions are not expected to be encountered and that could provide early indication that radiological conditions may challenge the establish controls. Workers should be cognizant of conditions that could compromise the effectiveness of established controls. The work crew should be able to acknowledge the impact of conditions pertaining to their radiological safety such as higher than expected dose rates or higher than expected contamination levels, the presence of discrete radioactive particles, or the generation of airborne contamination for example. In addition changing plant conditions (e.g., presence of water or liquid, loss of plant ventilation or portable filtration equipment provided for the task, and others) may pose significant radiological concerns and these issues should be understood by workers and the expected actions to be taken in such an event effectively communicated.

## 7.5.4 Organizational and Departmental Interface

Another complication that arises when coordinating work activities and communicating factually correct and understood information is the layers of interface that may be involved with a given task. Interfacing between various radiation protection groups, different departments, and organizations and the plant location at which these interfacing activities take place further contribute to complicating communications and work coordination activities. Each organizational interface point may also involve various worker classifications. These classifications may involve crafts and skilled technicians, supervisors, schedulers and planners. Worker classification should be taken into consideration when communicating the necessary details required of individuals at each interface point. A scheduler may be more concerned about the availability of needed resources to support the performance of a task at a given time while the individual worker performing the job may be more concerned about the radiological conditions and equipment needed to perform the task. It will serve little purpose if RP concentrates on providing the specific aspects of various radiological control measures to a planner or scheduler in lieu of providing this information to the work team.

A typical job performed within the radiological control area may involve multiple interfaces similar to those depicted in Fig. 7.3.

Information may flow from one box to another and back again. In fact interfacing may actually take place between boxes not adjacent to each other. Each interface point affords an opportunity to miss-communicate critical information which could result in poor work coordination and possible radiological incidents. Radiation protection personnel must be keenly aware of potential interfacing issues and

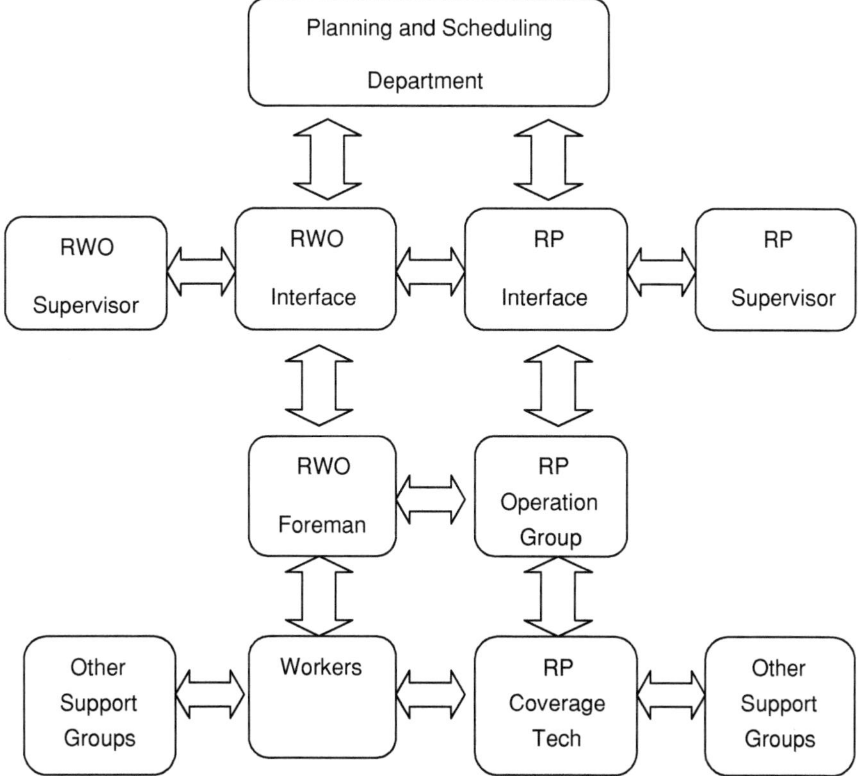



**Fig. 7.3** Interface points associated with radiological work activities. *RWO* responsible work organization

knowledgeable of what information is reqired at a given point or time to ensure the safe execution of radiological work activities. Consequently information should be communicated in a manner that takes into consideration the qualifications, skill, and knowledge levels of personnel dealt with at each interface point to ensure that information is clearly received, understood, and acted upon. It is crucial to provide the right information, to the correct individuals at the proper time for work activities to be safely controlled and coordinated. Even though communicating information is an activity people perform daily, effective transfer of verbal communications is not easily accomplished without the diligence of all people involved.

### 7.5.5 *Internal Departmental Communications*

Radiation protection serves a supporting role when it comes to maintenance and other work-related activities. As such RP receives information and direction from many sources. Various radiation protection sections within the RP organization

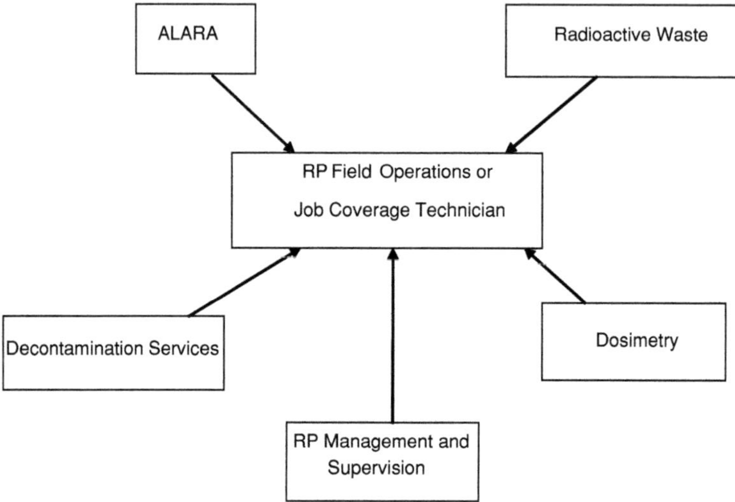


**Fig. 7.4**  Internal communication flow in RP organization

must coordinate their efforts to ensure proper execution of their responsibilities and job functions. Several interface points may be associated internally to radiation protection. The same concerns noted above pertaining to interface communications and coordination of work activities also apply here. It is essential that RP groups effectively communicate and coordinate activities which they are responsible, to ensure the successful accomplishment of tasks.

Though RP organizations may differ from one plant to the next, a core group responsible for performing radiological surveys and providing job coverage in the field is typically central to the organization. This group may be referred to as the "operations" or "surveillance" section within the RP department. For purposes of this discussion the RP section responsible for providing job coverage will be referred to as the "RP Field Operations" group. Assuming that this section is the lead group coordinating with other work organizations under most situations, the interface structure depicted in Fig. 7.4 could be associated with a given task.

Even though the RP field operations technicians are responsible for providing the RP support in the field and performing job coverage functions they receive support and assistance from other RP groups. These other groups share similar issues concerning the need to ensure that communication and coordination activities are effectively implemented. Considering the potential number and complexity of internal interfaces many opportunities to miss-communicate vital information may arise. Smooth coordination and accurate communications do not simply happen because people are in close proximity to each other or involved with the same task. Proper implementation of work activities and the radiological safety of workers require a conscientious effort on the part of each individual. If any of these communication chains or layers of interface are poorly executed the probability of initiating a sequence of events that may result in a radiological incident may quickly escalate.

It is essential that RP personnel understand the purpose and intent of communicating within their own department. The flow of information depicted in Fig. 7.3 is a daily process. Each group must clearly define the methods and processes used to support the flow of information and data from a particular RP group or section. This data must then be transmitted in the desired form to provide accurate information and to ensure that the information is understood by the receiver. Radiation protection personnel should guard against complacency when requesting and forwarding information internally within their department. Even though the gathering and processing of radiological survey data and associated information are a daily occurrence using common terms familiar to RP personnel it is vital to guard against informal or poor communication techniques and styles. When dealing with common terms and routine transfer of data it is still important to communicate in a formalized fashion to prevent incorrect assumptions being formed and acted upon to the detriment of worker radiological safety.

## 7.6 Planning and Scheduling Aspects

The mechanisms associated with the scheduling of work activities also play an important function in coordinating work activities and communicating key work-related details. It is important that RP personnel understand the scheduling process utilized at their facility and what type of information is available during the scheduling process. The earlier RP personnel are involved in the scheduling and planning activities the greater the opportunity to identify and provide more effective radiological safety controls for a given task. This aspect becomes more important for those activities involving significant radiological challenges and those that are complex in nature involving multiple work groups.

The specific details concerning the planning and scheduling of work activities may differ from one plant to the next. However, effective planning and scheduling programs have common attributes typically associated with these activities. The discussion that follows for planning and scheduling of activities performed during periods of power operation and for outage periods is generic in nature. An overview of those elements normally associated with a planning and scheduling department are presented. The details associated with the frequency of planning and scheduling meetings or how far in advance activities are placed onto a work schedule is not discussed. The primary aspect is to realize the role that a formalized planning and scheduling program plays with regards to the effective planning and coordination of radiological work activities. A formalized planning and scheduling program affords a unique opportunity to plan, schedule, and implement radiological work activities in a manner that fully utilizes dose reduction and contamination control measures. This discussion addresses those aspects of planning and scheduling that play a role in radiological work activities. The intent is not to present a detailed description of the work control process or the role of other organizations or groups that are integral to the overall process.

## *7.6.1  Power Operation*

A planning and scheduling program should provide a means for the major work groups and supporting organizations to meet and review upcoming work activities. Mechanisms are typically established that designate a period of time for which work is planned on a routine "look-ahead" basis. For instance work activities may be required to be identified several weeks or even months in advance as part of the initial planning process. As the schedule work date, for an activity progresses, specific tasks may be placed on a near-term work schedule perhaps a few weeks prior to the start date of a task. These near-term work activities are periodically reviewed by all the major work groups and support organizations needed to perform the activity. In addition to RP support many tasks may require support from other groups or departments such as plant operators to perform valve line-ups and place clearances on equipment or systems, quality control personnel to witness key aspects of a task, engineering support, industrial safety specialists and others. Therefore all these organizations should also be involved at some point during the planning and scheduling stages.

At some point a "final" work schedule is issued that details all the upcoming work activities for a given period (e.g., a 7-day work week). These are the work activities that have been approved to work. This approval should signify that all tasks on the "approve-to-work" schedule have been reviewed and approved by all groups involved in a work activity. Obviously this means that the major maintenance organizations have committed the resources, equipment, tools and parts, and have final approved work package documents to perform the task per the schedule. From the radiological safety perspective that means that RP is aware of upcoming jobs, has indentified the necessary radiological control measures required to support these jobs, and has coordinated and communicated the details of radiological control measures with work groups. In other words RP should have had sufficient time to plan and prepare for a given task. The length of the planning and preparation period should be commensurate with the complexity of the task and the degree of radiological hazards associated with the activity. Complex jobs requiring significant preparation time that may involve mock-up training or dry-runs to ensure workers are prepared and trained to perform a given task should be identified sufficiently far in advance during the planning and scheduling process to provide adequate preparation time. A work activity that involves the replacement of a highly contaminated pump or major repairs on a contaminated heat exchanger for instance should not be a "surprise" to RP or a last minute addition to the weeks' approved work schedule. Obviously allowance has to be made for the addition of emergent work or emergency type work packages to a work schedule. The process by which emergent work items are added to the work week schedule should be governed by the work control process. Emergent work or emergency type repairs represent those periods when communication and coordination efforts are in jeopardy of "short-cuts" being taken. The radiological aspects associated with these activities must be properly controlled and should not be compromised to the detriment of worker radiological safety just because it is an emergent task. A standard set of pre-established radiological control measures could be developed that address a range of radiological conditions and the nature of the emergent work activity. These measures may be overly conservative but could be implemented quickly to support emergency type work activities.

Scheduled work activities are often generated on some type of automated system that depicts the start and end date for a given task along with the task duration. These systems may be composed of a time line with specific steps broken down in duration together with the required craft resources. For example a mixed bed demineralizer in the Chemical and Volume Control System (CVSC) is scheduled to be replaced. This task may be broken down into distinct steps depicted in the work schedule. The first step may depict a step, with a duration, for operations to hang (or place) a clearance. The second step may be for RP to post and control areas of the plant impacted by the resin transfer. The next step, again with a specified duration may depict the time required to sluice the resin. This step may be followed with a step for RP to confirm post-sluice radiation levels and to update radiological sign postings. A final step for operations to remove the clearance may also be depicted. Assuming that RP is the only support group required for the task, a notation on the schedule may reflect the need for RP support required at a specific point in the task sequence. Obviously for more complicated jobs numerous steps may be required with much more detail reflecting the need for multiple support group coverage and perhaps different maintenance craft required for specific stages throughout the duration of the task that may require several days to perform.

Some planning and scheduling work schedules may include the person's name in the work control center responsible for coordinating changes or updates to the schedule as the job progresses. This person may also be responsible for verifying that all work groups and supporting organizations are prepared to support the task and have agreed with the planned schedule. Again the exact format of the automated work control planning and scheduling system may differ from plant to plant and oftentimes may be formatted to meet the specific needs of a plant. Lessons—learned may serve as input to revising the form and lay-out of the planning and scheduling tools to meet specific needs of a facility. The work control function should incorporate a means to monitor the status of jobs as they are performed. This may take the form of routine meetings whereby representatives of the major work groups and support organizations meet collectively to review progress and to address any issues that may have arisen (e.g., higher than anticipated exposures received by the work crew) during the performance of a task.

Notwithstanding the form or the degree of complexity that may be associated with a planning and scheduling process RP should be fully-integrated into the program. The sooner that RP is informed of radiological work activities the greater the opportunity will be for RP to identify, prepare and implement effective radiological safety measures for a given task. If a formal work control organization exists, RP should take advantage of the benefits to be gained by integrating radiological safety aspects of tasks at the planning and scheduling stages.

## *7.6.2  Outage Periods*

Plant refueling outages and maintenance outages represent the periods when the vast majority of personnel exposures are received and when the most challenging radiological jobs occur. These are the periods when major work activities are typically performed. Tasks associated with preparing the reactor vessel to support refueling activities often involve work in high radiation areas, high contamination levels or unknown amounts of contamination, and require multiple work groups and various support organizations. Overhaul and refurbishment of major components are performed during these periods. Surveillance testing and preventative maintenance activities are extensive during these periods and often involve components and equipment located in radiologically challenging areas, access to which is limited during power operation. Major plant modifications requiring extensive work hours in radiation areas and the involvement of numerous work groups are also performed during outages. Consequently close involvement of RP in outage planning and scheduling activities is crucial to the success of maintaining annual collective personnel exposures to a minimum and minimizing radiological incidents.

Depending upon the scope of the outage and scheduled maintenance and plant modification activities there may be hundreds to perhaps thousands of individual work packages or work steps that are scheduled to be performed during an outage. Scheduling and planning activities for an outage begin several months prior to the scheduled start date. Planning activities associated with complex work activities and extensive design modifications may start as early as a year in advance. Those plants that have been successful in executing well-planned outages with good schedule adherence rates often "freeze" the scope of outage work activities a year before the outage start date. Though beyond the scope of this book organizations are encouraged to develop strict guidelines and policies concerning controls placed on outage activities and the process by which work activities are added to an outage scope after the "freeze" date.

Due to the number and magnitude of outage-related work activities and based upon the maturity of the work control program RP may have to assign permanent or temporary resources to the work control group. Depending upon the length and scope of a given outage additional RP resources may be required in order to effectively review and plan radiological work activities.

## 7.7  Radiological Safety Aspects of Work Coordination

Based on the above discussion it can be seen that the RP representative assigned to the work control process plays a vital role in ensuring that radiological work activities are properly planned and scheduled. The RP representative should poses a practical knowledge of operational radiation protection and understand plant system and component radiological conditions as a function of plant operational status. To maximize the benefit of RP's participation in the work control process the RP representative should have a basic understanding of plant systems and their operational interfaces. For example if a work task involves draining contaminated water from the reactor cavity to a refueling water storage tank the RP representative should be cognizant of the potential radiological impact this evolution may have in various areas of the plant. Planning and scheduling activities should identify the need to review the schedule to ensure that non-related tasks that may have been scheduled concurrently have been evaluated with regards to the potential radiological impact. Bear in mind that the operations work control representative may be focused on the operational aspects of an evolution. The operations representative will probably be focused on such items as whether there is sufficient capacity in the receiving tank to accommodate the transfer of the anticipated volume of water, have valve line-ups been performed and are the necessary pumps available to support the transfer. The value that the RP represent brings to the work control function is that a dedicated person is evaluating and considering the radiological implications of work activities. These considerations may not be readily apparent to planners, schedulers, plant operators, or maintenance personnel assigned to work control who have limited knowledge of plant parameters effecting radiological conditions and who are focused on their specific areas of responsibility. Consequently the effectiveness of the RP representative in work control can be greatly enhanced if this individual has received training in basic system interrelationships. This training should focus on how contaminated systems are connected to other plant systems and system configurations that could pose radiological concerns. The content of the training would not have to involve an extensive knowledge of system operating parameters, component or system interlocks, valve actuation set points, or otherwise be trained to the level of that required of a plant operator. The level of training should be sufficient to allow the RP representative to be aware of operational conditions that could pose radiological safety concerns. Simply being able to recognize such conditions would bring the issue to the attention of other groups that could then be utilized to evaluate the situation. Even though other work control representatives may be able to identify these type concerns, they are not specifically trained as radiation protection specialists and tasks could go unidentified that have potential radiological safety significance.

A major aspect in coordinating work activities at a nuclear power plant involves the dynamic environment and complexity under which many of these tasks are performed. Contributing to this situation is the fact that radiological conditions are also dynamic and often times difficult to accurately anticipate. The very fact that work is performed within the RCA imposes further constraints and adds to the complexity of work coordination activities. Consider some of the aspects involved with performing radiological work:

- Use of protective clothing
- Work package and radiation work permit requirements
- Respiratory protection equipment
- Alarming dosimeters

- Hold points for radiological protection purposes
- Use of communication equipment such as head-sets or radios
- Quality of radiation worker practices
- Communication of radiological terms and conditions to workers
- Changing radiological conditions

Industry events have occurred that identified contributing factors associated with failures to perform accurate radiological surveys, individuals not recognizing the significance of unanticipated radiological conditions, failure to anticipate radiological conditions and plan accordingly, and not communicating work area radiological conditions in a timely or effective manner. Oftentimes these contributing factors arose or were compounded by the radiological control measures established for the task. For instance the use of full-face respirators may have contributed to miscommunications. Radiation protection personnel must be aware of these issues and the importance of properly communicating radiological safety measures to workers. During the planning and scheduling stages the RP representative should work closely with responsible work groups to evaluate the potential impact that radiological control measures may have concerning the successful completion of a task. The planning stage provides an opportunity to evaluate these issues and to consider alternate measures that may be less intrusive. The importance of adequate planning is crucial in this regard, especially considering other factors such as working in a hot environment or in a high noise area that may further influence the effectiveness of proposed radiological control measures.

### 7.7.1  Pre-Job Briefings

Many times the first formal setting in which RP has the opportunity to communicate radiological safety details to workers occurs at the pre-job briefing session. This is a crucial link in the work coordination process. This is the point at which the job either starts out correctly or the stage is set for a series of miscommunications or wrong assumptions that ultimately result in a radiological incident. Communications at this stage should be very deliberate, not hurried, and thorough and prescribed radiological requirements understood by members of the work crew.

Various guidelines should be followed to enhance the effectiveness of pre-job briefings. Some sort of pre-job briefing checklist should be utilized. The checklist serves a useful function to help ensure that key elements are covered in all pre-job briefings. In addition to the radiological safety topics that are included in pre-job briefings the checklist may include other standard elements or topics that are useful to review in a briefing. These other topics could include such items as emergency procedures, use of self-checking or peer-checks, use of three-way communications and repeat backs, a reminder to follow procedures and steps in the work package, and perhaps industrial-safety related aspects of the task as well. Discussion of these and other non-radiological topics that may be covered during the pre-job briefing should be lead by the responsible group or party. The RP representative should present the radiological portion of the pre-job briefing.

The responsible work group supervisor or foreman should have the overall responsibility for coordinating and scheduling the pre-job briefing. This person should also ensure that all necessary personnel are present for the briefing. This may seem a minor point, however; this serves the purpose of demonstrating to work crews that the work group has responsibility for the radiological safety of their workers. If the pre-job briefing session is seen to be an "RP" responsibility and RP assumes the lead role in the briefing then a culture could inadvertently result in which workers come to rely solely on RP for their entire radiological safety and work group supervisors assume a subservient role. If work groups assume a more active stance towards their own radiological safety then overall performance will be greatly strengthened. By having the responsible supervisor taking an active role in the radiological safety aspects of a task demonstrates to the work crew that their supervision takes radiological safety seriously. This approach should be no different than the role that supervisors play every day when stressing the importance of industrial safety with their employees.

Support groups and other key persons such as quality control, safety, engineering, and perhaps a management representative for instance should also be in attendance. It is essential that the major work groups and support groups are represented in the pre-job briefing to ensure that personnel are aware of common issues and key aspects of the task. The presence of the key parties also promotes discussion and minimizes the chance of conflicting work requirements that may have been overlooked during the planning and scheduling stages.

Depending upon the radiological significance of a given job it may be appropriate to have an ALARA specialist present and perhaps even lead the detailed discussion concerning the radiological control measures that will be applied during the course of the activity. The RP technician that will be providing job coverage for the task should normally be present. If more than one technician will be required to support the task then ideally all the RP technicians should participate in the pre-job briefing. Regardless of the approach taken concerning who represents RP at the briefing the RP representative should summarize the radiological aspects of the job. Particular emphasis should be placed on any crucial steps requiring specific surveys to confirm radiation dose rates, airborne concentrations, and contamination levels or otherwise would constitute a "hold point" for radiological reasons. If the program utilizes job-specific Radiation Work Permits then the applicable RWP requirements should be reviewed at this time. This is the last chance to ensure that all members of the work crew clearly understand the purpose and reasons for requiring various radiological control measures and provides an opportunity for workers to ask clarifying questions. An overview of the radiological conditions, both actual and anticipated, should be presented. If a formal ALARA review was performed for the task then details of anticipated exposure estimates that may have been developed for the task should be provided. Based upon the total dose commitment estimated for the job this portion of the briefing may provide specific details concerning individual dose limits established for the job and the associated dose and dose-rate alarm settings that may have been established for individuals if electronic dosimeters are utilized.

The pre-job briefing could also serve to address administrative tasks if not otherwise previously handled. This could include such tasks as confirming radiation worker training status of individuals, completion of any job-specific training required for the job, successful completion and participation in any mock-up sessions prescribed as a pre-requisite for participating in the job. Respirator qualifications of individuals as appropriate and details associated with the issuance of any special dosimetry (e.g., multi-badge packages) may also be addressed at this time. Many of these items may be incorporated into an automated RCA access control system (see Chap. 6) that verifies such items as worker training status, respirator qualifications, current whole-body count, and approval to work under a given RWP and perhaps additional administrative requirements. If such a process is utilized or if these type matters are handled by other means then it may not be necessary to cover these items during the pre-job briefing.

Table 7.3 presents the radiological safety topics that could be included in a pre-job briefing checklist. A brief explanation of the subject matter that could be associated with each element is also provided. The exact details that may be covered under any particular item should be tailored to meet the needs of a given program and may differ somewhat from those offered below.

**Table 7.3** Pre-job briefing checklist

1. Provide an overview of the sequence of events for the job

The responsible work crew supervisor or foreman should provide a brief summary of the tasks associated with the job. Particular attention should be given to key steps or points in the evolution whereby radiological conditions may be subject to change and identify any specific hold-points where surveys are required. Radiation protection could highlight radiological concerns, protective clothing requirements, and provide any additional radiological information required by the work crew

2. Provide an overview of radiological aspects associated with the job

Using the most current or best available data provide status of the current radiological conditions present in the work area prior to the start of the job

Provide details of anticipated radiological conditions and hazards as the job progresses and at key points during the job when conditions will be subject to change

Review the actions that workers are to take in the event that unanticipated radiological conditions are encountered.

Identify any radiological "hold points"

Review access routes to and from the work area. (During outage periods for entries into drywells and containment buildings or reactor buildings it may be beneficial to review access routes to the job site to ensure that workers do not traverse high radiation areas or airborne contamination areas or areas where ongoing work activities could be impacted. The purpose of this item is to minimize unnecessary personnel exposures while in transit to and from work locations. If a system is available that displays plant areas via a video or digital photograph type system this could be utilized to show workers the access routes)

(continued)

Emphasize any environmental conditions present in the work area that could impact communications or work performance (e.g., noise levels, heat stress conditions, available lighting, physical space limitations, etc.). If special communication techniques will be required (e.g., hand signals) then these details should be reviewed

Identify low-dose waiting areas and review expectations for use of these areas

Use of temporary shielding should be identified and reviewed

3. Review lessons-learned

Any lessons-learned from previous performance of the activity, either from the specific station or industry experience from other utilities should be reviewed. Personal experience from members of the work crew should be solicited at this point and discussed

4. Contamination Control Measures

Summarize the use of any equipment or methods to be employed to control the generation or spread of contamination or to minimize the potential for airborne radioactivity. Discuss as applicable the placement of portable HEPA ventilation units, the use of catch basins or enclosure devices to contain contamination. Identify any specific points during the evolution that may require decontamination measures

5. Housekeeping and System Cleanliness

Review critical steps that have the potential for introducing debris into systems and the applicable cleanliness controls. If the task requires the use of formal foreign material exclusion measures then these requirements should be discussed

6. Dosimetry Requirements

Describe any special dosimetry requirements associated with the job. This could include such items as multi-badging, extremities, neutron dosimetry and details associated with electronic dosimeter alarm set points

7. Protective Clothing and Equipment Requirements

Summarize the protective clothing and equipment requirements associated with the activity. For complex evolutions such as double dress-outs or if protective clothing requirements may be upgraded or downgraded during the job explain the details associated with such changes. If portable filtration units, local radiation monitoring devices, or respiratory protection equipment will be utilized describe the details and functions associated with these requirements as they apply to the specific job

---

The work control process should have provisions for addressing revisions to a work package or RWP as the job progresses. These provisions should detail guidelines that would require that another pre-job briefing be held based on the significance of changes to work packages or as necessitated by radiological conditions.

### *7.7.2 Job Coverage and Execution*

Once pre-job planning activities have been completed and the work crew properly prepared to perform a given task it is essential that work be performed utilizing sound radiological work practices. Strict adherence to good radiological work practices by properly trained and qualified individuals are cornerstones for ensuring that radiological work is performed in a safe and competent manner. Pre-job planning efforts, training, and preparation activities could be for naught if good radiological work practices are not followed in-the-field. Adherence to basic radiological work practices should include such items as proper use of protective clothing, including donning and removal techniques; contamination control practices; basic ALARA principles and techniques; and routine verification of in-progress personnel exposures.

A considerable amount of time and effort is expanded by LWR radiation protection groups in providing services in support of maintenance and operations. These periods of direct interfacing afford an opportunity for developing worker awareness of the radiological aspects of their duties and for promoting a good working relationship between RP and other departments at the working level, which is so essential in maintaining an effective radiological safety program. Radiation protection personnel should constantly be on the alert to spot poor radiological work practices and to assist workers in the correct methods. If work procedures or governing documents do not adequately address associated radiological conditions then they should be revised as necessary. Specific hold points may prove useful at critical steps in procedures for radiological monitoring purposes that may not have been previously identified. Caution statements to inform workers that contaminated fluids may be encountered at certain stages or that dose rates are likely to increase as a result of a specific step, or similar type warning statements, may need to be incorporated into working procedures based upon experience. Radiation protection personnel should be cognizant of these type issues and assist in revising associated procedures and work documents. The opportunity for assisting workers in practical radiological safety work techniques should be taken advantage of whenever the opportunity arises. These efforts will help to promote lower personnel exposures over the long run and improve overall performance of radiological work activities.

As noted above the crucial interface between RP and work groups occurs at the working level. The job coverage RP technician can oftentimes initiate actions to ensure the radiological safety of workers. Even if all else fails there is one last opportunity to prevent radiological incidents from occurring. Personnel providing job coverage should not assume anything and should always be prepared for the worst. The same degree of effort afforded to workers in preparing for the job must also be given to RP personnel. Details associated with the functions and responsibilities of the job coverage technician must be clearly communicated. Abbreviated communications of job preparation and coverage details among RP personnel could result in serious consequences. Personnel must guard against incorrect assumptions being formed as a result of their familiarity of common terms used within RP or having covered the job on previous occasions.

The following activities should be taken into consideration in order to assist with job coverage technician preparations and responsibilities.

1. The job coverage technician should have participated in the pre-job briefing if appropriate. If the job coverage technician did not attend the pre-job briefing then mechanisms should be in place to ensure that the job-specific details are provided to the technician.
2. Verify the job location, component tag number and unit as applicable. Even though this may be the responsibility of the work group lead, the job coverage technician should also be aware of the component and plant system involved and the work area location.
3. The job coverage technician should be aware of any radiological safety hold points or special survey requirements associated with the task.
4. If pre-staging activities were required (e.g., the placement of temporary shielding, contamination control barriers or catch containments) for radiological purposes, the technician should verify that these activities have been completed and the area properly prepared prior to start of work.
5. When assigned to job coverage the technician and supervisor making the assignment should use repeat backs during the communication process. This may be vitally important when communicating those facets of the job that are crucial to ensuring radiological safety.
6. If unforeseen circumstances arise the job coverage RP technician should not hesitate to implement necessary measures, which could include stopping the job, to ensure the radiological safety of workers.
7. If job duration will necessitate the need for breaks the job coverage technician should ensure that these breaks occur at points when radiological work conditions allow and conditions are otherwise stable. Breaks should not be taken during crucial stages such as when breaching a contaminated system, when air sample data is being collected for assessment of airborne exposure to workers, unstable radiological conditions exist due to the current stage of the task, or when breaks could result in additional re-work or redundancy resulting in additional exposure.
8. If job coverage activities are transferred to another technician while the task is still in progress then the turnover should be accomplished at the work location. The turnover must be sufficiently detailed to ensure that the relief RP technician fully understands the nature and scope of work and the status of radiological safety measures.

The job coverage technician is a key player associated with ensuring the proper performance and implementation of radiological work activities. If for whatever reason workers make it to the field and do not understand RWP requirements, hold points, radiological conditions or some other radiological aspect of the job then the job coverage technician has an opportunity and responsibility to address the issue before a problem arises. Consequently it is essential that RP technicians assigned to jobs have been properly and thoroughly briefed, understand the radiological concerns associated with the job (both actual and potential), and are aware of the actions to take in the event that unanticipated radiological conditions are encountered.

Radiation protection job coverage does not end upon completion of work activities. Radiation protection personnel should ensure that the work area has been properly secured to prevent the spread of radioactive contamination. System components should be properly sealed or placed into an operational or standby mode and not left in a dismantled configuration. Formal work control programs may have requirements that the responsible supervisor ensure that work areas are properly secured upon completion of an activity. The work area should be decontaminated to the maximum extent practical with due consideration given to ambient radiological conditions. For instance, in may not be practical to decontaminate a work location to clean area levels if there is an active contaminated leak in the area that would soon render the area contaminated. Additionally, if the work was performed in a high radiation area, extensive decontamination efforts should be delayed until such time as ambient radiation levels are reduced. Under these conditions efforts to reduce gross contamination levels may be all that is warranted. Again, the extent of any post-job decontamination should be determined on a case-by-case basis. The importance of following good contamination control practices during the performance of a task is an important element in minimizing personnel exposures during post-job recovery efforts and for subsequent operation and maintenance activities. Programs that emphasize the importance of minimizing the extent and magnitude of radiological zones and take steps to implement an aggressive policy in this regard will reap long-term radiological safety benefits.

Tools and equipment should be monitored for contamination prior to leaving the work area. Contaminated items should be packaged and labeled and transferred to a designated radioactive material storage area or to a location for decontamination, as appropriate. A post-job survey should be performed as necessary, signposting updated, and zoning materials re-established. In general, good housekeeping practices should be followed and the area left in a clean and orderly state.

### 7.7.3 Post-Job Debriefings

The best planning efforts may still not identify every possible measure that could have been utilized to lower exposure for a given set of circumstances. Additionally, work processes or procedures may reveal areas needing improvement that were not previously considered or identified, or up till now were otherwise effective. Control measures that may have appeared to be practical during pre-job planning and review sessions may in fact prove counterproductive from an exposure management perspective. For instance, the work area may be encumbered with too much equipment resulting in space limitations that required additional time to relocate equipment and tools during the performance of the task. The configuration or operating status of nearby equipment may increase noise levels in the work area, impacting communication among members of the work crew, increasing the time required to complete the task. Consequently work control programs should include provisions for assessing the effectiveness of pre-planning efforts during and after the performance of an activity. Depending upon the nature of the job this evaluation could be performed by dedicated individuals, specially trained in observation techniques, or delegated to a member of the work crew, or workers could be required to provide feedback upon completion of the job.

The purpose of a post-job debriefing is to provide an opportunity for the work crew to collectively reflect on the performance of the task. Synergies may occur as a result of having a "team" debriefing. To ensure that appropriate and accurate information is obtained upon completion of the job, the post-job debriefing should be performed as soon as possible. This is especially important for activities in which significant radiological safety issues were encountered or activities did not go as planned. If a post-job critique is held several days after the performance of a given job, opportunities to solicit recommendations or areas for improvement may be lost. This may be particularly important during outages if contractor personnel were members of the work crew. Contractors may only be available for a short period of time after completion of a given job. Therefore mechanisms should be in place to ensure that debriefings are conducted for contractors and temporary employees before they depart the site.

To ensure that effective post-job debriefings are performed certain guidelines should be followed.

- Ideally the debriefing should be done in one setting and include all members of the work crew. This will assist with the accurate identification of issues and a chance for all workers to contribute.
- An ALARA specialist should facilitate the discussion and make note of items and lessons-learned that may contribute to dose savings when the activity is next performed.
- If appropriate other personnel who may not have been directly involved with the field work, may also participate in the debriefing. This could include supervisors, support personnel, planners and schedulers and related groups.
- Some method should be utilized to formally record and collect the information obtained during the debriefing. This should include a process to assign responsibilities and track action items to completion.

It is essential that lessons-learned and potential improvement items be identified for later evaluation. Post-job debriefings play a vital role in preventing and minimizing future incidents. Oftentimes the job supervisor may be responsible for conducting these debriefings. Radiation protection personnel should ensure that workers are afforded an opportunity to identify improvements, especially if unforeseen problems were encountered during the performance of the activity. Lessons-learned should be actively solicited and captured to ensure continual improvement in radiological safety performance. Workers should be encouraged to identify possible improvements regardless of the magnitude of the radiological conditions or dose savings. It should be recognized that the identification of lessons-learned, even for those activities that may not require a formal post-job debrief, may result in significant radiological safety improvements. The thoroughness of the post-job critique is an important element in achieving continued improvement in radiological safety.


## 7.8  Radiation Work Permits

Radiological safety requirements must be identified and appropriate measures implemented to control and minimize personnel exposures during the performance of radiological work activities. Throughout this chapter reference has been made to a document referred to as a Radiation Work Permit (RWP). This is the document commonly used within the industry to communicate the radiological controls required for a specific task. An RWP may be issued by the RP department to specify the activity or task to be performed and may include any necessary protective equipment and clothing, special precautions, the expected and anticipated radiological conditions, and the degree of radiation protection job coverage required. Additionally the names and signatures of individuals who are to perform the activity, along with their available dose margins may also be provided on the RWP or a companion document. The purpose for associating individuals to a given RWP is to signify that these individuals have attended any pre-job briefing that may have been required, are trained and qualified to perform the task, and are prepared to perform the task. Signatures signify that individuals have read and understand the RWP and associated radiological safety measures that have been prescribed.

To meet the specific needs of a given utility or station, the format and content of the RWP may differ somewhat from one facility to the next. However, for an RWP to be useful it should specify a minimal amount of essential information. This basic information may be common to most LWR radiation protection programs. Figure 7.5 depicts an example of an RWP. The RWP is an important document that should be completed and issued by suitably trained and qualified individuals. The RWP in Fig. 7.5 is a basic RWP and the contents and format of RWPs should be designed to meet the needs of a given RP program. The RWP depicted includes those elements generally associated with a radiation work permit. In addition there may be addendum sheets as noted previously that may include the listing of authorized personnel and signature places for workers.

The discussion above applies to the situation whereby RP essentially prepares and issues the RWP. As noted in Chap. 6, computerized access control systems afford flexibility pertaining to the authorization for entry into the RCA. Work control centers also provide opportunities for incorporating the development of the RWP into the planning and scheduling process. In some cases RWP attributes may be provided directly in the work package, eliminating the need for a stand-alone RWP document. The computerized RCA access control process may incorporate steps that require individuals to acknowledge that they have read and understand the RWP requirements,

**Number:**________

| | |
|---|---|
| Date: __________  Time: __________  Valid Through: _____________ | |
| Work Order No. ________________  Work Order Task No. _______________ | |

Work Description:

Location:

### Radiological Conditions

| -Dose Rates- | -Smearable Contamination- |
|---|---|
| Gen. Area __________ μSv/hr | General Area __________ Bq/100 cm$^2$ |
| Work Area __________ μSv/hr | Maximum __________ Bq/100 cm$^2$ |
| Maximum __________ μSv/hr | |

### Required Dosimetry and Protective Equipment

-Dosimetry-

| | TLD | ED | | TLD | ED |
|---|---|---|---|---|---|
| Whole Body | | | Extremities | | |
| Chest | __________ | __________ | Upper | __________ | __________ |
| Head | __________ | __________ | Lower | __________ | __________ |
| Back | __________ | __________ | | | |
| Upper Arms | __________ | __________ | | | |
| Upper Legs | __________ | __________ | | | |

-Electronic Dosimeter-

ED Alarm Setpoints:     Dose: __________          Dose Rate: __________

-Protective Clothing-

| Hands : | Body: | Feet: |
|---|---|---|
| ☐ Cotton inserts | ☐ Coverall | ☐ Booties ☐ 2 pair |
| ☐ Rubber gloves ☐ 2 pair | ☐ Disposable | ☐ Shoecovers |
| ☐ Work Gloves | ☐ Labcoat | |
| ☐ Surgeons glove | ☐ Wet Suit | Head: |
| | ☐ Top | ☐ Skull cap |
| | ☐ Bottom | ☐ Hood |
| | ☐ Top and bottom | ☐ Face shield |

-Respiratory Protection-

☐ None Req. ☐ Full-Face Air Purifying ☐ Full-Face Supplied Air ☐ Hood    ☐ SCBA

Special Instructions:

| | |
|---|---|
| Job Coordinator: __________________ | Phone: ________ |
| Approved by: __________________ | Date: ________  Time: ________ |

**Fig. 7.5**  Radiation work permit

thus eliminating the need for a hard-copy sign-in or acknowledgement form. Additionally, the automated RCA access control data base may be utilized to verify that all prerequisites for entry into the RCA for the given RWP (or work package) for an individual are satisfied as part of the authorization process. Based on the philosophy pertaining to radiological safety and the maturity of the work force, work procedures could include requirements pertaining to radiological safety measures associated with the task covered by the procedure. Appropriate steps in work procedures could include such items as the following:

- Contact RP for survey prior to performing the following step
- Work area dose rates may increase at this step of the procedure
- Prior to breaching the system contact RP
- Don respirators prior to performing the next step
- Continuous RP coverage required at this stage

Obviously these are generic examples, in actual practice specific radiological safety measures would be more detailed and could be incorporated as procedure steps warning of changing radiological conditions. The use of caution statements prior to the performance of a task or evolution to provide warning of potential changes in radiological conditions may be beneficial. In some cases various generic-type radiological safety measures may be more appropriately provided under the precautions or prerequisites sections of work procedures. This could include items such as the need to obtain an RWP prior to the job or a reminder to follow specific ALARA work practices.

Regardless of the method employed to develop and issue RWPs, and whether or not separate RWP documents are utilized or applicable requirements incorporated into work documents, the details of the task must be known in order to prescribe effective radiological safety requirements. The job description may be provided by a RP representative or preferably by the job supervisor. The details of the job description should include such information as the start time and date, job location, number of workers, the expected duration of the activity, the system(s) involved and a detailed summary of the task to be performed. It is essential that the job description be sufficiently detailed to allow RP personnel to understand the scope of the job. Job descriptions such as "repair valve" or "inspect gauge" are of little value in determining radiological safety measures for the job. Will work involve a contaminated system, what type of repair or inspection work is involved, is breach of a system required, are just a few of the parameters that must be known in order to complete an RWP. In fact an RWP should not be initiated until an adequate job description has been obtained.

A convenient method of identifying those activities that typically pose radiological concern is to provide a checklist on the RWP or in the work package for such activities as welding, cutting, grinding, system breach, use of air powered tools and similar activities that by their very nature could impact radiological conditions during the performance of an activity. The individual initiating the RWP could simply check off the tasks involved in the work activity providing a convenient mechanism to inform RP personnel during the planning and scheduling stages that certain activities posing special radiological concerns will be involved with the activity. Once an adequate description of the task is provided then appropriate radiological safety requirements may be established for the RWP.

Under certain circumstances a pre-job radiological assessment of the work area may be required. This may be necessary due to the nature of the job, or lack of current survey data for the work area, or the need to obtain more detailed dose rate data to perform pre-job dose assessments if significant personnel exposures are involved. The pre-job radiological data should be provided as an addendum to the RWP or otherwise communicated to the work crew. Depending on the circumstances and location of the job, pre-job radiological survey data may include any, or all, of the various radiological hazards including: gamma, beta, and neutron radiation levels as appropriate, and surface and airborne contamination levels.

## 7.9  ALARA Work Plans

Depending upon the complexity of a given job and the estimated collective dose that will be received a comprehensive work plan may be developed based on the radiological significance of the task. Guidelines may be established that provide dose thresholds requiring more extensive pre-job evaluations and planning. These thresholds could include values for the total collective dose received by the work crew as well as a value for the highest individual worker exposure. Pre-job collective dose estimates exceeding 10 mSv for example, could trigger the need for a formal pre-planning review of the given task. Additionally if a member of the work crew is estimated to exceed 5 mSv, regardless of the estimated collective dose to the entire work crew that may also trigger the need for more thorough pre-planning efforts. The thresholds established for initiating more extensive pre-planning efforts should be based on historical performance and the specific needs of a given plant. A graded approach may also be useful triggering the need for additional planning and preparation efforts based on pre-job dose estimates. Regardless of the approach taken the program should result in the initiation of what is commonly referred to as an "ALARA work plan."

The establishment of formalized "ALARA Committees" is an important tool in minimizing station exposures over the long run. Guidelines noted above could serve as formal triggers for initiating review of projects and work activities that may exceed a certain dose threshold. The ALARA Committee should be comprised of representatives from key discipline areas. Representatives from plant management, engineering, operations, the various maintenance disciplines, industrial safety, chemistry and perhaps others in addition to radiation protection should be assigned to the ALARA Committee. Representatives should be at a high enough organizational level to have the authority to approve the expenditure of funds to implement dose reduction measures. The committee should have the authority to direct changes in the work scope, to require certain provisions that they deem necessary to minimize exposures and have overall authority to approve or disapprove a project based upon the committee review.

The ALARA Committee should not be perceived to be a solely radiation protection function. Nor should the committee chairperson be a representative of the radiation protection organization. The cross-discipline representation of the committee at a relatively high organization level demonstrates to site personnel that the control of station exposures is a site-wide responsibility. Obviously RP will play an important role and will be closely involved with the major work groups in the development of ALARA work plan packages. The ALARA representative assigned to the ALARA work plan will most likely play a key role during formal presentations to the ALARA Committee.

The degree of involvement of the ALARA Committee often is based on a graded scheme. This scheme should specify the role of the committee based on estimated person-sievert exposures. Tasks or projects with estimated exposures between 10 and 50 mSv could require that committee members review the ALARA work plan individually without the need for a formal presentation to the collective committee. Tasks exceeding 50 mSv may require a formal, detailed presentation, of the ALARA work plan to the ALARA Committee. The ALARA Committee chairperson may have to formally approve these ALARA work plans. The dose threshold that triggers ALARA Committee reviews may be different than those noted here. The involvement of the ALARA Committee should consider historical dose performance. If the plant is a perennial high dose plant then it may be appropriate for the ALARA Committee to formally review projects at the 10 mSv or even the 5 mSv level. On the other hand plants that are industry leaders in dose performance may elect to have the ALARA Committee review projects that would not otherwise rise to the level requiring a formal review. For instance if 50 mSv is the "formal" threshold for initiating ALARA Committee involvement and there are few or no projects exceeding this threshold for a typical outage then the committee should have the authority to designate specific projects for review regardless of the dose estimate. The important point is to guard against the establishment of high dose thresholds that either minimize or eliminate the involvement of the ALARA Committee. The ALARA Committee affords an opportunity for station management to become more aware of those activities that contribute significantly to station dose. If the opportunity to engage cross-discipline managers in reviewing key dose-significant tasks is not fully utilized then the identification of important dose reduction initiatives may be lost.

ALARA work plans should address several key elements. Tasks requiring review by the ALARA Committee should have a formal ALARA work plan developed and submitted to the committee for review and approval. Those items that should be addressed by an ALARA work plan are summarized below. The key elements are presented along with a summary of their purpose and function.

## ALARA Work Plan

**Job or Task Title:** ________________________________________________

**ALARA Work Plan Number:** ___________________  **Revision:** __________

**Radiation Work Permit:** ____________  **RWP Work-Hour Estimate:** ________

**Dose Estimate (mSv):** __________

| |
|---|
| Prepared by: ___________________ Date: ___________ |
| Approved by: ___________________ Date: ___________ |
| (Project Manager) |
| Approved by: ___________________ Date: ___________ |
| (RP or ALARA Coordinator) |
| Approved by: ___________________ Date: ___________ |
| (ALARA Committee Chairperson)* |
| *Note:  Required for dose estimates $\geq$ 50 mSv |

1. Job or Task Description
   Provide brief description of the job or task
2. Historical Radiological Performance
   If the task has been performed previously (such as repetitive outage activity) provide a summary of previous dose totals accrued for the job. The associated historical RCA-work hours should also be provided.
3. ALARA Measures and Radiological Safety
   This section of the ALARA work plan should describe in detail those radiological measures established for the job to minimize worker exposures and to support the overall radiological safety of the crew during the performance of the activity. This section may include such items as the need for specialized training including mock-up training for individuals prior to performing the task. The details associated with the use of any specialized tooling or equipment should be provided. The use of experienced workers may be specified for certain steps or evolutions.

   The establishment and location of low-dose waiting areas in support of the job should be specified. The use of engineering controls, temporary shielding, portable filtration units and any other radiological control measures should be detailed. The establishment of dose and dose-rate alarms for electronic dosimeters for specific stages of the task should be described.

   If specific system or equipment configurations are required to support various stages of the job then details should be included here. This could include such elements as the need to perform a system flush prior to starting the job or verifying that components are drained. Components in the vicinity of the job (e.g., heat exchangers and waste hold-up tanks) that may contribute to radiation levels at the job site may be required to be filled with "clean" water to reduce radiation levels due to radioactive material content of nearby components.

Any other radiological control measures identified for the task should also be included here. This could include the use of respiratory protection equipment, measures to minimize the spread of contamination, airborne contamination control measures, basic time-distance-shielding controls to be followed at various stages of the activity, and protective clothing requirements.

4. Previous Lessons-Learned

   Previous lessons-learned should be compiled and listed in this section of the ALARA work plan. Mechanisms should be available to track and record previous lessons-learned to a specific task, work package, or ALARA work plan. During the development of the ALARA work plan the lessons-learned data base should be reviewed and related events identified.

   Previous lessons-learned should be reviewed and discussed with the work crew. Emphasis should be placed on ensuring that workers understand the causes and corrective actions stemming from previous performance of the activity that contributed to higher than estimated exposures. Lessons-learned play a key role in supporting continued improvement in dose reduction efforts. The effective implementation of corrective actions identified in conjunction with lessons-learned is vital in preventing the repeat of events that contributed to unnecessary personnel exposures or radiological incidents.

5. Industry Operating Experience

   This area is similar to those items addressed under "Previous Lessons-Learned" above. Whereas the items covered under Point 4 pertain to in-house experience this section summarizes any pertinent industry "lessons-learned". Industry operating experience may contain a wealth of information pertinent to the task covered by the ALARA work plan. This section should summarize industry operating experience related to the task. Oftentimes an ALARA work plan may be identical in scope and coverage to similar jobs performed routinely at other nuclear plants. It is essential that this experience be incorporated into the ALARA work plan to improve the overall radiological safety performance of the task and to prevent repeat occurrence of previous industry events.

   Specific industry incident reports should be identified and reviewed with the work crew. Any measures incorporated in the ALARA work plan as a result of industry operating experienced should be clearly communicated and the importance of effectively implementing the corrective actions emphasized.

6. Project Dose Goal

   The time and effort required to develop the work plan dose goal will depend upon the complexity and scope of the task. The basis for the dose goal may be detailed in a separate document attached to the ALARA work plan. The key aspect is to ensure that the dose goal is developed in close coordination with the work crew. Details involving the number of workers required, the types of craft involved, the equipment and tooling utilized that may impact dose estimates, the time required for craft to perform each task for which a dose estimate will be developed all require input from the members of the work group. The availability of historical dose records and previous performance records for the task may prove vital in developing an accurate dose estimate. The dose goal should include all the major steps to complete the project. A matrix may be developed that depicts the number of workers required for various steps of the project, the total person-hours to complete the step and the dose rate figures that were used in estimating exposures for each separate task. Based on a detail review of the work activities an overall dose goal is assigned.

Depending upon the estimated dose for the project and the duration of the project, daily dose goals may be established and tracked on a daily basis, or more frequent, to provide early indication of adverse dose trends. The tracking of daily dose trends could trigger the need for an in-progress review as discussed in Point 7 below.

The project dose goal should be reviewed with the work crew. Each individual should be aware of the dose assigned for the specific tasks they are to perform and the actions to take in the event that task-specific dose goals are in jeopardy. A key aspect in developing the dose goal for a project is to obtain "ownership" on the part of the work crew. Dose goals developed in isolation by the ALARA group or RP that do not have the support of the work groups are less likely to be met. If individual workers are consulted and have had a role in developing dose estimates for the tasks they will be performing then the chances of meeting established dose goals will be greatly improved. The achievement of the overall dose goal for the project requires a team effort. Equally important the likelihood of identifying lessons-learned and suggestions for future improvements will be greater when workers have ownership of their activities.

7. In-Progress Review

To ensure the timely identification of an adverse trend in dose performance for the work plan and to allow sufficient time to identify and implement corrective actions the need to perform an in-progress review may be specified. The frequencies at which these in-progress reviews are to be performed are typically predicated on the overall dose estimate. The number of in-progress reviews is usually greater for those projects with higher dose estimates.

The in-progress reviews are usually assigned based on the percent completion of the project. These reviews could be triggered at 50% or 75% completion for instance. Additionally reviews could be performed during key stages of the task upon completion of one of the more dose-significant phases. These reviews should be performed by representatives from the ALARA group and the key work group. The purpose of in-progress reviews is to verify that dose estimates are accurate and that any deviation or unforeseen event that may impact the basis for the initial dose estimate are identify. If necessary the results of the in-progress review may require the ALARA Committee to review and approve revisions to the ALARA work plan and dose estimate.

## 7.10  Scaffold Management

A major area that has a significant impact on outage exposures relates to the installation and removal of scaffolds in support of outage activities. This key area should not be overlooked concerning a stations' exposure management program.

Effective "scaffold management" is a key element in minimizing outage exposures. Proper coordination of scaffold management activities may also eliminate outage exposures resulting from repetitive installation and removal of scaffolds.

Many of the LWRs constructed over the period from the 1960s through the late 1980s and early 1990s were not optimally designed to support routine maintenance activities. Perhaps the same situation can be said for units recently constructed? Issues such as the location of valves and components requiring routine inspection, testing, or maintenance were placed in confined locations or overhead areas or otherwise located in areas not conducive to support maintenance activities. An area that posed unique radiological challenges was the need to build and install numerous work platforms during outages to support maintenance activities. Oftentimes the installation of ladders, scaffolds and other work platforms took place in high radiation areas and at a time during outages that limited opportunities to employ dose reduction initiatives. Compounding this issue was the fact that these temporary access and work platforms had to be installed and removed during each outage. Thus these activities represented a repetitive dose contributor for each maintenance or refueling outage. As industry annual dose totals decreased over the last 30 years, it has become more apparent that exposures relating to the erection and removal of temporary work platforms often represent a major contributor to outage dose totals.

Though this discussion is primarily focused on the dose aspects associated with the installation and removal of temporary work platforms the financial costs associated with these activities may also be considerable. Dose expenditures, costs of maintaining an inventory of scaffold supplies, and the resources required to install and remove temporary work platforms during outage periods should not be overlooked. The installation and removal of scaffolds and work platforms must be properly planned and scheduled. These activities should be coordinated by the work control center. Poor coordination could result in work crews showing up to perform a task only to find out that the required scaffold has not been installed. Inadequate planning and scheduling could result in scaffolds being erected for one task that interfere with another task or the repetitive installation and removal of the same scaffold. Obviously all these situations may result in additional radiation exposure to workers. Many work control centers assign resources specifically to the coordination of the installation and removal of scaffolds. A common practice is to designate each work package requiring the use of a particular scaffold to that scaffold. A prerequisite for removing the scaffold would be verification that each work package requiring use of the scaffold has been verified field-work-complete.

Focus on shortening the length of outages over the last 10–15 years has also contributed to industry process improvement efforts pertaining to the overall management of scaffolds and temporary work platforms. Several tens of millisieverts of dose may be expended on these activities during an outage. These doses could be higher for outages with significant design modification work or major projects (e.g., steam generator replacement). To continue the long-term downward trend in annual dose totals it was recognized that improvements in the management of "vertical access" would be necessary. The term "vertical access" is now commonly used within the industry and encompasses those activities associated with temporary work platforms, lifts, and other temporary measures required to gain access to components or equipment.

Each plant should evaluate its vertical access needs required to support outage activities. The aim of the evaluation should be multi-targeted and include such initiatives as:

(1) Reducing the number of scaffolds required;
(2) Erecting seismically-qualified permanent platforms to eliminate the need to erect scaffolds in the future;
(3) Designing easier to install scaffolds; and,
(4) Reducing the time required to erect and remove scaffolds.

Obviously if the number of scaffolds can be reduced this will result in lower dose totals for an outage. Secondary benefits may include fewer workers required to install and remove outage scaffolds and perhaps reduce the time required to perform a given task. The next logical step would be to evaluate the possibility of reducing the time needed to install and remove those scaffolds that are required. Conventional "tube and clamp" scaffolding requires the use of tools and for multi-platform scaffolds could require significant work hours to install and remove. Measures to reduce the time required to erect these scaffolds would have resultant dose savings. This would be especially important for those activities requiring the installation of scaffold in high radiation areas. Scaffolds supporting outage activities erected in such locations as inside the biological shield wall, inside the drywell, or for inspection activities on recirculation piping, and other components located in high radiation areas would be prime candidates for exploring ways to reduce associated work hours.

The use of modular scaffolds or work platforms could reduce the time required to install and remove scaffolds. Another option would be to use "quick-erect" scaffold. A scaffold design that has gained widespread acceptance in the LWR industry is the Excel Modular Scaffold System. The advantages of this system include quick connect locking mechanisms that do not require hand tools to connect pieces of the scaffold thus reducing the time to install and remove scaffold units. Figure 7.6 depicts a close-up view of the Excel locking mechanism that engages and locks onto the tubular support sections. A significant savings in the time required installing and removing an Excel scaffold versus a tube and clamp scaffold can be achieved. The Excel scaffold system is also seismically qualified for use in permanent installation packages. The benefits of installing permanent scaffolding whenever possible are discussed later in this section.

The modular scaffold system is quicker to install and remove resulting in less time spent in radiation areas. The ease of assembly also requires fewer workers to install the same size scaffold arrangement as compared to the tube and clamp design.

It should be emphasized that the vertical access program also pertains to the installation and use of work platforms that are not necessarily true scaffolds. In fact dose savings associated with the installation of work platforms to gain access to such areas as the reactor vessel head, steam generator platforms, or recirculation

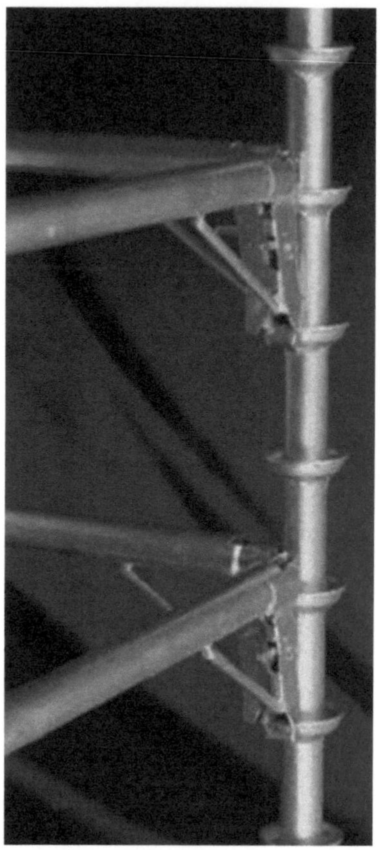

**Fig. 7.6** Quick connect fittings utilized in the Excel modular scaffold system (Courtesy of BHI Energy www.bhienergy.com)

piping may also represent significant dose expenditures that accrue for each outage. Specially designed platforms utilized for a specific purpose that facilitate installation and removal of work platforms and access walkways would have similar benefits as those noted above. The supporting structure for these work platforms should maximize the use of quick-erect scaffold components. Figure 7.7 shows a reusable work platform specifically sized to fit around a reactor head. Note the use of Excel scaffolding for the safety handrail. This feature reduced the amount of time required to install and remove the handrail.

Temporary shielding packages are often installed by hanging lead blankets or shielding materials from racks built for that purpose. Scaffold materials may be utilized to construct these frames or racks from which temporary shielding may be hung. Use of the scaffold frames eliminates the need to place the shielding material directly on components or system piping. This may be necessary since the added weight of the lead shielding may render a system inoperable for the period in which the shielding will be placed or the additional weight may pose seismic or other system-operational constraints. The use of the quick-erect or modular scaffold systems could also prove beneficial in these cases also. Many utilities have

**Fig. 7.7** Reactor vessel head work platform for use during outages. The platform was specifically designed for the work location and individual pieces fabricated to facilitate installation and removal (Courtesy of BHI Energy www.bhienergy.com)

performed engineering analyzes to install shield frames inside reactor and containment buildings permanently thus negating the need for repetitive installation and removal of the shield racks each outage. The temporary shielding will still have to be placed each outage but often times the installation of the shield racks represents the primary dose contributor associated with the placement of temporary shield packages.

For these reasons locations that require extensive placement of temporary shielding during outages should be evaluated for the installation of permanent shield racks. Even though lead shield blankets may not remain in place during power operation the installation of a permanent rack greatly facilitates the placement of temporary shielding material during subsequent outages. The presence of permanent shield racks allows temporary shielding to be installed in a sequence to maximize shielding to workers as the installation progresses. While installing the racks to hold the shielding no temporary shielding is present during the installation of the racks. The time required to install the shield racks is often much greater than the time required to place the shielding itself. However once the racks are in place lead blankets may be hung in a sequence to take advantage of the shielding already installed. This "shield-as-you-go" installation process results in additional dose savings. Permanent shield racks not only save exposures associated

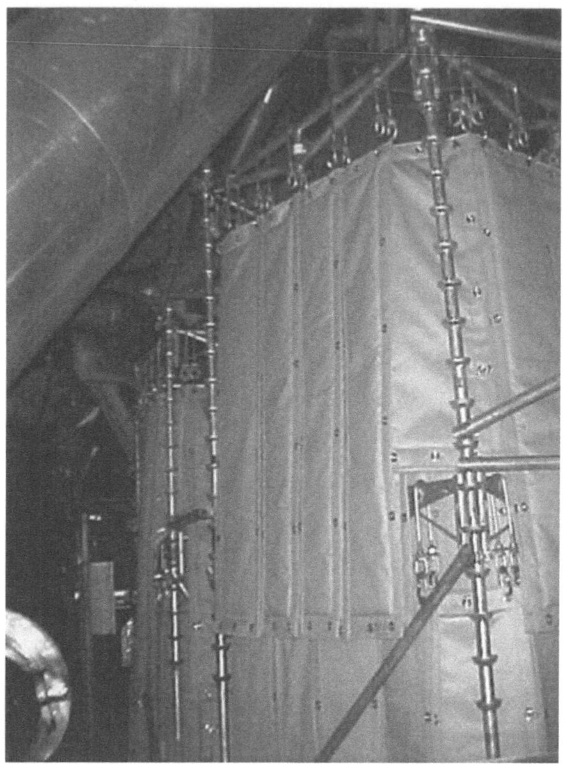

**Fig. 7.8** Use of quick-erect Excel scaffolding as a rack to hold temporary shielding (Courtesy of BHI Energy www.bhienergy.com)

with the installation and removal of the racks themselves but once in place also decrease exposures received in subsequent outages since the shielding may be hung in a more efficient, dose-saving sequence.

The dose savings may justify the costs to perform these engineering evaluations. In some cases the dose savings coupled with costs associated with temporary shield packages that impact critical path time may be sufficient to support permanent installation of shield racks. Figures 7.8 and 7.9 depict the use of Excel scaffolding as frames to support temporary shielding. The use of hooks to hang the lead shield blankets from the frame minimizes the time required to install and remove the shielding.

The number and complexity of scaffolds required to support outage activities can represent a significant source of radiation exposure received during outages. Utilities should ensure that programs are established to minimize exposures associated with this program. The Electric Power Research Institute, Scaffold Program Optimization and Dose Reduction Guide, is an excellent resource addressing this important area. EPRI-member utilities and organizations seeking additional guidance in this area should find the report useful. Each plant should perform an assessment of their vertical access program and incorporate the results of this assessment into their long-range dose reduction initiatives. This is an area where

**Fig. 7.9** Use of quick-erect Excel scaffolding as a rack to hold temporary shielding (Courtesy of BHI Energy www.bhienergy.com)

significant process improvements may be achievable and should not be overlooked. Even though the emphasis here is directed towards the dose savings aspects of a vertical access program, as noted previously, cost savings and overall productivity and industrial safety improvements may be equally, if not more, significant.

## 7.11  Summary

Exposure management involves numerous inter-connected activities all of which must be properly coordinated to achieve maximum dose reduction results. These activities include proper planning and scheduling of radiological work activities and comprehensive radiological preparations and execution of work activities. Additionally the important role that organizational relationships and interdepartmental and intradepartmental communications play in ensuring the effective implementation of dose control measures must be recognized. Often times the success of dose reduction initiatives depends more on the training and qualification of workers, how well a specific task is coordinated, communicated and implemented versus the effectiveness of measures implemented to actually minimize radiation levels at the work location.

## Bibliography


1. Electric Power Research Institute, Scaffold Program Optimization and Dose Reduction Guide, EPRI Report 1021102, Palo Alto, CA, 2010
2. International Atomic Energy Agency, Safety Guide No. RS-G-1.3, Assessment of Occupational Exposure Due to External Sources of Radiation, Vienna, 1999
3. McKenna P., Planning & Scheduling to Minimize Refueling Outage, Nuclear Plant Journal, 27(3):20–22, 2009
4. National Energy Agency, Organization for Economic Co-Operation and Development, Optimisation in Operational Radiological Protection, A Report by the Working Group on Operational Radiological Protection of the Information System on Occupational Exposure, 2005
5. National Energy Agency, Organization for Economic Co-Operation and Development, Occupational Exposures at Nuclear Power Plants, Eighteenth Annual Report for the ISOE Programme, NEA No. 6826, 2008
6. States Nuclear Regulatory Commission, Occupational Radiation Exposure at Commercial Nuclear Power Reactors and Other Facilities 2001, Thirty-Ninth Annual Report, NUREG-0713, Washington, D.C., 2006
7. States Nuclear Regulatory Commission, Occupational Radiation Exposure at Commercial Nuclear Power Reactors and Other Facilities 2001, Forty-First Annual Report, NUREG-0713, Washington, D.C., 2010
8. U. S. Nuclear Regulatory Commission, Regulatory Guide 8.19, Occupational Radiation Dose Assessment in Light-Water Reactor Power Plants Design Stage Man-Rem Estimates, Revision 1, June 1979
9. Wolge E., Integrated Exposure Reduction Plan, 39-41, Nuclear Plant Journal, May–June 2008

# Chapter 8
# Source Reduction

## 8.1 Overview

The radiological environment encountered at a given LWR facility can be governed, to a large extent, by the approach taken pertaining to the management of contamination control and source term reduction programs. Depending upon the approach taken the results may be wide ranging. Where management expectations are less stringent the use of protective clothing and radiological control measures may be more prevalent. These controls may extend to such routine activities as operator rounds and inspections and other non-work type tasks. On the other hand at those facilities where aggressive contamination control and source reduction programs are implemented the need for radiological control measures will be minimized. This could range to facilities that afford entry into the containment building (at PWRs) for minor maintenance or inspection type tasks, with minimal or no protective clothing. The point to emphasize is that the degree of radiological control measures required at a given facility is more often predicated by the existing culture that management establishes when dealing with the presence and the control of contamination and source terms. If the attitude is one of complacency then the radiological environment will probably be one where large areas of the RCA are routinely zoned contaminated and higher than average radiation levels may be present in the vicinity of various components and systems. Higher annual collective radiation exposures, a high number of contamination events, increased radioactive waste volumes (both solid and liquid), and an increased number and severity level of radiological incidents may be indicative of those facilities that do not aggressively pursue excellence in radiological safety.

Over the years, especially in the USA, the industry has experienced a marked change in both the effort and attitude towards the importance of controlling contamination at the source. Contamination was accepted as an expected nuisance encountered during operation and maintenance activities associated with nuclear power plants. It became a "rite of passage" that one was not considered a nuclear power plant worker until they had the occasion to don the full-set of protective

R. Prince, *Radiation Protection at Light Water Reactors*,
DOI: 10.1007/978-3-642-28388-8_8, © Springer-Verlag Berlin Heidelberg 2012

clothing, including a full-face respirator and an outer set of plastics for good measure! The need for such protective measures often went unquestioned and little thought was given to implementing or establishing controls that perhaps would lessen or eliminate the need for the use of protective clothing and equipment. Thankfully those days have been left behind. The industry has come to realize that contamination can be controlled to a large extent, minimizing radiological safety issues, personnel exposures and improving the productivity of workers no longer encumbered with multiple layers of protective clothing or respiratory devices.

Corrosion products deposit throughout the primary and auxiliary systems causing radiation levels to increase in the affected areas. The corrosion products of interest along with their principle production modes were described in Chap. 4. Corrosion products are formed in various materials and transported through the core where they become activated. These soluble and insoluble activated corrosion products are then transported to ex-core locations where they can be deposited. Industry efforts employed to minimize the amount of those materials that are precursors to activated corrosion products resulted in significant source term reduction. These and other efforts have resulted in a decreasing trend in the total collective dose at LWRs over the past two or three decades. A secondary benefit of source term reduction efforts is that it minimizes the amount of contamination encountered during system breaches and maintenance activities. The establishment of effective source reduction and contamination control programs is key elements in the industry's efforts to decrease personnel exposures and in improving the economic performance of nuclear units. This principal has been emphasized throughout this text.

## 8.2  Contamination Control

Contamination control measures are most effective when targeted at the source or as close as possible to the source. Consider the advantages of placing a drip bag under a leaking valve and routing the contaminated flow to a nearby floor drain or other suitable collection device versus allowing contaminated water to run uncontrolled onto the floor. The result may be the difference between controlling an entire room or otherwise large area as contaminated or allowing access to the area with no protective measures required. Figure 8.1 displays a typical drip bag, or catch containment, and a drip funnel. Routine contamination surveys play a vital role in the early detection and control of contamination. Whenever practical, measures should be established to contain the source of contamination and min-imize the area affected by the presence of contamination. Ideally control measures should be directed towards eliminating the source (e.g., fix the flange leak) and should not end once the source has been controlled (e.g., drip bag installed). The following benefits are associated with an effective contamination control and minimization program:

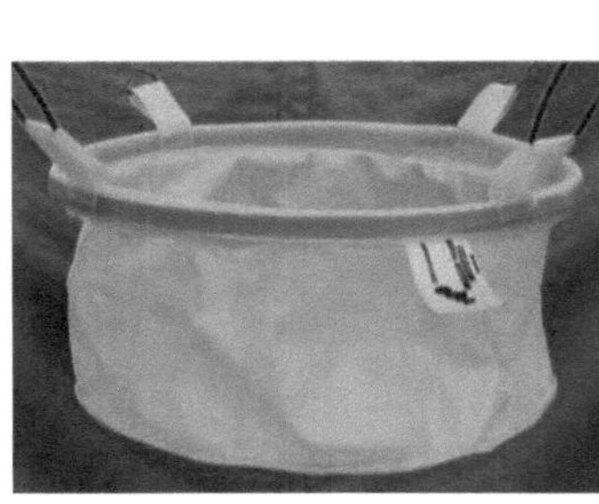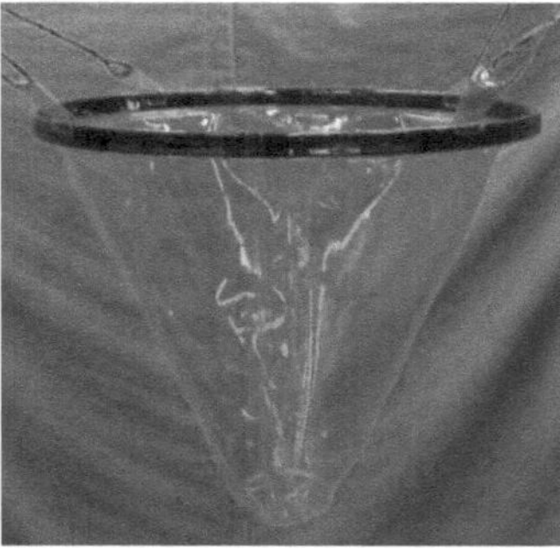

**Fig. 8.1** Contamination control devices commonly used to minimize the spread of contamination (Courtesy of Nuclear Power Outfitters, www.nuclearpoweroutfitters.com.)

- Plant areas are more accessible to plant personnel
- Less stringent radiological control measures are required
- Plant material condition is more easily maintained
- Reduced administrative and overhead costs associated with access controls and the performance of work in contaminated areas
- The volume of contaminated waste water requiring processing, monitoring and associated effluent controls is reduced
- Stay times associated with entry and exit from contaminated areas and the time required to perform tasks in contaminated areas are reduced resulting in lower radiation exposures to personnel

These and other benefits result in significant improvements in other important areas, most notably, productivity enhancements in addition to intangible benefits such as maintaining a culture that does not tolerate the unnecessary presence of contamination. A comprehensive contamination control program should include the monitoring and control of contaminated leaks (i.e., leak control), measures to minimize the spread of contamination, personnel monitoring, and the importance of following good work practices, training and the use of protective clothing among others. Methods should be established to track and trend the status of contaminated leaks. Ideally the work control process should have guidelines for assigning priorities to repair leaks based on their radiological significance. Effective contamination control programs aggressively identify and schedule repairs in a timely manner to address contaminated system leaks.

Routine in-plant contamination surveys should be directed at identifying potential as well as actual sources of contamination. Particular attention should be given to valve flanges, valve stems, pump seals, drain lines, and any contaminated system components when performing surveys or while making routine tours of the RCA. Radiation protection technicians should be knowledgeable in the aspects of system operations and parameters, as noted in Chap. 4 that may contribute to potential contamination issues. Radiation protection personnel should be diligent in their efforts in identifying sources of contamination and aggressively pursue the implementation of corrective measures. Complacency with regards to accepting the presence of contamination must be continuously guarded against to ensure the maintenance of a strong contamination control program.

## 8.3 Source Control Techniques

Lowering of in-plant radiation fields is obviously a key component in reducing personnel exposures. Plant radiation fields can be reduced utilizing various techniques. The techniques employed to reduce in-plant radiation fields have the most impact during periods of reactor shutdown, when the majority of the worker exposure is received during the performance of refueling and maintenance activities. Some of the more easily implemented measures involve the use of shielding, either temporary or permanent, that may be installed around contaminated system components to reduce radiation levels in nearby areas. Purification system filters can be sized to minimize the inventory of insoluble radioactive species present in the reactor coolant system and other systems containing radioactive fluids. However as with contamination control it is more effective to control the source of radiation exposure at the source or in this case to prevent the formation of corrosion products in the first place. Corrosion products cannot be totally prevented however techniques can be employed to minimize the corrosion process and to reduce the inventory of activated corrosion products after they are produced. Water chemistry plays a vital role in achieving these objectives.

### 8.3.1 Water Chemistry Fundamentals

Water chemistry plays a major role in determining the magnitude of plant radiation levels, especially over the long run. Since water is utilized as the coolant and heat transfer medium its chemistry must be maintained within established operational limits. Water Chemistry programs are designed to maintain efficient plant operations and to minimize the long-term buildup of corrosion products. Since the PWR and BWR nuclear power plant designs operate on different concepts the water chemistry control programs must of necessity differ somewhat for each type of reactor. Water chemistry in a PWR is complicated by the use of boric acid in the primary system for reactivity control purposes while in BWRs the goal is to maintain the primary water in as pure a state as possible. Chemistry programs for PWRs are designed to control pH and conductivity and concentrations of dissolved oxygen, hydrogen, chlorides, total gas, boron and radioactivity.

If water chemistry is not properly controlled, system components in contact with this water may be subject to accelerated corrosion mechanisms, releasing fine particulate matter that may become activated and subsequently be deposited in out-of-core components. These so-called crud[1] deposits not only increase plant radiation fields but also decrease the operational efficiency of various plant components. Over the long run resultant operational problems may require additional or more frequent maintenance, thus contributing to increased personnel exposures. Since these mechanisms have a negative impact on radiological safety, RP personnel should have a basic understanding of water chemistry fundamentals in order to evaluate radiological issues that may result due to water chemistry control issues encountered during plant operations. Chap. 4 summarized the principle production modes for the common activation products many of which result from corrosion products whose inventories are influenced by the quality of plant chemistry. There are various parameters associated with LWR chemistry programs. Radiation protection personnel should have a basic understanding of these parameters and their radiological implications. Additionally, various chemical techniques may be utilized to reduce or control the amount of crud buildup to reduce plant radiation levels. Numerous studies have been conducted to evaluate the effectiveness of various chemical regimes in reducing plant radiation fields. This section provides an overview of the fundamentals of plant water chemistry. Several chemical treatment techniques employed to manage in-plant radiation fields at LWRs are presented later in this chapter.

Many substances are normally present in water. These substances may be present in either a suspended or dissolved state and are generally referred to as impurities. When impurities are dissolved in water the resultant mixture is referred to as a solution. The concentrations of these dissolved substances are often reported in terms of parts per million, abbreviated ppm or parts billion (ppb). This is a convenient unit to use when the amount of a substance dissolved in a solution is small when compared to the amount of the solution. One ppm simply indicates that there is one part of a substance present to every million parts of the solution. In solutions the dissolved substance is usually present in small amounts and is called the solute while the dissolving medium is termed the solvent.

Substances dissolved in water may dissociate into positively and negatively charged species referred to as ions. The positively charged ions are called cations while the negatively charged ions are referred to as anions. This dissociation process is referred to as ionization and, depending on the chemical conditions, does not always go to completion. Ions play a key role in the corrosion process and in the buildup of crud. They are the principle participants, which promote the chemical reactions, both in the water itself and the structural materials, which lead to corrosion.

The pH is a term used to express the hydrogen ion (H+) concentration of a solution and is measured on a scale from zero to fourteen. When the hydrogen ion concentration equals the hydroxyl ion (OH−) concentration, the solution has

---

[1] Crud is a term that refers to the accumulation of radioactive corrosion products on the surfaces of plant components. The term was derived from Canada's Chalk River site as an acronym for Chalk River unidentified deposits.

a pH of 7 and is considered to be neutral. When the hydrogen ion concentration increases the solution becomes acidic and has a pH value less than 7. If the hydrogen ion concentration decreases, then the solution becomes basic and has a pH value greater than 7. When acids ionize in water they liberate hydrogen ions. When a base ionizes in water, hydroxyl ions are liberated. The pH range is controlled to inhibit corrosion. Protective films, which form on the inner surfaces of system piping and components, are removed by the action of H+ ions, which accelerate the rate of corrosion. Iron oxide, a common species present in LWR systems, is soluble in highly acidic environments. High pH conditions are favorable for the promotion of caustic stress corrosion. Consequently, pH values are maintained within an optimal range based on plant conditions to minimize corrosion concerns.

Studies have been performed to investigate the effects of coolant pH on plant radiation fields. The Electric Power Research Institute (EPRI) sponsored one such study that studied this effect at two PWR stations. One plant maintained constantly low lithium hydroxide concentrations (0.3–0.7 ppm) that allowed the pH to gradually increase during the fuel cycle. The second unit involved in the study maintained a constant pH by coordinating lithium hydroxide concentration reductions with decreasing boric acid concentrations during the fuel cycle. The coordinated lithium-boron approach allowed for relatively high lithium concentrations (>1 ppm). Even though this study was based upon data from only two stations several conclusions were drawn from the results. In general operating with low lithium concentrations yielded relatively high Co-58 and Co-60 activity levels throughout the primary circuit, predominantly on cold legs, which resulted in increased radiation fields. Effects on crud and impurity disposition patterns were also noted during this study. In the intervening years, based on these and other studies, and operational data, many PWR stations are now operating on a coordinated lithium-boron platform to minimize the amount of crud deposition.

As previously mentioned when substances are dissolved in a solvent they will dissociate to form ions. These ions are capable of conducting an electric current. Pure water is a poor conductor of electricity due to the small concentrations of hydrogen and hydroxyl ions. As the amount of dissolved impurities is increased the concentration of ions increases and the ability of a solution to conduct an electric current improves. Conductivity is the term used to measure the ability of a solution to conduct a current. The conductivity of a solution is an indicator of the degree of purity of a given solution. Conductivity is measured in units of mhos and is the reciprocal of electrical resistance. Hence the term "mho"—ohms spelled backwards. As the ionic concentration increases, so does the conductivity of a solution. As a general rule as the conductivity increases the more effective a solution becomes in promoting corrosion.

## *8.3.2  Water Purification*

A unique property of water is its ability to dissolve essentially every substance found in nature. For this reason water is commonly referred to as the universal solvent. Due to this characteristic all sources of water found in nature will contain some amount of impurities. If not removed, or at least reduced to acceptable concentrations, the presence of impurities could be detrimental to nuclear power plant operations. Additionally, to maintain the required purity level of water for use in various LWR systems, water must be continuously conditioned. This conditioning usually consists of some combination of filtering and demineralization steps. A major factor in reducing annual collective radiation exposures in the LWR industry, along with greatly improved plant capacity factors, over the last couple of decades has been associated with improvements made in maintaining extremely low water impurity levels.

Filtration is a process used to remove suspended solids from a liquid or process stream. Filtration removes suspended solids via a combination of adsorption, inertial impaction, straining and interception processes. The process stream is routed through a filter medium that is contained within a vessel referred to as the filter housing. Various filter types are employed in the LWR industry, including precoat and cartridge type filters and deep-bed filters. Filters may be constructed from various materials the most common type being a fiber material. Many process systems as noted in Chaps. 2 and 3 are filtered on a continuous basis. The removal of suspended solids in these systems results in the collection and concentration of activated corrosion products, fission products that have escaped from fuel assemblies in addition to non-radioactive solids. The capture of radionuclides on process filters lowers plant radiation levels. While in service, as the amount of radioactive material builds up on these process filters, radiation levels will gradually increase. Radiation levels on many system filters will eventually increase to perhaps several Sv/h (hundreds of rem/h) depending upon the amount of time they are in service. Filter housings are located within shielded housings access to which usually requires the removal of a heavy shield plug.

Cartridge filters consist of a perforated metal core around which a yarn type medium is wound. These cartridges are cylindrical in shape and may be several centimeters in diameter and 20–30 cm in length, depending on their function (e.g., pre or post filter or debris type filter). A filtering unit may contain several or more of these filter cartridges. The filter housing holds the cartridges in place and provides a flow path for the process fluid through the cartridges. Various cartridge fibers are utilized depending upon the filtering application and are sized, or rated, to remove suspended particulates exceeding a minimum size. Over the years filter sizes have been reduced in an attempt to minimize the concentration of suspended solids in the various systems. This reduces the amount of material present that may later become activated and decreases the resultant source term present in crud levels. Filter sizes in the sub-micron range, down to 0.1 $\mu m$, are in common use in the LWR industry.

Pre-coat filters consist of a filter housing containing perforated stainless steel tubes called septums. These septums are usually a couple of centimeters in diameter and up to a meter in length. A pre-coat filter unit may contain hundreds of septums. The pre-coat filtering medium consists of finely divided fibers that are introduced as a slurry. The filter fibers deposit on the septums and held in place by the pressure of the water. When the pre-coating process is completed the filter is ready to be placed into service. Cellulose or other substances are used as pre-coat filter mediums. When the filter is changed, the spent fibers are backwashed from the septums and processed for disposal. The unit can then be pre-coated and placed back into service.

Cylindrical stainless steel mesh filters are often employed as a screen-type filter to remove large sized particles and debris from the process stream. These can be used as a pre-filter to extend the useful life of downstream filters. The filters are removed from service periodically for cleaning and re-used.

Filtration is primarily effective for removing suspended, non-dissolved impurities. Dissolved solids are not removed by filtration.

The process of ion exchange removes dissolved impurities and sub-micron sized materials. Ion exchange is a chemical process that involves the exchange of ions between the process stream and another substance (i.e., the demineralizer bed). Certain materials have a preferential affinity for specific ions. These materials are known as resins and usually consist of hydrogen-based compounds. Resins are essentially small bead-like materials. Under the proper conditions resins will readily exchange one ion type for another. Each resin (or bead) may have many ion exchange sites, which can be saturated or loaded with suitable ions capable of being displaced by the impurity ions present in the process stream.

Placing a volume of fine resin beads into a tank or other suitable container or vessel forms a "resin bed". This unit is commonly referred to as a demineralizer. Depending on their purpose, demineralizers encountered at LWR's may be several cubic meters in volume. The process stream enters at the top of the demineralizer tank and flows down through the resin bed. The resins release their ions from the exchange sites into the water as the impurity ions are retained at these sites on the resins. Impurity free water exits from the bottom of the resin bed. When all available exchange sites have been utilized the resin is exhausted and must be regenerated or replaced. The regeneration process involves a chemical treatment of the spent resins to remove the undesirable ions, replacing them with a suitable ion. The regenerated resin bed can then be placed back into service. Alternatively the spent bed may be sluiced directly to the spent resin storage tank. New resin is then loaded into the demineralizer.

Two types of demineralizers are used for purification purposes. These two types are cation and anion demineralizers, which remove cations and anions respectively. Cation and anion resins may be mixed together in the same tank to yield a mixed bed demineralizer that is a common practice in LWR's. Calcium and magnesium are examples of undesirable cations while chloride is an undesirable anion. In addition to removing dissolved impurities, demineralizers can also be used to control pH and to minimize the inventory of radionuclides in contaminated systems.

## 8.4 Corrosion Processes

The excessive formation of corrosion products ultimately contributes to the magnitude of radiological problems. It is therefore important that RP personnel understand those water chemistry parameters pertaining to corrosion mechanisms. Several corrosion mechanisms encountered in LWR environments either directly or indirectly are of radiological concern. In order to minimize the degree of corrosion it is essential that water chemistry programs include actions to control the conditions and the formation of those chemical species that promote corrosion processes. Corrosion is the process whereby the integrity of a metal is compromised as a result of chemical and electrochemical reactions with its environment. There are various types of corrosion mechanisms that can occur when certain conditions exist. Corrosion can occur due to the presence of impurities in base metals or in the water itself, localized stresses, temperature and pH affects, the presence of oxygen as well as for other reasons. Several types of corrosion are encountered at LWR's including stress corrosion cracking (SCC), intergranular attack (IGA), general corrosion, galvanic corrosion and pitting.

**Stress corrosion cracking**: Stress corrosion cracking may occur in some metals when they are subjected to stress in the presence of a corrosive environment. Metals may be subject to stress during construction as well as during their operational lifetime. Austenitic stainless steels are susceptible to SCC at high temperatures, particularly when in the presence of chloride and oxygen ions. Hence one of the primary reasons why oxygen scavenging is performed to minimize oxygen concentrations in the primary system and why strict controls are placed on maintaining chloride ion concentrations in the range of a few ppb or less. SCC is most likely to occur in crevices and those locations where water velocities are low or restricted. Oftentimes IGA occurs in conjunction with SCC and involves a localized attack at the boundaries between metals. This type of corrosion could be encountered in stainless steels that were not properly heat-treated during fabrication.

General corrosion occurs over large surface areas of a metal and involves the reaction of water or oxygen with the surface of the metal. Corrosion results in the formation of metal oxides that in LWRs may include oxides of iron, nickel, chromium, and zirconium among others. The rate of general corrosion is heavily dependent upon the oxygen concentration in the water. In general as oxygen concentrations increase the rate of corrosion also increases. Oxygen concentrations are maintained as low as possible. Introducing hydrogen gas or the addition of hydrazine into the primary system of PWRs promotes the scavenging of oxygen thereby reducing the concentration of free oxygen. The pH of water also influences the rate of general corrosion. As previously noted a major effort associated with water chemistry programs involves control measures to minimize oxygen concentrations and to maintain pH values within acceptable limits.

**Galvanic corrosion**: Galvanic corrosion may occur whenever two different metals are in electrical contact in water. The magnitude of the potential difference between the two metals determines the rate of galvanic attack and varies depending upon the particular metals involved. The potential difference between the two metals causes ions of one metal to migrate to the other. This migration causes pits in one of the metals and deposits on the other. Under these conditions the metals act as two electrodes, allowing an electrical current to flow between them via the water medium connecting the two. One metal will act as the anode (whose surface will become pitted) while the other will become the cathode (whose surface will contain deposits). The metal comprising the anode will lose material. Maintaining high water purity to minimize the concentration of impurities lowers the ability of water to conduct a current thus minimizing galvanic corrosion. Hence conductivity is maintained as low as possible.

**Crevice corrosion and pitting**: The crevice corrosion mechanism is similar to that of galvanic corrosion even though only one metal need be involved. Gaps or crevices located between two adjoining metal surfaces of a component may lead to different concentrations of impurities. These gaps or crevices could be present as a result of microscopic imperfections during fabrication. The formation of pits causes the water solution within the pit to become highly corrosive since it is isolated from the main water stream. Locations where water velocity is low (e.g., pits and crevices) are susceptible to crevice corrosion. Once the crevice forms a localized area may result whereby impurities may be preferentially trapped resulting in higher oxygen concentrations in the crevice area then in the process stream. The presence of chloride ions will enhance these corrosion mechanisms. Historically crevice corrosion often plagued steam generators, as well as other system heat exchangers, where the heat exchanger tubes connect to the tube sheet.

The various corrosion processes described above can produce cracks and pits and cause general corrosion of plant materials that may ultimately lead to component failures. Consequently corrosion phenomena have been investigated for many years to evaluate the effect of various chemical parameters on the corrosion rates of LWR structural materials.

The extent and magnitude of corrosion processes have wide-ranging affects on the radiological environment of LWR facilities. Chapter 4 summarized various corrosion processes and the resultant activated corrosion products along with their nuclear characteristics. These corrosion products not only contribute to the ambient radiation environment but the production and transport of corrosion products may result in increase rate of fouling of plant components such as valves, heat exchangers, filters and resin beds resulting in increase maintenance and inspection activities. The undesirable aspect of accelerated corrosion is that it results in higher radiation levels in the vicinity of those major components and systems that are subject to routine maintenance and inspection activities.

Steam generator maintenance and inspection activities provide valuable insight into the detrimental impact the presence of activated corrosion products may have on radiological safety performance. Though this discussion applies to steam generators specifically the basic concepts apply to valve repairs, other heat exchangers, impact on pump seals and any other situation whereby the presence of activated corrosion products are encountered. Even though filter and resin bed change-out may be accomplished by means of remote operations and do not result in significant radiation exposures, if accelerated corrosion results in more frequent replacement of process filters and resins beds there is an economic cost involved.

Steam generator tubes are subjected to a severe operating environment that may ultimately lead to tube failures. Steam generator tube failures result for several reasons, the repair and maintenance of which may result in considerable dose expenditures. The major causes of tube defects have changed over the years as experience was gained in understanding the underlying causes of steam generator tube degradation. Design improvements, use of more corrosion-resistant stainless steel alloys, and improved primary system and feed water chemistry controls have all played a key role in improved steam generator performance. In the early 1970s stress corrosion cracking was a major cause of tube defects while in the mid 1970s phosphate wastage was a major contributor. By the late 1970s denting had become a leading cause of tube defects. Since the steam generator tubes in a PWR serve as the primary barrier between the primary and secondary sides the integrity of these tubes must be maintained at all times to prevent radioactive contamination being introduced to secondary side systems and components and the turbine building in general. To minimize the probability of severe tube failures while at power, which could result in a radiological release offsite, the integrity of these tubes is subject to a comprehensive in-service inspection program. Personnel exposures associated with routine steam generator inspection and maintenance activities are often significant contributors to outage exposure totals.

## 8.5  RCS Oxygenation (Hydrogen Peroxide Addition and Crud Burst Promotion)

Over the last couple decades a major effort has been put forth by the industry in analyzing and quantifying source terms at LWRs. These efforts highlight the significant contribution of Co-58 and Co-60 as primary contributors to the overall source term. In some cases these two species may comprise over 80% of the radiation fields in excore areas of LWR facilities. Consequently source term reduction efforts have been targeted towards reducing cobalt inventories. One of the more common techniques employed in this regard has been the implementation of shutdown chemistry measures. The primary measure utilized has been the intentional promotion of crud bursts brought about by RCS oxygenation techniques.

During the cool down phase as a unit is entering shut down and oxygen concentrations increase in primary and auxiliary systems crud deposits become more soluble. This solubilization results from the temperature reduction and the increase in boron concentration levels associated with PWR plant shutdown operations that cause the pH to decrease. The increase in solubility leads to an increase in Co-58 and Co-60 concentrations in the primary system. These increased concentrations subsequently migrate to several auxiliary systems (e.g., RHR, CVCS and others), which impact radioactivity levels. The major source of activated corrosion products are released from the surfaces of fuel assemblies and reactor vessel components and other system locations and transported throughout various systems. This phenomenon causes dose rates to increase in those systems that are in direct communication with the reactor vessel and core, particularly in the RHR train that is providing shutdown cooling or decay heat removal. Additionally, a few days after shut down when the reactor coolant system and auxiliary systems are opened to atmosphere there is an increase in the dissolved oxygen concentrations in the primary circuit. The increased oxygen concentrations produce a "burst" of activated corrosion products above and beyond the existing source term that results in higher radiation levels in many plant locations. These concentrations typically peak several days after shutdown, leading to increases in system radiation fields that may result in higher personnel exposures during maintenance activities. If the process is allowed to proceed naturally dose rates will be increasing in many plant locations at the time that systems and components are just becoming available for maintenance. Obviously this is not an ideal situation from a dose perspective. Compounding this situation is that normal clean-up systems may not be running at 100% capacity a few days after plant shutdown. Flow rates through demineralizers may be lower based on plant conditions thus minimizing clean-up capacity. Consequently at a time when optimum clean-up is needed the ability to do so is limited.

Processes to promote earlier oxygenation of the RCS to allow removal of released activated corrosion products prior to performing maintenance activities would prove beneficial. A process that has gained wide-spread use in the industry for PWR units is to add hydrogen peroxide (i.e., oxygenating the RCS) during the early stages of plant shutdown to solubilize crud layers at an earlier time to promote the release of crud containing these activated corrosion products. The addition of hydrogen peroxide increases the oxygen concentration in the RCS and results in a "crud burst". The intent is to promote release of the crud layer, that is relatively loose, and the concomitant source term (i.e., Co-58 and 60) during cool down, prior to maintenance activities. The letdown system is used to remove the radioactive species, via filtration and demineralization. This process can result in the release of a few tens of terabecquerels (several hundred curies) of activity from fuel surfaces and RCS components. The vast majority of this activity is due to the presence of Co-60 and Co-58. It is essential that plant chemistry parameters be maintained within prescribed limits to optimize the crud burst and to keep released species in a soluble state for an extended period to maximize removal of the released corrosion products. The solubility of cobalt is highly dependent on pH values requiring strict control on boron and lithium concentrations during the promotion and subsequent removal of the released corrosion products upon initiation of a crud burst. Failure to maintain required conditions could result in the increased plate-out and deposition of activated corrosion products in ex-core locations resulting in higher than normal radiation fields.

The crud burst will take place over a period of time (typically 12–48 h or longer) depending on plant conditions. It is essential that plant purification and clean-up systems remain in service while the crud burst proceeds. Sufficient resin bed capacity should be available to handle the anticipated crud release. The selection of resin material that optimizes the removal of the released ionic species increases the removal and clean-up of corrosion products. If a resin type different than that used during power operations will be utilized then the demineralizer (resin bed) that will be in service during the crud burst should be pre-loaded and available prior to promoting a crud burst. Additionally system purification filters may require frequent change-out during this period depending upon the particle sizes encountered. Typically particle sizes are sub-micron and are removed primarily by the resin beds minimizing the build-up on filters. In any event a sufficient supply of letdown and RCS system filters based on previous experience should also be available to support removal of corrosion products.

The progress and effectiveness of a crud burst may be monitored by obtaining RCS samples from an appropriate letdown line or by measuring the changes in dose rates on the RHR train that is in operation. The use of area radiation monitors with remote readout capability or electronic teledosimetry systems are ideally suited for this purpose. Soon after the initial hydrogen peroxide addition dose rates may increase by orders of magnitude in the vicinity of the letdown heat exchanger and the RHR heat exchanger and RHR pump and associated piping on the train in service. Assume that contact dose rates on the RHR system that will be in service during the crud burst are several tens of $\mu$Sv/h (several mrem/h) before the addition of hydrogen peroxide. Dose rates of a few hundred $\mu$Sv/h (hundreds of mrem/h) would be indicative of a crud burst. This increase may be seen within tens of minutes after the addition of hydrogen peroxide not hours. These dose rate values could be significantly different under certain situations, most notably for units that have undergone several refueling cycles before implementing shutdown chemistry programs. Under these conditions the first-time crud burst could result in the removal of several tens of terabecquerels or more (greater than a thousand curies) of cobalt. Several liters of hydrogen peroxide may be added to the RCS to promote the crud burst. The total amount of hydrogen peroxide may be added in increments as determined by RCS radiochemistry data or based on changes in radiation levels at the selected monitoring locations. Typically several system or area locations will be monitored to evaluate the progress of the crud burst.

Since the initiation of an intentional crud burst occurs during the initial stages of an outage the time required to achieve a successful crud burst is most likely a critical path activity. As such the radiological benefits achieved must be balanced against the "costs" associated with critical path time. Many health physicists have probably experienced the "joys" of having plant management ask "when" will the crud burst be over? While the crud burst is ongoing purification flow rates should be at a maximum. In fact some utilities have performed engineering evaluations to support increased flow rates through the CVCS filters and demineralizer and associated purification piping and equipment to shorten the cleanup time and thus saving critical path hours. If hydrogen peroxide additions have been made during previous outages then experienced gained during those evolutions should be used to establish clear criteria for determining when the maximum results have been achieved. This may be predicated on area dose rates, the estimated Co-58 or Co-60 activity removed over a specified time period, or some other radiochemical parameter.


## 8.6  Techniques to Reduce Source Terms

Various techniques and methods have been developed and utilized by the LWR industry in an attempt to minimize the long-term buildup of source terms. Numerous technical reports, industry symposiums, research organizations, and individual utilities have devoted much time and effort in developing methods to reduce and minimize the build-up of source terms. This section highlights some of the efforts utilized by the LWR industry to address source term reduction. This is not an exhaustive review of the various techniques available to the industry. Based on the needs of a specific nuclear station individuals may wish to consult the available technical reports and appropriate source term reduction techniques in greater depth to evaluate their applicability to a given situation.

Though not strictly a source term reduction technique for the case of BWRs, where the primary objective is to maintain high purity feed water to the reactor vessel, various systems serve a vital function in controlling the inventory of activation and corrosion products. Strict adherence to primary coolant chemistry controls is an essential element in controlling these corrosion mechanisms. The reactor water cleanup system (RWCU) plays a vital role in maintaining the purity of the primary coolant. The condensate system is equipped with several full-flow filters and demineralizers to maintain feed water chemistry. A sufficient number of demineralizers are available to allow a resin vessel to be taken offline to be regenerated without impacting plant operations. These systems should be maintained at optimal performance levels at all times.

Intergranular stress corrosion cracking (IGSCC) of austenitic stainless steel and Inconel materials has been one of the more significant corrosion issues encountered in BWRs. The presence of oxygen in the feedwater will accelerate this type of corrosion. If hydrogen is injected into feedwater, dissolved oxygen concentrations can be greatly reduced. Pilot projects and tests in the early to mid 1980s have showed this process to be effective in reducing (IGSCC). Hydrogen injection (or hydrogen water chemistry) is a common practice employed in today's fleet of operating BWR's. Hydrogen water chemistry; however, increases the amount of N-16 carried over in the steam due to the production of more volatile forms of nitrogen. The additional N-16 results in an increase in dose rates in the vicinity of the main steam lines and associated equipment and components due to the increased carry-over of N-16 in the steam. The increased dose rates and their impact on station exposures will vary from one plant to the next and must be evaluated on an individual basis to determine the overall radiological impact of operating with hydrogen water chemistry. However, IGSCC of recirculation lines can result in replacement, additional inspection and repair activities with associated exposures that greatly exceed the additional dose resulting from operations due to hydrogen water chemistry.

Noble metal chemical addition (NMCA) is another process utilized at BWR units to minimize IGSCC of boiling water reactor internal components. Use of NMCA improves the overall efficiency of HWC for mitigating potential IGSCC with lower amounts of hydrogen. This reduces the amount of N-16 carried over in the steam. The use of NMCA basically provides a more corrosion resistant layer on internal surfaces mitigating the IGSCC process.

Another process employed to lower Co-60 concentrations in crud layers utilizes zinc injection. Initially this process was applied to BWR units but now is used in both PWR and BWR units. Early studies had shown a correlation between the amount of Co-60 contained in crud layers and the amount of ionic zinc in reactor water. Subsequent laboratory testing and industry pilot projects (e.g., various EPRI sponsored projects) indicated that oxide film formation could be suppressed if ionic zinc concentrations in reactor water were increased. Depleted zinc oxide is introduced via the reactor feedwater. Decreasing the formation of the oxide film layer resulted in lower amounts of Co-60 present in the crud and thus, a smaller radiation source term. Additionally, zinc replaces some of the cobalt in the oxide layer that does form on internal surfaces of piping and components. The majority of BWR units in the USA use zinc injection primarily to lower dose rates emanating from recirculation piping. This in turn lowers shutdown radiation dose rates in the drywell. The number of PWR units implementing zinc injection is rapidly increasing.

## 8.7 Summary

Industry efforts in identifying and evaluating the mechanisms that cause and promote corrosion of materials contained in primary system components have been instrumental in reducing annual exposures in the nuclear power industry. Understanding the underlying corrosion processes has resulted in the establishment of improved water chemistry guidelines for both BWR and PWR nuclear plants. The applications of operational controls to minimize power operation time when chemistry parameters are not within specification have also minimized corrosion processes and the formation of crud. Continued industry efforts in this regard will undoubtedly result in further improvement in dose reduction efforts. It is vital that RP and plant management stay abreast of current and future industry developments in this key area. Those nuclear plants in the top exposure quartile should ensure that they evaluate, and whenever possible, implement industry accepted practices to support long-term dose reduction initiatives and continued program improvements.

# Bibliography


1. Deligiannis A., Master's Paper, Comparative Analysis of Source Term Removal at Cook Nuclear Plant, Department of Nuclear Engineering, University of Illinois at Urbana-Champaign, 2009
2. Dionne B., and Baum J., Discrete Radioactive Particles at Nuclear Power Plants: Protection, Mitigation and Control, Radiation Protection Management, 8:66–79, 1991
3. Electric Power Research Institute, BWR Radiation Assessment and Control Program, Assessment and Control of BWR Radiation Fields, EPRI Report NP-3114, Palo Alto, CA, 1983
4. Electric Power Research Institute, BWR Source Term Reduction-Estimating Cobalt Transport to the Reactor, EPRI Report TR-1018371, Palo Alto, CA 2008
5. Electric Power Research Institute, BWR vessel and Internals Project, BWR Water Chemistry Guidelines, EPRI Report 1016579, Palo Alto, CA 2008
6. Electric Power Research Institute, Cobalt ContaminationContamination Resulting from Valve Maintenance, EPRI Report NP-3220, Palo Alto, CA, 1983
7. Electric Power Research Institute, Coolant Chemistry Effects on Radioactivity at Two Pressurized Water Reactor Plants, EPRI Report NP-3463, Palo Alto, CA, 1984
8. Electric Power Research Institute, Effects of Cold Shutdown Chemistry on PWR Radiation Control, EPRI Report NP-3245, Palo Alto, CA, 1983
9. Electric Power Research Institute, Effects of Shutdown/Control Techniques on Radiation Fields in PWR Primary Coolant Loops, EPRI Report NP-3461, Palo Alto, CA, 1984
10. Electric Power Research Institute, PWR Operating Experience with Zinc Injection and the Impact on Plant Radiation Fields, EPRI Report TR-1003389, Palo Alto, CA, 2003
11. Electric Power Research Institute, PWR Primary Shutdown and Startup Chemistry Guidelines, EPRI Report TR-101884, Palo Alto, CA, 1993
12. Electric Power Research Institute, Radiological Effects of Hydrogen Water Chemistry, EPRI Report NP-4011, Palo Alto, CA, 1985
13. Hettiarachchi S., Miller W.D., Cowan, R.L., and Diaz T.P., Noble Metal Chemical Addition for IGSCC Mitigation of BWRs—Field Successes, Corrosion 2000, NACE International, Orlando, FL, March 26–31, 2000
14. Lin C., Radiochemistry in Nuclear Power Reactors, National Academies Press, 1996
15. Messier C., and Lane E.J., Radiation Protection Consideration for Boiling Water Reactors Using Zinc Injection, Radiation Protection Management, 8:40–56; 1991
16. National Council on Radiation Protection and Measurements, Dose Control at Nuclear Power Plants, NCRP Report No. 120, Bethesda, MD, 1994
17. National Energy Agency, Organization for Economic Co-Operation and Development, Occupational Exposures at Nuclear Power Plants, Eighteenth Annual Report of the ISOE Programme, 2008
18. US Nuclear Regulatory Commission, NUREG-0713, Volume 30, Occupational Radiation Exposure at Commercial Nuclear Power Reactors and Other Facilities 2008: Forty-First Annual Report
19. Wagner, D.S. and Banks T., Virginia Power's Source Term Reduction Efforts, Radiation Protection Management, 6:37–43; 1989

# Chapter 9
# Protective Clothing and Equipment

## 9.1 Overview

Ideally a LWR environment would be one in which workers are not encumbered with protective clothing and equipment. The effectiveness of controlling contamination at the source and the minimization of the magnitude and extent of radiologically contaminated areas are key elements in eliminating, or at least minimizing the use of radiological protective equipment. The radiological environment oftentimes is governed to a large extent, on the management approach towards contamination control and source reduction measures. Based upon the approach taken results may be wide ranging. One facility may require the use of protective clothing and radiological control measures for the performance of routine, non-work type, activities; while another facility may aggressively control the magnitude and extent of contamination to allow entry into plant areas with minimal or even no protective clothing. When contamination control measures are less stringent the use of protective clothing and need for more extensive radiological control measures may be more prevalent. These controls may even be applied for entries into plant areas that contain relatively low-levels of contamination or where radiological conditions are "mild". These controls may extend to such routine activities as operator rounds, supervisor tours, inspection activities and other non-work type tasks. On the other hand at those facilities that take an aggressive approach towards contamination control and source reduction, corresponding use of protective clothing and the need for more-extensive radiological control measures will be lessened. A point in case would be those PWR facilities that maintain clean (or "street clothes") containment buildings that allow general entry with no special protective clothing requirements versus those that treat the containment building as a "contaminated area" requiring protective clothing for all entries to the containment building, regardless of the nature of the task (i.e., inspection or entry to perform maintenance). This chapter describes the various types of radiological protective clothing and equipment commonly used in the LWR industry. Guidelines for the proper selection, use and maintenance of protective clothing and respiratory protection equipment are described.

R. Prince, *Radiation Protection at Light Water Reactors,*                                    211
DOI: 10.1007/978-3-642-28388-8_9, © Springer-Verlag Berlin Heidelberg 2012

As noted above the degree of radiological control measures required at a given facility may often be determined by the culture that management has established (either intentionally or unintentionally) when dealing with the presence and control of contamination and source terms. Ideally the need for radiological control measures should be based on the actual radiological conditions encountered in the field. If the attitude is one of complacency then the radiological environment may be one where large areas of the RCA are zoned as contaminated areas with higher than average radiation levels present in many areas of the plant. Higher annual collective personnel radiation exposures, greater number of contamination events, increased radioactive waste volumes (both solid and liquid), and a higher number and severity-level of radiological incidents may typify those facilities that do not aggressively pursue a comprehensive contamination control program.

Elements of a contamination control program were presented in Chap. 8, some of the key elements of which are listed below:

- Contamination survey program
- Source identification and control
- Equipment and material movement controls
- Leak identification, management and control
- Personnel contamination monitoring and use of hand-held portable contamination survey meters and PCMs
- Contamination control work practices including such measures as hose control, tool control, and availability of designated "hot tools"
- Protective clothing
- Training and qualification of workers
- Contamination control measures including a decontamination program, use of barriers, enclosures, ventilation and fixatives

Aggressive contamination control programs, comprehensive efforts in minimizing and controlling activation and corrosion product source terms, and good operating performance will not necessarily eliminate the need for radiological protective clothing or equipment during the performance of all activities. Various types of protective clothing and equipment have been developed for use in radiological environments, oftentimes uniquely designed for a specific purpose. Due consideration should be given in the selection and use of protective clothing and equipment to maximize protection of the individual, while providing the best possible level of comfort to minimize the impact on worker productivity. Established protective measures should not be cumbersome as to be counter-productive pertaining to the impact such measures may have on the expenditure of collective dose associated with a given task. Therefore when prescribing the proper level of protective clothing and equipment a balance must be struck between providing adequate protection against contamination and the impact on collective dose to the entire work crew. For instance when evaluating the use of portable filtration equipment to minimize or control the spread of airborne contamination during the performance of a given task the dose required for installing, testing, operating and removing the equipment should be evaluated against the anticipated benefits to be gained. If the dose received as a result of setting-up and maintaining the equipment exceeds the potential dose savings then the control measure should be reconsidered.

## 9.2 Protective Clothing

Protective clothing (PC) includes those articles utilized to protect specific areas of the body from radioactive contamination. The primary function of protective clothing is to afford protection against contamination. Even though under certain circumstances, the use of protective clothing will afford some protection against beta radiation exposure, the use of protective clothing is primarily predicated on the physical type and level of contamination present in the work area. A wide range of protective clothing designs and types are available for use in radiological work areas. Protective clothing designs are available for use when dealing with dry or dusty conditions or when dealing with wet environments. Specific clothing materials have been developed to minimize migration of contamination through the fabric to prevent contamination of the underlying skin area. Additionally lightweight, "breathable" type materials are available if heat stress factors are a contributing concern when prescribing the use of protective clothing.

Once the need for protective clothing has been determined the next step is to prescribe the specific set of articles required for the task. Protective clothing requirements may be specified on an RWP or other work related document, perhaps within the work package itself. A detailed evaluation of the task to be performed, the nature of the work activity, expected and anticipated radiological conditions, the potential for unforeseen circumstances that could dramatically change radiological conditions and other factors are considered when determining the required protective clothing and equipment. This evaluation may be an integral part of pre-job planning activities discussed in Chap. 7.

As previously noted a wide-range of protective clothing types to choose from is available. Various parameters are considered when selecting protective clothing and may include such items as those noted below.

- Magnitude of radioactive contamination present
- Physical and chemical form of the contamination (e.g., wet, dry, corrosive)
- Physical nature of the work (e.g., heavy lifting, climbing, crawling)
- Degree of dexterity required (e.g., mechanical or electrical work)
- Environmental conditions in the work area (e.g., hot, humid, cold)
- Nature of work process and potential for generating airborne contamination
- Tools and equipment required to perform the task (e.g., power tools, air movers, grinders, welding)
- Nature of work process and potential to spread contamination
- Duration of the task (e.g., will individuals be required to work extended periods in hot or humid areas posing perspiration or heat stress concerns)?

**Table 9.1** Full set of protective clothing

| Location | Protective clothing article |
| --- | --- |
| Body | A whole-body coverall (or a protective coverall or anti-contamination coverall commonly referred to as an "anti-C") |
| Hands | Light-weight, cloth insert or glove liner worn under the outer protective glove |
| Hands | Outer protective gloves typically rubber or latex type fabric |
| Feet | An inner shoe cover that extends above the ankle covering the lower calf |
| Feet | Outer shoe cover (or bootie) typically a heavy-weight rubber or cloth material that provides non-slip footing |
| Head | Skull cap (or surgeons cap) to cover the top and sides of the head above the ears |
| Head | Hood that covers neck area in addition to the sides of the face and top of the head to the shoulder region |

- Type of work to be performed (e.g., electrical, mechanical)
- Relative strength of the beta/gamma component comprising the contamination

A standard set of protective clothing may be designated for use covering a specified range of contamination levels and the nature of the task. For instance work involving dry contamination not exceeding 100,000 dpm/100 cm$^2$ (or $\approx 1,700$ Bq/100 cm$^2$) that does not involve the potential to generate airborne contamination may require the use of a standard set of "full-PCs". A set of "full" protective clothing basically consists of those clothing articles that provide a degree of protection for the entire body. A set of full-PCs would typically include the items presented in Table 9.1 with perhaps minor deviations uniquely tailored to meet the needs of a given program.

For purposes of this text, shoe covers (often low-cut) are considered to be the outer primary shoe cover, usually of sturdy design, to provide secure footing.

Full-body coveralls are available in a variety of fabrics and materials. The different fabrics have their own advantages and disadvantages. Such factors as the ability of the fabric to absorb moisture, breathability of the fabric, effectiveness as a barrier against the absorption and migration of contamination, the durability of the material in everyday use and how many laundry cycles can it go through before the fabric starts to wear out, are all factors to consider when choosing a particular fabric. A synthetic fabric that has gained wide spread use in the industry is the Frham Tex II offered by Frham Safety Products. The material has a unique combination of characteristics that allows it to be used in numerous applications. The fabric has excellent breathability properties, is light-weight, when compared to cotton fabrics, provides a water resistant barrier, minimizes sweat-through contamination issues, is launderable, and provides added cooling capacity to minimize heat stress concerns. The material is also used for hoods and other articles of protective clothing. Figure 9.1 depicts a worker in Frham Tex II protective clothing. Note the use of Velcro straps to secure gloves and booties to the coverall. The use of these type straps eliminates the need for tape. By eliminating tape whenever possible, radioactive waste volumes are reduced, this is a primary factor in the USA concerning the use of Velcro for these applications.

**Fig. 9.1** Worker in a full-suit of Frham Tex II protective clothing (Courtesy of Frham Safety Products, Inc., www.frhamsafety.com)

Figure 9.2 displays examples of the types of gloves that may be associated with a full-set of PCs. Note the varying weights of material that are available for gloves.

A less extensive set of protective clothing could be prescribed for activities meeting certain criteria. For instance, work in contaminated areas not involving heavy physical activity or breach of a contaminated system could be safely performed with a less-stringent set of protective clothing. Activities such as operator rounds to obtain instrument readings, electrical and instrument and control work

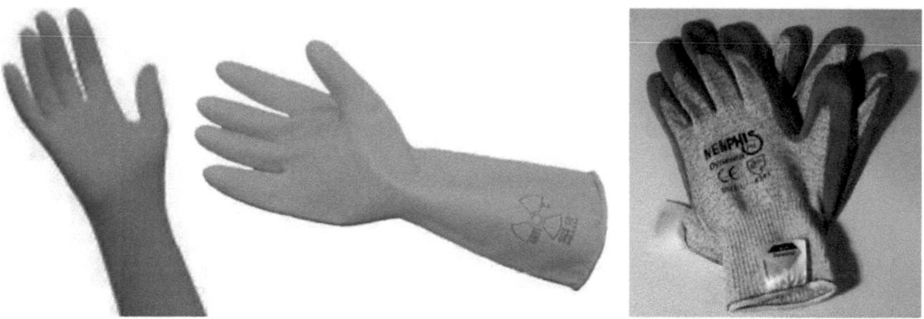

**Fig. 9.2** Various types and designs of gloves often used with a full-set of protective clothing (Courtesy of Frham Safety Products, Inc., www.frhamsafety.com)

activities, inspection type tours and general plant observation type activities could be performed with a less robust set of protective clothing. Additionally the use of less cumbersome protective clothing could prove more practical for these type tasks. Electrical work performed inside instrument panels requiring a high degree of dexterity may be difficult or impossible to perform while wearing a set of rubber gloves. Consequently it may be beneficial to establish a prescribed set of protective clothing that may be worn to facilitate the performance of various tasks under less demanding contamination control conditions. A less stringent set of protective clothing could consist of the following:

- A lab coat
- One set of shoe covers
- Surgeon gloves or a fine latex type glove
- Surgeons or skull cap or perhaps no head covering

## 9.2.1 Donning and Use of Protective Clothing

Obviously for protective clothing to serve its intended function it must be worn correctly and removed in a manner that does not cause the inadvertent spread of contamination from articles of protective clothing to individuals during the removal procedure. Though slight deviations may be encountered from what constitutes an acceptable dress-out method, the basic procedure for donning protective clothing is essentially uniform. The primary objective is to ensure that a given article of protective clothing is donned in such a fashion to serve its intended function and removed in a manner that minimizes the probability of spreading contamination to the wearer. An acceptable sequence for donning a full set of single PC's is provided in Table 9.2.

**Table 9.2** Sequence for donning full set of protective clothing

| Step | Clothing article |
| --- | --- |
| 1 | Don inner shoe cover (or bootie) over personal footwear |
| 2 | Don single coverall (or anti-c) with inner booties tucked inside coverall pant legs |
| 3 | Don outer shoe cover over inner bootie |
| 4 | Don set of inner gloves (glove liners) |
| 5 | Don skull cap or hood |
| 6 | Don outer set of rubber or latex gloves (or otherwise whatever constitutes the outer work glove) |
| 7 | Ensure gloves, booties, coveralls, are properly fastened and secured |

Various methods to secure protective clothing openings are acceptable. Coveralls may be equipped with Velcro straps, snaps or bands to secure openings at the wrists and ankles. Alternatively protective clothing may be taped at the cuffs and ankles to prevent contamination from entering at these locations. Regardless of the specific sequence established for donning protective clothing, the important aspect is to ensure that the articles are properly secured and in good physical condition prior to entering a contaminated area. Obviously the exact sequence in which head coverings, or outer shoe covers and gloves are donned is not necessarily of overriding concern. The primary consideration is to ensure that prescribed protective clothing has been correctly donned in accordance with established procedures by individuals prior to entering radiological work areas. As general practice individuals should be instructed in the importance of inspecting each article of protective clothing to ensure it is in good material condition. Obviously the use of protective clothing that is in poor physical condition could result in needless personnel contaminations. The presence of tears, holes, missing Velcro straps or fasteners or other imperfections and overall general integrity of the fabric should be inspected prior to use. Any condition that may compromise the effectiveness of protective clothing to serve its intended function should be cause to discard the article.

In addition to the above sequence for donning protective clothing, various practices may be established requiring the use of modesties or under garments prior to donning PC's. The use of under garments may be considered an integral component of the protective clothing program. Consequently, prior to donning PC's, individuals are assumed to have already removed their outer personal clothing and donned any required modesties. This practice also implies that the proper facilities (e.g., change rooms) are available whereby individuals may remove their personal clothing and have a location where modesties may be donned.

Protective clothing's primary purpose is to afford protection to workers during the performance of work activities in contaminated areas. This protection could be nullified if proper practices and procedures are not followed during the removal of PC's when exiting a contaminated area. During the course of a given task protective clothing may become highly contaminated. Particular attention should be given to the fact that outer gloves and shoe covers will most likely become contaminated during the course of work activities. Depending upon the physical nature of a task, if such activities as crawling, kneeling or climbing were performed during the course of the task, then other areas of PC may be highly contaminated. Protective clothing covering the knees, elbows, back or other body locations that may have been in intimate contact with contaminated surfaces or components should also be considered as prime suspect areas.

The proper removal of contaminated protective clothing often represents one of the primary challenges for the inexperienced radiation worker. Since the spread of contamination is not inherently obvious when removing PC's, training and the need to follow strict contamination control techniques are essential elements in preventing inadvertent personnel contamination during the removal of PC's. As experience is gained and individuals become more proficient (or accustomed) to dealing with radioactive contamination these skills will improve. Consequently, diligent contamination control practices must be maintained while removing potentially contaminated protective clothing when exiting a contaminated area.

To minimize the chances of spreading contamination from articles of protective clothing to an individual's skin or to under garments, not to mention to adjacent clean areas, a prescribed sequence for removing protective clothing should be established. The sequence essentially entails starting with the removal of those PC articles most likely to be contaminated and finishing with those items that should be non-contaminated or minimally contaminated. For instance if an individual were to remove outer work gloves and then proceeded to remove outer shoe covers with just glove liners on, then the glove liners could possibly become highly contaminated. Since the inner glove liners do not serve as the principle barrier against contamination, an individual could cross-contaminate under-lying skin areas on the hands after removing the outer shoe covers.

Even though slight deviations in the sequence in which PC's are donned may not result in any radiological consequences, failure to follow the proper procedure when removing PC's could result in personnel contaminations. Under certain circumstances potentially significant radiological safety concerns could result. Though the exact sequence of steps may differ somewhat within the industry, the process detailed in Table 9.3 follows the basic practice of removing the most highly contaminated items first (or those with the greatest potential of being contaminated) progressing to the least contaminated items.

### 9.2.2 Double Sets of Protective Clothing

Oftentimes it may be beneficial to designate contaminated areas based upon the magnitude of contamination present. To minimize the chance of personnel contamination or the spread of contamination within areas of the RCA it may be customary to designate highly contaminated areas from areas of lower contamination. This practice could consist of a graded approach whereby various

**Table 9.3**  Sequence for removing full set of protective clothing

| Step | Clothing article |
| --- | --- |
| 1 | Remove outer head covering—skull/surgeons cap, hood or head covering |
| 2 | Remove outer shoe covers while outer work gloves are still on |
| 3 | Remove outer work gloves being careful not to touch the inner glove liners with any portion of the outer work glove |
| 4 | Depending upon the design of the coverall (e.g., Velcro, zipper, or other closure mechanism) undo the coverall while the inner glove liners are still worn. Again being careful not to touch outer surfaces of the coverall with the gloved hand |
| 5 | Slip coverall down over the shoulders to the lower legs not allowing outside portions of the coverall to come into contact with undergarments. Slip the coverall over the inner shoe covers/booties being careful not to allow outside areas of the coverall to come in contact with personal shoes/footwear |
| 6 | As the inner shoe cover/bootie is removed from each foot the person places his foot onto the clean step off pad, repeating the process for the other foot. During this step it is essential that the bootie be pulled off the foot towards the contaminated side of the step off pad to prevent contamination from spreading to the clean side of the contaminated area boundary. |
| 7 | Remove inner glove liner or surgical gloves and dispose in designated container, ensuring that no cross-contamination occurs at the contaminated area boundary line of demarcation |

classifications of contaminated areas are established. This may include such nomenclature as Level 1 and Level 2 contamination areas (with perhaps additional levels) or a color scheme or some other suitable designation. Alternatively, a value may be designated as a lower bound for what constitutes a "high" contamination area. A convenient value for the designation of a high contamination area could be greater than 100,000 dpm/100 cm$^2$ or approximately 1,700 Bq/100 cm$^2$ (beta–gamma). This value may serve as a threshold for the establishment of a so-called double step-off pad area (discussed below) or for prescribing additional protective clothing as well as more extensive contamination control measures. The actual value used to designate a highly contaminated area may be predicated on the specific needs or conditions encountered at a given facility. Primary consideration is the recognition that for work in highly contaminated areas a value to serve as a threshold to ensure proper review of contamination control measures and protective clothing requirements may prove beneficial.

When working in highly contaminated areas additional protective clothing may be required. In addition to the single dress-out noted above, an additional coverall and extra set of outer gloves and shoe covers may be prescribed. A common practice is to establish an outer contamination zone around an area that contains significantly greater amounts of contamination. Oftentimes a double step-off pad arrangement may be established. Individuals exiting from the more highly contaminated area would remove outer shoe covers, gloves and coverall before exiting the inner step-off pad area. The outer shoe covers are removed in a manner similar to that described above when exiting a contaminated area, as the individual proceeds to the outer step-off pad area and prior to stepping onto the inner step-off pad. In actuality the individual will probably have two sets of booties and one set of shoe covers. Booties are typically made from a less rugged material such as thin plastic or cloth material that fit loosely over shoes, and usually extend up and over the ankle area. Booties do not usually serve as the outer working shoe cover. Consequently when exiting the inner step-off pad area an individual may have two sets of booties on after removing the outer shoe cover. Again, this procedure may differ somewhat from facility to facility. In some cases a double set of shoe covers may be utilized with one set of inner booties. Due to the tight-fitting nature of shoe covers extra effort may be required when removing the outer set of shoe covers while wearing an inner set due to the design of shoe covers. This may not be a problem at those facilities that utilize a physical barrier or provide benches at contaminated area exit locations. These arrangements allow individuals to be seated as they proceed from the contaminated area to the clean area. Though these type arrangements may prove beneficial in certain circumstances, contamination control and monitoring programs must be in place to ensure that any seating surface (e.g., chairs or bench tops) are surveyed on a frequency to minimize the potential for cross-contamination.


## 9.2.3 Disposable Protective Clothing

The utilization of single-use only PC's affords some advantages when dealing with residual contamination levels present on laundered articles of protective clothing. Under certain circumstances the use of disposable PCs may be a suitable alternative. Lightweight, low cost fabrics are available that offer an effective barrier against contamination and may be practical in certain situations versus the use of launderable protective clothing. Over the last several years many new fabrics have been developed and are ideal for use in certain environments. Lightweight, breathable fabrics that are water repellant are ideally suited for activities involving heavy physical work or when heat stress issues may be of concern. These fabrics also have an added advantage in that they provide a more effective barrier against the absorption of contamination relative to cotton or composite fabrics. Personnel contaminations may result due to profuse sweating of an individual during the course of an activity. Profuse sweating and the resultant moisture may saturate the protective coveralls, promoting the absorption of contamination through the fabric weave. This is often referred to as the "wick affect". During the course of an activity under conditions, where the outer PC has become saturated with moisture, and PC's subsequently come in contact with contaminated surfaces in the work area, contamination may leach through the fabric to the underlying skin area. These contaminations are often referred to as "sweat-through" personnel contamination events and could be prevalent during outage conditions. In fact residual contamination remaining after laundering could leach through the fabric of wet PC's also resulting in personnel contamination events.

The costs involved with the laundering and handling of protective clothing along with the processing and disposal of any radioactive effluents stemming from the laundering process may be comparable to the costs associated with the use of disposable protective clothing. The focus of this discussion pertaining to single-use and disposable protective clothing deals primarily with coveralls or anti-C's. However, it is recognized that it may be advantageous to utilize other articles of protective clothing such as hoods and skullcaps from these disposable materials. Also when one considers the administrative costs of dealing with personnel contaminations resulting from sweat-through events, not to mention the potential radiological safety aspects of these events, single-use PC's may be cost beneficial. As the price of disposable fabrics decrease and the quality of these materials improve while the costs of radioactive waste processing and disposal increases, the use of single-use or disposable PC's may become more advantageous from both a radiological safety and overall cost perspective.

Several types of disposable fabric materials are available that are incinerable or dissolvable after use, that minimize handling and disposal costs. One material that has gained widespread use in the industry is the OREX® brand of disposable, one-time use articles of protective clothing. The OREX® material provides an effective barrier against the migration of radioactive particles through the fabric. A patented film layer forms an impermeable barrier to contamination. The combination of these properties greatly minimizes sweat-through contamination events. This material is ideally suited for those activities that involve a high degree of physical work in hot areas with high levels of contamination. The material is light-weight and breathable providing extra comfort to workers especially in high heat environments. The specially designed material also provides an effective moisture barrier to minimize sweat through contamination events. Figure 9.3 depicts examples of an OREX® coverall and modesty garment. Figure 9.4 shows a work crew in OREX® protective clothing working inside a containment building.

The OREX® protective clothing products are fabricated with specially treated fabric that has unique solubility properties. These properties allow the clothing articles to be dissolved utilizing a proprietary process. The main advantage of the dissolution process is the significant reduction in radioactive waste volumes requiring processing and disposal. Due to the limited availability of radioactive waste disposal facilities and the high disposal costs in the USA this feature often proves cost beneficial. Essentially all that remains after the dissolution process are the coverall zippers and a small amount of non-compatible material.

### 9.2.4 Wet Work Protective Clothing

Certain activities may involve exposure to wet surfaces. This could consist of residual water or moisture inside components that may be encountered by workers during maintenance activities. Conditions may exist where the possibility of being sprayed by water may be encountered. This could arise as a result of residual pressure within a system or component or simply head pressure if the work location is at a system low point. Under these conditions an outer garment to protect individuals

**Fig. 9.3** The OREX® Ultra coverall and modesty protective clothing products (Courtesy of Eastern Technologies Inc., www.orex.com)

from exposure to contaminated liquids may be required. Again, a balance must be achieved between affording the desired protection and not unnecessarily encumbering workers. To serve their intended function as a barrier against liquids, water resistant protective clothing will contribute to heat stress concerns. Therefore close monitoring of work conditions and individuals while wearing these items should be performed to minimize the probability of heat stress. As a minimum provisions should be established to ensure that water resistant protective clothing is removed as soon as work conditions allow. Many situations may only require the use of these garments during the initial breach of a system or component. Once the residual water is drained or it is established that liquid is not present arrangements should be made to allow for removal of these garments.

**Fig. 9.4** Work crew dressed-out in OREX® protective clothing (Courtesy of Eastern Technologies Inc., Incorporated www.orex.com)

### 9.2.5 Staging and Maintenance of Protective Clothing Inventories

Consideration should be given to maintaining an adequate inventory of protective clothing. Additionally the manner in which protective clothing is staged and made available to workers may present logistical issues if adequate consideration is not given to these aspects of the program. The administrative aspects of providing protective clothing and various radiological safety supplies and equipment, will require sufficient resources to maintain the program. Items associated with the staging and location at which supplies will be made available to workers, stocking, storage and handling aspects, and facilities to maintain these supplies must be properly managed. If protective clothing is laundered at a given plant then obviously arrangements must be made to ensure that adequate laundry facilities are properly equipped to support the needs of the plant. The maintenance and inventory of protective clothing must also be capable of handling increase workloads encountered during refueling outages and extended maintenance outages when the demand for these supplies may increase by orders of magnitude as compared to routine operating conditions. The administrative aspects of these tasks should not be underestimated.

Many existing nuclear units were not designed with sufficient change room or locker facilities, and in some cases no allowance was made for female radiation workers. Consequently change room facilities may be undersized with designs that do not allow for ideal placement of personnel contamination monitors or floor layouts that do not facilitate contamination control during egress and exit. These factors should be taken into consideration when establishing protocols concerning the use of modesties. The location (or locations) where individuals will don protective clothing either in a designated facility or specific areas of the RCA where there is sufficient space to stage supplies may also be other aspects to consider. The availability and location of personal decontamination shower facilities should also be evaluated. Centralized change facilities and "satellite" facilities may be required. Satellite or temporary change facilities may be established during heavy work load periods to support a large influx of outage workers and the increase in the number of entries made into the RCA during these periods. The complexity of these activities will be compounded at multiple unit sites, especially if interconnecting centralized facilities are not available. Necessary resources must be available to collect, stock, launder, monitor, and to otherwise maintain the inventory of protective clothing necessary to support plant operations.

Ideally protective clothing should be free of any residual contamination after laundering. However, since this may not be practical (see discussion below) post laundering screening values should be established for what constitutes "acceptable" levels of residual contamination. Based on practical considerations an upper limit is typically established for contamination levels allowed to be present on articles of laundered protective clothing. One is confronted with a tradeoff, namely; how much time and effort to devote to the laundering process to minimize residual contamination levels versus the cost and radiological safety benefits to be gained? Obviously the amount of residual contamination should be minimized commensurate with affording adequate protection to the wearer while not posing a direct contamination concern due to the levels present.

It is common practice, at least in the American LWR industry, to launder protective clothing at an offsite facility. Many articles of protective clothing are re-used and include coveralls, shoe covers, booties, hoods, skullcaps, gloves and other protective clothing items. Ideally, laundered protective clothing should not result in the transfer of residual contamination from the article itself (e.g., primarily coveralls) to the user. However, as any Health Physicist would testify, previously contaminated protective clothing will contain some amount of residual contamination after being laundered. As the detection sensitivity of whole-body personnel contamination monitors used at RCA exit points has improved, the ability to detect lower levels of contamination on individual's exiting the RCA has dramatically improved. Consequently it is essential to recognize the need to provide effective laundering of protective clothing and establish residual contamination limits that are compatible with personnel contamination monitor (PCM) release limits. For instance if laundered protective clothing is allowed to be re-stocked below a certain monitoring value (e.g., 10,000 dpm/100 cm$^2$ or $\approx 170$ Bq/100 cm$^2$), low-levels of residual contamination may be transferred to individuals while wearing the protective clothing. Upon exit from the RCA, with PCM alarm values set at a lower alarm threshold, the presence of "contamination" may be detected. Upon investigation the cause of the contamination may be discovered to be nothing more than the fact that the use of protective clothing was the source of the contamination and not the result of poor work practices or an operational incident. Therefore the laundering and subsequent monitoring of protective clothing must be afforded a comparable level of effort as that directed towards personnel contamination monitoring. Various measures may be implemented to address this issue.

Clothing may be worn under the coverall to serve as a barrier between the wearer and the "contaminated" PC. The use of modesty garments, or surgical scrubs, or some type of light-weight under garment, may be employed for this purpose. Modesties could be treated as clean clothing monitored in a similar fashion as that for street clothes worn within the RCA. Contamination monitors used to measure residual contamination levels on laundered PC's should be highly sensitive, state-of-the-art systems to ensure accurate monitoring of residual contamination levels. Various measures can be taken during the laundering process to improve the effectiveness of washing. For instance more-highly contaminated clothing (e.g., booties and gloves) could be laundered separately either in a separate wash cycle or perhaps even in dedicated laundry machines or facilities to minimize the chance of cross-contaminating lesser-contaminated items. Articles of protective clothing exceeding an upper contamination screening value for washing could be discarded. Multiple wash cycles could be employed to achieve the desired residual contamination limits. Notwithstanding the approach taken regarding the laundering and re-use of protective clothing it is essential that procedures be established to address monitoring limits that are deemed acceptable and controls prescribed for the laundering process.

## 9.3 Respiratory Protection Program

A respiratory protection program is a key element in the control and minimization of internal exposures at LWR facilities. Even though exposures resulting from the inhalation or intake of radioactive material at LWRs are usually insignificant and represent a small fraction of the collective dose, an effective respiratory protection program plays a vital role in ensuring the overall radiological safety of employees. The need for respiratory protection equipment (RPE) may be necessary when responding to plant operational occurrences, involving radiological safety concerns when dealing with the presence of airborne material of unknown con- centrations. Under these circumstances personnel entry may be required into affected areas, to place equipment in a safe configuration, manipulate valves, or take other measures to mitigate an event, without an opportunity to pre-evaluate airborne concentration levels. Consequently an effective respiratory protection program should be established with initial responders trained in the proper use and function of the various types of RPE available for use at a given facility.

The goal of a respiratory protection program is one whereby the primary focus is to minimize the need for RPE. To accomplish this goal or objective engineering controls and strict contamination control measures should be utilized to preclude the need for RPE. The proper utilization of these and other techniques in combination with an effective airborne monitoring program can play a key role in minimizing the use of RPE, improving worker productivity, providing a higher level of comfort to workers, and minimizing collective radiation exposures. The intent is not necessarily to eliminate the need for RPE but rather to ensure that appropriate focus is maintained on addressing conditions that warrant the use of RPE. It may be relatively easy to prescribe the use of full-face respirators for a given task versus establishing controls to eliminate a potential source of airborne contamination. However, by not controlling the possible presence of airborne activity the situation could unwittingly contribute to a slow but gradual degradation of radiological conditions in plant areas. For instance low-level amounts of contamination may be deposited on plant systems, structures and components, in overhead areas or on floor areas, that otherwise would be free of contamination. Over time the radiological requirements for entries into affected plant areas for routine tasks may become more burdensome. If a lax approach is taken regarding the use of RPE then opportunities to evaluate measures that could possibly eliminate the need for RPE may be missed. Even though the one-time use of respirators for a given task may be inconsequential, over the long-run, if measures are implemented to eliminate the need for RPE the long term savings in dose, improved worker productivity, and lower maintenance and operational costs could make significant contributions to safe and reliable plant operations. Consequently one of the primary objectives of a LWR radiological safety program should be to minimize the need for the use of RPE.

A contributing consideration concerning the use of RPE is the need to evaluate the impact of RPE on the overall radiological safety aspects associated with the performance of a given task. This stems from the desire to maintain worker comfort, minimize impact on worker productivity, and the fact that under certain situations the use of RPE may result in an overall increase in the collective dose received by either an individual or the entire work group when performing tasks involving the use of RPE. This latter concern is typically addressed by an "ALARA/TEDE" evaluation. This evaluation basically consists of determining the dose the individual would receive to complete the task while wearing a respirator and without a respirator. This essentially results in determining the extra time required to perform the task with a respirator. This may be relatively straight forward when only one individual is involved. However other considerations often come into play when dealing with an entire work crew. The task may involve a step requiring one worker to don a respirator for a length of time to complete an evolution. The evaluation should factor in the exposures received by the other members of the work crew while the one individual completes the step. A comparison should be made between the exposure saved by wearing a respirator for the one person and the additional exposure received by the other members of the work crew waiting for the step to be performed. If the other members have access to a low-dose waiting area this may not be a concern. On the other hand if the other members of the work crew are required to stay in close proximity of the work area due to the short-duration of the step or otherwise will remain in an area resulting in measurable dose to the other crew members then this should be factored into the dose evaluation. The purpose of the ALARA/TEDE evaluation is to focus on the collective dose received by the entire work crew while performing tasks requiring RPE. By performing the evaluation other options may be identified to minimize exposures that may have otherwise been overlooked. Determining the extra time required to perform a given task in a respirator due to the impact on worker productivity, efficiency, communication issues, and limitations on visibility and assigning values to these items may prove the most difficult part of the evaluation.

Even when an ALARA/TEDE evaluation is not performed due consideration should be given to such items as the obstruction of worker vision, ability of members of a work crew to communicate, as well as the physiological and psychological aspects associated with the use of respirators when prescribing RPE for a given task. The use of RPE must be properly assessed with an evaluation of the overall hazard adequately analyzed prior to prescribing RPE. The use of RPE should not be a prescriptive-based program but one in which thorough consideration is given to eliminating the source of the airborne contamination whenever possible. A proper balance should be struck between the risks associated with the use of RPE and the radiological safety benefits to be gained.

### 9.3.1 Elements of a Respiratory Protection Program

A comprehensive LWR respiratory protection program may include such elements as the maintenance and inspection of RPE, a training and qualification program for users, medical surveillance of respirator users, and administrative aspects to address the storage, handling, and issuance of RPE. Various types of respirators may be required as part of an overall respiratory protection program. The types of respiratory protection devices may include full-face respirators, powered air purifying respirators (PAPR), fresh air or air-supplied bubble hoods, self-contained breathing apparatus (SCBA), and bubble suits. Though not strictly a respiratory protective device, face shields are often used to protect against facial contamination. Under certain circumstances face shields serve as an effective protective device with the added advantage that communications and worker comfort are not impacted to the same degree as when compared to the use of a full-face respirator.

To ensure the safety of users of respiratory protection equipment certain requirements should be established when using respirators. This section presents those requirements that are closely related to respiratory protection programs implemented at LWRs in America. However, many of these elements are integral to the establishment of an effective respiratory protection program.

The employer or utility is typically responsible for providing respiratory protection devices for workers. The employer has the responsibility for ensuring that approved respiratory equipment has been certified by the competent authority (e.g., the primary regulator or international safety organizations as appropriate) and that equipment is maintained and utilized in accordance with the manufacturers' recommendations in order to maintain the designated certification. The NRC requires that respiratory protection equipment utilized in nuclear power plants in the USA be tested and certified by the National Institute for Occupational Safety and Health (NIOSH). There are exceptions to this requirement that allow a licensee to submit an application to the NRC requesting the use of a specific device. The performance characteristics and protection attributes of the particular device would have to be supported by appropriate test data. The training and qualification program should be sufficient to ensure that individuals are knowledgeable of techniques and procedures required in order to properly utilize respiratory protection equipment.

## 9.3.2  Types of Respiratory Protection Equipment

A full range of RPE devices is available from various suppliers. Each device offers unique advantages and disadvantages. The full-face, negative demand, respirator is probably the most common type of respirator utilized in the LWR industry. A negative-demand respirator is designed to maintain positive pressure inside the respirator face piece during exhalation. While the wearer inhales (air drawn through the air purifying filter or filters) the pressure inside the face piece is negative compared to the outside air pressure. These respirators are relatively easy to maintain, offer an effective protection factor that is adequate to provide the degree of protection sufficient for the majority of activities involving exposure to airborne radioactivity. These devices afford a relatively long service life with proper maintenance and care.

Figure 9.5 depicts the Ultra-Twin® full-face respirator commonly used by the industry offered by Mine Safety Appliances. Various filter cartridges are available for use with full-face respirators. Often times a combination filter cartridge is used at LWRs to provide protection against both particulates and organic vapors. Since airborne particulates (e.g., Co-58 and Co-60 and other airborne activation products) represent the primary constituent of airborne radioactivity it is essential to provide an appropriate filter cartridge designed for this purpose. A combination cartridge that protects against organic vapors offers added protection when airborne radio-iodine species are also present. A combination filter cartridge is also depicted in Fig. 9.5.

Powered air purifying respirators (PAPRs) are equipped with a blower that delivers filtered air to the breathing zone inside the respirator. The face piece is essentially identical to that of a full-face respirator equipped with a filter cartridge. The units are equipped with a battery powered blower that routes the air through a filter and then directs the air via a breathing tube to the full-face respirator breathing zone. As with other RPE devices a PAPR has distinct advantages and disadvantages. Figure 9.6 depicts the OptimAir 6A PAPR offered by Mine Safety Appliances. This unit may be used with several different NIOSH approved face pieces.

**Fig. 9.5** The mine safety appliances Ultra-Twin® full-face respirator and combination particulate-filter cartridge (Courtesy of MSA, www.msanorthamerica.com)

**Fig. 9.6** The mine safety appliances OptimAir 6A powered air-purifying respirator (Courtesy of MSA, www.msanorthamerica.com)

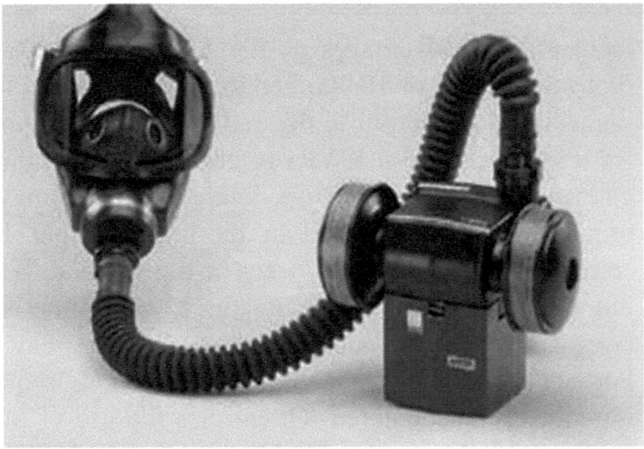

Advantages of a PAPR are the higher protection factor of 1,000 versus that of a full-face air purifying respirator. (Protection factors for various types of RPE devices are discussed below). The forced air flow affords limited cooling for the wearer and minimizes fogging of the respirator lens. The powered blower reduces breathing resistance that could be an important factor when respirators are worn for extended periods of time.

The self contained breathing apparatus (SCBA) may be the respiratory protection device of choice when entering areas of unknown airborne radioactivity concentrations. This situation may be encountered as a result of an operational

**Fig. 9.7** The mine safety
appliances AirHawk II SCBA
unit with an Advantage 4000
face piece (Courtesy of MSA,
www.msanorthamerica.com)

incident. The advantages of the SCBA under these circumstances is the high
protection factor of 10,000 and that it is approved for use in atmospheres that are
immediately dangerous to life and health (IDLH atmospheres). The use of SCBA's
for drywell entries at BWRs or containment building entries at PWRs to investigate operational occurrences while at power may be the preferred RPE. The
primary disadvantages include the heavy weight, limited air supply and limited
communication capability when using these devices. However; for rescue and
initial emergency response situations, when time is not available to evaluate airborne radioactivity concentrations, SCBAs may be the only viable option.
Fortunately in the LWR industry, situations requiring the use of SCBAs are
infrequently encountered. Oftentimes the use of SCBAs may be based more on
conservative reasons to take advantage of the high protection factor when entering
areas with unknown airborne concentrations. Figure 9.7 depicts a SCBA unit
offered by Mine Safety Appliances.

The use of SCBA units may be primarily driven by the need to maintain fire
brigades and emergency response teams. Emergency response personnel must be
trained in the proper operation and use of these units. Training should provide an
opportunity for emergency response team members to don an SCBA under
simulated conditions where the user can become familiar with the breathing time
limitations, escape procedures, as well as the overall operation and use of an
SCBA.

Air-line respirators are another type of RPE commonly used at LWRs. These devices offer more comfort then that of a full-face respirator. They provide a constant flow of air to the user which maintains a slight positive pressure of air inside the face piece. Air-line respirators require a source of breathing air which could be supplied by a plant-wide breathing air system or by bottled-air. A manifold arrangement is often used that allows up to several air-line respirators to be used at the same time. To maintain the proper airflow to the end user it is vital to ensure that the breathing air pressure is maintained within an acceptable range. The required air pressure may be a function of the number of users connected to a given manifold or primary source of breathing air. To maintain proper air flow and pressure in the air-lines manufacturers provide specifications regarding the maximum length of airline that may be deployed to an individual user.

Air-line respirators offer advantages when working in hot environments or for long-duration jobs requiring respiratory protection. The higher protection factor of 1,000 allows the use of these respirators in high airborne radioactivity areas, where a full-face air purifying respirator may not be sufficient. Air-line respirators are not approved for use in atmospheres immediately dangerous to life and health. A precaution associated with the use of air-line respirators is to ensure that the breathing air sources, including the manifold airline connections are maintained free of contamination. Breathing air manifolds located inside reactor or containment buildings or other potentially contaminated areas should be protected from contamination and surveyed prior to connecting airlines to the system.

### 9.3.3 Selection of Respiratory Protection Equipment

The primary consideration when selecting a given type of respirator for radiological safety purposes is to limit the potential intake of inhaled radioactive material by the wearer. Ideally, once the airborne radioactivity concentration is determined the applied protection factor should reduce the actual concentrations to which the wearer will be exposed, to less than a DAC, and preferably lower. Under normal operating conditions and for the vast majority of maintenance and refueling activities airborne concentrations exceeding tens of DACs are seldom encountered. Additionally the diligent use of engineering controls and measures to restrict the time duration of airborne exposures exceeding concentrations of several DACs presents a situation whereby a full-face air purifying respirator offers sufficient protection in the vast majority of situations requiring the use of RPE. Assigned protection factors (APF) for various types of RPE based on their operational mode are provided in Appendix A of 10CFR20. The APFs for the commonly used RPE are summarized in Table 9.4.

To gain the afforded protection RPE devices should be used in accordance with the manufacturer's instructions or recommendations. Oftentimes the

**Table 9.4** Assigned protection factors for respirators

| Device | Protection Factor |
| --- | --- |
| Full-face air purifying respirator (particulate cartridge respirator) | 100 |
| Powered air purifying respirator (full-face piece or hood) | 1,000 |
| Air-line respirators, continuous flow or pressure demand (atmosphere supplying protects against particulates, gases and vapors) | 1,000 |
| SCBA pressure demand mode | 10,000 |

approval of an RPE device includes not only the respirator face piece itself but could include any connecting airlines, fittings and attachments. This is often the case for airline, supplied air, respirators. For example the use of airline respirators may require the use of specific type hose fittings and associated attachments to ensure that prescribed flow rates are maintained to provide the assigned protection factor. Governing regulatory authorities or national safety organizations may also specify requirements associated with the approval and use of a given respiratory device in order to use assigned protection factors.

The use of RPE, regardless of the design or type, results in some obstruction of vision, makes communications more difficult, and restricts freedom of movement (e.g., air-supplied hoods). Safety considerations that should be considered when prescribing the use of RPE, in addition to those pertaining to the radiological conditions, is ease of access to the work location and exit from the location. Work areas accessible only by means of a vertical ladder may pose unique safety concerns due to limited visibility while individuals are ascending and descending ladders. In fact, under certain situations industrial safety concerns and increased risk of trips or falls for example, may be the primary concerns versus exposure to potential airborne contamination. Obviously that's not to say that the use of RPE for radiological safety purposes is not important, rather proper evaluations are necessary when prescribing the use of RPE to ensure the overall protection of individuals has been properly considered. Under most circumstances the use of engineering controls or the confinement or elimination of the source of the airborne contamination to eliminate the need for RPE may be the most effective option.

### 9.3.4 Maintenance and Inspection

Respiratory protection equipment must be properly maintained and repaired in order to afford the intended protection. Persons performing these tasks should be trained and qualified to perform these activities. Full-face respirators are washed, sanitized and inspected after each use. Persons performing these activities should be trained to identify any defects in the full-face respirator that may prevent a proper seal from being achieved. The condition of exhalation valves and the sealing area for the filter cartridges, and respirator straps should be inspected for any signs of damage or defects. Typically a method is employed to inform the wearer that the respirator has been inspected, sanitized, and tested and is otherwise available for use. Respirators that have been made available for re-use should have an inspection sticker or tag placed inside the bag that holds the individual respirator. The inspection tag should include the date of the inspection especially if filter cartridges with a given shelf-life are provided with the respirator.

Maintenance and repair of SCBA units require manufacturer-certified training. Maintenance and repair on components such as the regulator and low pressure alarm are required to be performed by specially trained and certified individuals. Manufacturers of SCBA units identify those components essential to maintaining the NIOSH certification of a specific device. Repair of these components require manufacturer-certified training. SCBA units available for use are typically stored in separate containers or wall-mounted storage units. A break-away security seal may be affixed to the handle of each SCBA storage unit cover. Individuals authorized to use SCBA units should be thoroughly trained on the inspections to be performed prior to donning an SCBA unit. In addition to verifying the condition of the face piece as discussed above users should be trained and qualified in the operational readiness checks of the SCBA unit. These checks should include verification that sufficient air cylinder pressure is available, verifying that the breathing air cylinders' hydrostatic test is current, that the low pressure alarm is functional and inspections associated with the regulator and SCBA harness.

### 9.3.5 Training and Qualification

Respirator users must be trained in the correct use of RPE to ensure the safety of individuals while using RPE. Training requirements are typically incorporated into an initial respirator qualification program. Respirator qualification should include specific training required that covers all aspects of respirator usage. Training and qualification efforts should also address an evaluation of an individuals' "comfort" level while wearing a respirator. Training should include some type of performance evaluation whereby users are observed while demonstrating the proper donning, use and removal of RPE. Training may include mock-up sessions or other elements to mimic physical stress levels that may be encountered while wearing a respirator. During the mock-up sessions it may be beneficial to evaluate physiological and psychological stress factors also.

Training should include instructions in the proper donning and removal of the respirator. Training should address the responsibilities of respirator users with regards to ensuring that they verify the suitability of a respirator for use. Requirements associated with inspection of the respirator prior to use and the need to perform a field check to confirm that a proper seal has been obtained after donning the respirator should be included in the training program. Individuals should be trained to inspect the physical condition of a respirator prior to use. The sealing surface of the face piece should be checked for cracks or tears and that straps are in good condition. Exhalation and intake valves should be checked for the presence of foreign material or debris that may interfere with their function. Verification that the respirator has been properly donned is often accomplished by simply covering the filter(s) intake with the palm of the users' hands and inhaling while observing an inward deflection of the respirator face piece.

Respirator users should be instructed in the procedure to exit an area while wearing a respirator in the event of an emergency. Typically for a filter cartridge full-face respirator this may entail nothing more than instructing wearers to simply remove the full-face respirator whenever discomfort is experienced and leave the area immediately. An exposure of a few DAC-hours may be of no consequence when compared to the safety aspects that could result if an individual experiences problems (whether physical or psychological) while in a respirator and does not take immediate actions. When using an SCBA the situation is more complicated and potentially life threatening. Assuming that the SCBA was prescribed as a result of entering an area with a dangerous atmosphere the option of immediately removing the face-piece may not be available. These situations could include entries into areas with low oxygen levels (<19.5%), the presence of a hazardous gas or toxic substance or entry involving a confined space. Under these circumstances approved emergency exit procedures should be reviewed prior to entry. Preparations in the event of an emergency may include the staging of a rescue team, equipped with the necessary respiratory protective equipment, and prepared to enter the work area immediately in the event that rescue is necessary due to discomfort, injury, or incapacitation of a member of the entry team. The exit or escape route should be described to members of the entry group prior to entering the area. Standby rescue personnel should be properly trained in their duties and responsibilities.

The training and qualification for full-face respirators may consist of a medical screening and confirmation of a suitable fit test. USA programs require respirator users to complete a medical screening to evaluate pre-existing conditions that could impact users of respirators. These conditions may include heart conditions, stress factors, history of smoking and perhaps other parameters, and a pulmonary function test. In addition the need for prescription eye glasses may also be evaluated. Depending on the individuals job classification prescription eye wear may be crucial for an individual. This may be particularly important for welders for example.

A quantitative fit test is often conducted to ensure that a suitable seal is obtained and can be maintained during the course of normal work conditions. In the USA quantitative fit tests are performed on an annual basis. The fit test may involve having the wearer demonstrate an adequate fit while talking, bending, and having the subject move their head in various directions. A quantitative fit test measures the percent ambient particles inside the respirator verses particles outside of the respirator while the subject is wearing the respirator. The test respirator is fitted with an adaptor that monitors the presence of the ambient particles inside the face

**Fig. 9.8** The TSI Incorporated PORTACOUNT® PRO respirator fit tester 8030 (Courtesy of TSI Incorporated, www.TSI.com)

piece during the quantitative fit test exercise. The adaptor is connected to a device that analyzers the air inside the respirator.

The PORTACOUNT® respirator fit tester offered by TSI Incorporated offers a convenient and easy to use method for performing quantitative fit tests. Unlike earlier qualitative testing methods that used such challenge agents as irritant smoke, isoamyl acetate (banana oil) or other agents the PORTACOUNT® system utilizes microscopic particles in ambient air. This method eliminates the need for fit test booths or other enclosures when determining fit test factors. Fit tests may be performed in any location convenient for an individual to perform the required fit tests movements. Fit test software is utilized with the unit to determine fit factors and when connected to a PC provides automated fit test reports. Figure 9.8 depicts a respirator fit tester offered by TSI.

## 9.4 Portable Air Filtration Units

Mechanisms to control the spread of airborne contamination are key ingredients in minimizing the need for respiratory protection equipment. Supplemental filtration units provide a means to locally capture and direct the airflow through filters. Two techniques are commonly employed to achieve this end; namely, the use of air movers and filtration units. Air movers utilize various apparatus to collect airborne contamination at its source and route the contamination to a remote location not occupied by workers. These devices consist of a vacuum pump to move the airborne contamination and transmit it through an enclosed duct work or confinement device. The air is then exhausted to a local intake of the installed plant ventilation system or directly through a portable filtration unit staged for that purpose.

Portable filtration units are often self-contained equipped with an air pump, couplings to direct air flow through filters and an exhaust unit. Units consist of one or more HEPA filters, a pre-filter, and perhaps a charcoal impregnated filter (for iodine removal). The filter section is enclosed in an air-tight housing equipped with an intake and outlet side. The filter housings are equipped with manometers to measure the differential pressure across the various filter stages. Filters may be changed-out based on differential pressure across a filter or a filter bank. Routine replacement of pre-filters can extend the life of the more expensive HEPA and charcoals filters on those units with multiple filter stages. The entire unit is mounted on a chassis often equipped with wheels to allow the units to be transported to the work location. Various sized units are available that can be selected based upon the anticipated airborne contamination levels and constituents associated with a given task. Units capable of providing air flow capacities as high as 3–6 $m^3$/min (or approximately 100 or 200 $ft^3$/min) are common. Units used for extended periods such as during outages that may collect significant quantities of radioactive material on filters should be subject to routine radiological surveillance. Depending on the location at which the actual filter housing is located the filters could become a source of radiation exposure to individuals in the immediate vicinity of the filtration unit. Radiation levels emanating from the filter housing should be checked periodically. Figures 9.9 and 9.10 depict two different size portable filtration units. The smaller unit in Fig. 9.9 has a capacity of 1,000 cfm (28 $m^3$/min) and the larger unit in Fig. 9.10 has a capacity of 2,000 cfm (57 $m^3$/m).

Vacuum cleaners may be available for use within the RCA for general housekeeping purposes and to vacuum radioactive material spills. If the operation and maintenance of these units is not properly controlled then their use could pose radiological concerns. Since the purpose of these vacuums is to collect radioactive contamination they will contain varying amounts of radioactive material over time. Just the simple act of emptying these vacuum cleaners could result in the spread of contamination and possibly internal uptakes if proper procedures and adequate contamination control measures are not followed. To minimize the chances of these type events administrative controls associated with the use of these units should be established.

Vacuum cleaners used for the collection of radioactive material should have certain design features. Vacuum cleaners used for these purposes should be equipped with a filtered exhaust (typically a HEPA filter) to prevent the spread of contamination. Manometers may also be installed on these vacuums to provide indication when the filters need to be changed-out. The units should be designed to prevent access to the collection housing by unauthorized personnel. This is usually accomplished by equipping the units with a lockable housing feature. Additionally administrative controls to prevent unauthorized access could include affixing labels to the vacuum cleaners informing individuals of special handling requirements. Chances that a vacuum cleaner could be used for an extended period of time without being monitored for radioactive material content and left unaccounted for should be guarded against. Vacuum cleaners used for an extended period could collect sufficient amounts of radioactive material within the housing

**Fig. 9.9** The Bartlett AP-1000-B portable filtration unit (Courtesy of BHI Energy www.bhienergy.com)

**Fig. 9.10** The Bartlett AP-2000 portable filtration unit (Courtesy of BHI Energy www.bhienergy.com)

to pose a radiation hazard of its own. A good practice is to designate a centralized storage location for vacuum cleaners approved for use in contaminated areas within the RCA. Utilization of a centralized location in conjunction with a formalized inventory program ensures that these units will be properly maintained and greatly minimizes the chance of a radiological-use vacuum cleaner being left unattended in an unauthorized location.

As noted above the units should have a lockable housing to prevent access to the internal waste collection area. A program to ensure that vacuum cleaners are surveyed and inspected on a routine basis should be established. The availability of a centralized storage area together with a process that requires vacuum cleaners to be signed-in and out by users facilitates the proper control of these units. Figure 9.11 depicts a HEPA equipped vacuum unit used in RCA areas.

**Fig. 9.11**  HEPA vacuum
unit (Courtesy of BHI Energy
www.bhienergy.com)

## 9.5  Temporary Shielding

Oftentimes the need arises to provide temporary shielding to reduce personnel exposures during the course of maintenance activities. Localized hot spots that may exist in such areas as pipe bends, valves, system low spots and various other locations may represent the primary contributor to worker exposure for a particular task. Oftentimes the localized source of the radiation may be adjacent to the primary work location and not directly involved with the task. Under these conditions strategic use of temporary shielding may result in significant dose reductions to the work group. For these and other reasons an inventory of shielding supplies should be maintained for use in work locations to reduce general area dose rates while performing work activities.

Various configurations of lead may be utilized for shielding depending on the purpose. Lead bricks are ideally suited for shielding radiation detectors utilized at local counting stations established to support outage activities. Sheets, rolls, and various sizes of lead blocks may also be utilized. Sheets of lead could be affixed to existing walls or incorporated into a custom made rack to provide additional shielding in plant locations. One of the most convenient forms for utilizing lead as a shielding method is by incorporating various densities of lead into a blanket design. This design is used extensively in the LWR industry due to its' versatility. These blankets may be hung or wrapped around pipes and valves to shield localized hot spots. Lead blankets are available in essentially any width or length and various sizes and shapes. For practical reasons the sizes of lead blankets usually do not exceed a thickness of a few centimeters or a length of about 2 m. A compromise must be achieved between the ease of handling and installing a lead blanket in the field versus the shielding factor afforded by a given blanket. The weight of an individual lead blanket should facilitate installation in order to ensure that the dose received while placing the shielding is minimized.

Depending on the density and thickness of lead blankets dose rate reduction factors for Co-60 in the range of 20–30 may be achieved. Additional blankets may be added to obtain the desired overall dose reduction factor for a task. Due to the high density of lead and depending on the physical arrangement of the installed shielding lead blankets must be properly fabricated. The fabrication should ensure that no shifting of the lead matrix occurs while in use. The lead is often incorporated into a cake-type mixture and secured within an inner wool or specialized high-strength fabric. The quilting must be such as to prevent shifting of the lead when it is hung from racks or draped over pipes or components. Depending on the quality and design of the stitching the lead within the blanket could sag or shift after repeated handling or extended use. Obviously if the lead matrix is subject to shifting during use the attenuation factor could be reduced resulting in unexpected radiation levels. Lead blanket design should also incorporate features to facilitate installation. A common design feature is the incorporation of grommets along the outer edges of the blanket. These grommets provide a convenient mechanism to hook blankets in place and subsequent removal. Applications involving the long-term use of lead blankets should be periodically inspected to ensure that the afforded shielding protection has not been compromised. These inspections should include verification of dose rates and a visual inspection of the material condition of the lead blankets and that the blankets are still secured in their original configuration. Figure 9.12 depicts a series of lead blankets arranged in a shield wall configuration and a close-up of the grommets to facilitate placement of the blankets on hooks or a rack.

Temporary shielding during outages may be required to be placed around large components such as heat exchangers, around the bottom of waste collection tanks, or along extended suctions of piping. Personnel exposures associated with maintenance activities conducted in the vicinity of these areas may be significantly impacted by these radiation sources. Under these circumstances the use of a "shadow-shield" arrangement may prove beneficial, especially if multiple jobs may be impacted by a common radiation source. Specially designed racks to accommodate hanging lead blankets that can then be rolled into place are ideally suited for these applications. Dose savings may be realized by placing the lead blankets on the racks in a low dose area then moving the racks into position. Depending on the total amount of lead to be placed on a given rack the ultimate weight of lead may make movement of the rack burdensome. Under these conditions a minimal amount of lead shielding may be arranged on the rack to minimize exposure to individuals installing the remaining amount of lead blankets once the shielding rack is placed into position. Figure 9.13 depicts a sample of a moveable rack designed to hang lead shield blankets. Lead shield blankets may be placed on either side of the rack.

The use of temporary shielding often needs to be evaluated for any potential impact on plant systems and components. Depending upon the circumstances and

**Fig. 9.12** A standard lead blanket shield wall configuration using lead blankets equipped with grommets (Courtesy of Nuclear Power Outfitters, www.nuclearpoweroutfitters.com)

specific locations of placement of shielding some type of evaluation may be necessary to ensure that the weight of the shielding has not produced unacceptable loading on system structures. The primary impact stems from the weight of lead shielding applied to system piping or components. Auxiliary impacts that may also need to be considered, even though not as obvious as those relating to the weight of the shielding, may be due to the physical location of the applied shielding. It is essential to ensure that temporary shielding does not impact the movement of valve stems for example. Even though the system may be out of service valves and other components may be required to operate if called upon. The manner in which the shielding is secured should not interfere with any component operation. A valve whose stem may be in a fully closed position when the shielding is initially applied should still be capable of operating through the entire range of its valve stroke.

Often times the need arises to provide form-fit shielding. This application is highly suited when the source of radiation is highly localized or when large amounts of shielding would be counterproductive. Tungsten is a relatively high dense, malleable metal that may be fabricated in sheets or ribbons. Nuclear Power Outfitters offers tungsten shielding products under their trade name of T-Flex®.

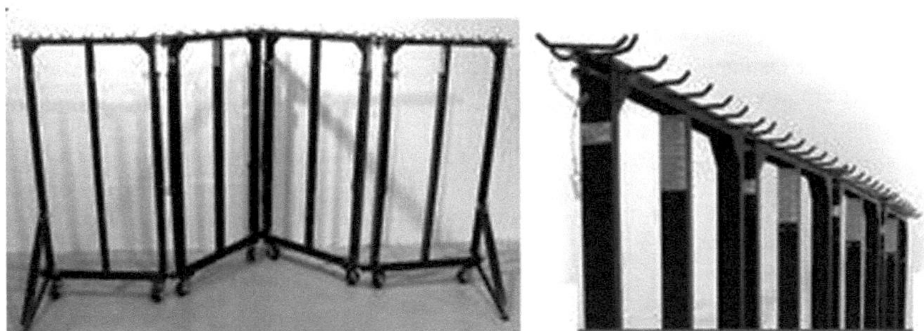

**Fig. 9.13** Portable lead shield racks (Courtesy of Nuclear Power Outfitters, www.nuclea rpoweroutfitters.com)

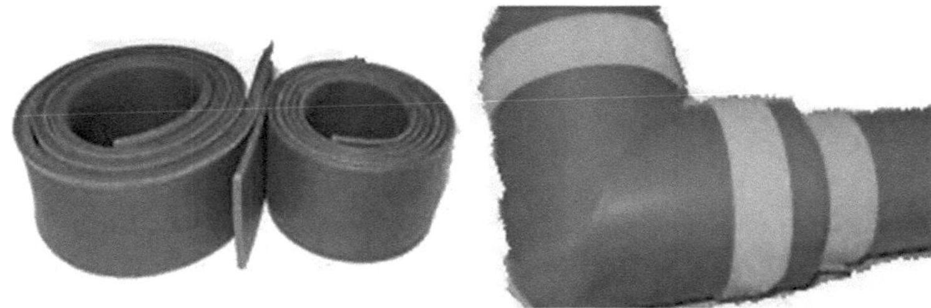

**Fig. 9.14** The NPO tungsten T-Flex® shielding product and example of its application to wrap a small diameter section of pipe (Courtesy of Nuclear Power Outfitters, www.nuclearpower outfitters.com)

The T-Flex® material is a blend of tungsten and iron to maximize the dose reduction and cost of the material. The T-Flex® products are useful for many shielding applications involving small diameter piping, small valves, pipe elbows and low-point crud traps. Figure 9.14 depicts the T-Flex® shielding product.

Typically the primary impact when applying temporary shielding is to ensure that the weight of the shielding does not overstress pipe supports or hangers or impact the performance of snubbers. Since it is usually desirable to apply the shielding directly on the source of the hot spot (e.g., a valve body or crud trap on a length of pipe) the additional stresses associated with the weight must be evaluated, especially when dealing with safety related systems. Alternatively a temporary frame structure not physically connected to the length of pipe or component to be shielded could be employed if space is available.

Another means of providing temporary shielding is the use of water shields. Water shields are convenient, easily installed and removed. The primary advantage of water shields is that the containers may be placed in the location to provide the shielding and then filled with water remotely. Exposures received installing and removing water shields is often much less than that required installing the equivalent amount of

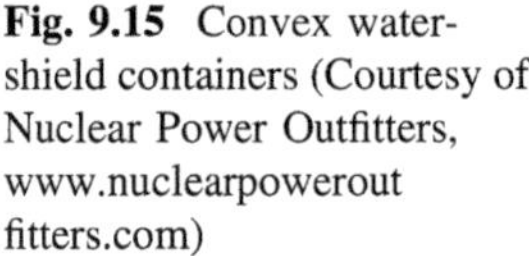

**Fig. 9.15** Convex water-shield containers (Courtesy of Nuclear Power Outfitters, www.nuclearpowerout fitters.com)

lead shielding. Water shields also offer a distinct advantage for those applications in which floor loading may be an issue. The primary disadvantages are that water shield configurations do not readily lend themselves for use in tight areas and the water-filled containers require a larger floor area to provide the same amount of shielding reduction of that of lead. Therefore the opportunity to use water shields may be limited. However under certain circumstances when space is not a limiting factor, the use of water shields may be the preferred method. Water shields are only limited by the physical size and shape of the container itself. Large water shields arranged in a shield wall configuration have a convex shape to prevent streaming between adjacent containers. Figure 9.15 depicts a series of water shields arranged as a shield wall (note the convex shape on the end unit), while Fig. 9.16 shows a specific application of a water shield in a wall configuration.

The vast majority of temporary shielding is installed based primarily on the need to reduce exposures from gamma source terms. On occasion the need may arise to provide shielding against neutron radiation. Neutron radiation is not a concern during outage periods; however, areas of the plant that may need to be accessed during power operations where neutron exposures may pose a concern may require supplemental neutron shielding material. Neutron shielding materials consist of low-Z materials such as water or light weight composite materials with high boron content. For example certain areas of the containment building outside

**Fig. 9.16** Water shield utilized to provide long-term shielding in general plant walkway area (Courtesy of Nuclear Power Outfitters, www.nuclearpowerout fitters.com)

the biological shield wall may contain flow transmitters, pressure gauges or other instrumentation that may need to be accessed to evaluate the operational status of components. If pipe penetrations in direct line of site to the reactor vessel are in the vicinity of instrumentation neutron exposure rates could contribute to personnel exposures while making entries at reactor power.

Since neutron shielding has to remain in place during periods of reactor power evaluations are required to ensure that no operational concerns are introduced. Such considerations as seismic concerns, fire loading, impact on adjacent structures and related issues due to the weight and location of the proposed shielding would have to be addressed. Any required supplemental neutron shielding is typically installed under a permanent design modification package. The need for any additional neutron shielding would most likely be an isolated occurrence.

## 9.6  Summary

A wide range of protective clothing and equipment is available to the LWR industry. The proper use of protective clothing plays an important role in protecting workers from radioactive contamination. Radiation workers should be trained in the proper donning and removal of protective clothing in order to afford the required protection when working in contaminated areas. Radiation protection personnel should be knowledgeable of the available types of protective clothing and prescribe the most appropriate articles of clothing for a specific task based on work area conditions. The use of respiratory protection equipment by suitably trained and qualified individuals is a key element in minimizing exposures due to the presence of airborne radioactivity. Various types of respiratory protection equipment should be available and maintained to provide protection against the range of airborne radioactivity species encountered in a LWR environment.

The use of portable filtration equipment and temporary shielding materials is important in controlling the spread of airborne contamination and reducing worker exposures during the course of maintenance activities. Pre-job preparation stages for radiological challenging work activities should identify the need for these measures. The radiation protection organization should ensure that programs are established that maintain an adequate inventory of protective clothing and equipment to support the needs of the facility.


## Bibliography


1. American National Standard Practices for Respiratory Protection, ANSI Z88.2, 1980
2. U.S. Nuclear Regulatory Commission, NUREG/CR-0041, Manual of Respiratory Protection Against Airborne Radioactive Material, January 2001
3. U.S. Nuclear Regulatory Commission, Regulatory Guide 8.15, Acceptable Programs for Respiratory Protection, Revision 1, October 1999

# Chapter 10
# Personnel Dosimetry (Monitoring of Personnel Exposures and Bioassay Programs)

## 10.1 Overview

A personnel dosimetry program includes those activities associated with the measurement, monitoring and assessment of worker exposures, and retention of related exposure records. The personnel dosimetry program also serves a vital function in maintaining documentation to demonstrate that individual exposures are in compliance with regulatory exposure limits. Personnel dosimetry programs include those aspects associated with the monitoring and evaluation of personnel exposures at LWR facilities. These programs include whole-body monitoring, the use of extremity dosimeters and multiple dosimeters (often referred to as "multi-badging"), provisions for neutron dosimetry and bioassay processes for the assessment of internal exposures. This chapter discusses the various elements commonly associated with such a program. The focus is oriented towards the operational and maintenance aspects of a LWR personnel dosimetry program. Details associated with the theory and principal of detection of dosimetry devices is presented in a level of detail necessary to evaluate the application of a specific dosimeter to a LWR dosimetry program. The use, advantages and limitations of dosimeters as they apply to a LWR personnel dosimetry program are also discussed. The references for this chapter provide more detailed information concerning the detection and operational principles of the dosimeter types presented.

The fission process produces either directly or indirectly a multitude of radionuclides. Neutrons and gamma rays produced during fission, while the plant is at power, contribute to relatively intense radiation fields. High radiation levels will be present in the vicinity of the reactor vessel and systems and components closely situated to the core. Additionally, a host of activation products that are produced contribute to both long term and short term radiation fields. Consequently, personnel exposures resulting from the presence of these radioactive species must be monitored. The accurate monitoring of worker exposure is a central element of a LWR Radiation Protection program. Various international standard setting organizations along with governmental regulatory agencies have established personnel

R. Prince, *Radiation Protection at Light Water Reactors*,
DOI: 10.1007/978-3-642-28388-8_10, © Springer-Verlag Berlin Heidelberg 2012

radiation exposure dose limits. Maintaining worker exposures below limits established by these agencies serves to minimize any long-term health detriment to workers.

Personnel dosimetry programs must maintain strict standards and ensure the accuracy of monitoring data and be implemented in a manner that affords a high level of confidence on the part of radiation workers that their personnel exposures are adequately assessed. Radiation workers typically take a keen interest in knowing the amount of radiation exposure that they have received. Accurate monitoring of personnel exposure is not only important in demonstrating regulatory compliance but is also important in tracking exposures to specific tasks to identify jobs that may be candidates for dose reduction initiatives. This allows exposure management efforts to be targeted to areas with the greatest potential for dose reduction benefit.

The primary contributor to the collective dose at LWRs is from gamma-emitting radionuclides and most notably from the two predominant cobalt isotopes encountered, Co-58 and Co-60. Exposure to neutron radiation typically represents a small percentage ($<5\%$) of annual exposures. Significant levels of neutron radiation are only present during power operation and confined to such locations as inside drywells and containment buildings, access to which is administratively controlled while at power. Therefore, the opportunity for individuals to be exposed to neutron radiation fields is limited. Even though beta-emitters are prevalent and oftentimes comprise a significant fraction of the activity present in contamination, the use of protective clothing effectively serves as a means to limit exposures due to beta radiation. Under unique circumstances beta radiation may be a limiting factor when evaluating extremity dose, skin dose and dose to the lens of the eye for examples. The evaluation and control of beta radiation exposure under these conditions is discussed in more detail in Chap. 6.

## 10.2  Dose Limits

The primary organization that recommends exposure limits is the International Commission on Radiological Protection (ICRP). The ICRP was founded in 1928. The ICRP is an advisory body that provides recommendations that may be used by regulatory and standard-setting organizations throughout the world. Reports are drafted by task groups comprised of internationally recognized experts. The ICRP develops recommendations and guidance for protection against the risks associated with exposure to ionizing radiation. The ICRP evaluates relevant information to periodically develop recommended limits for radiation exposure based on a review of current data concerning the biological effects of radiation exposure. The most recent ICRP recommendations were adopted by the ICRP in March 2007 and presented in ICRP Publication 103. This report updates the previous recommendations that were issued in 1990 in ICRP Publication 60. The previous dose limits recommended by the Commission are maintained in the ICRP 103 report. The

**Table 10.1** Occupational dose limits

| Type of dose | 10 CFR 20 annual dose limit—mSv (rem) | ICRP-103 dose limit (mSv) |
| --- | --- | --- |
| Whole Body | 50(5) | 50 (in any year)<br>100 (over 5 years)<br>20/year (5-yr average) |
| Organ | 500(50) | Controlled by the above stochastic effects dose limits |
| Lens of the Eye | 150(15) | 150 |
| Skin | 500(50)<br>(over 10 cm$^2$) | 500 (over 1 cm$^2$) |
| Hands and Feet | None specified | 500 |

recommendations update tissue weighting factors utilized in calculating total effective dose. The revised factors are based on the latest available scientific information of the biology and physics of radiation exposure.

The ICRP applies three fundamental principles regarding protection against ionizing radiation hazards. Exposure to ionization radiation should be: (1) justified; (2) optimized; and, (3) controlled by the application of dose limits. The principles of justification and optimization are embodied in elements of exposure management and ALARA programs as detailed in Chap. 7. The ICRP recommendations for dose limits are established to limit both stochastic and nonstochastic effects. Stochastic effects are those for which the probability of occurrence is a function of dose without threshold and include genetic effects and cancer. The severity of stochastic effects is not a function of dose. Nonstochastic effects include erythema and other biological effects with no threshold dose and whose severity is a function of dose.

Even though the ICRP recommendations serve as guidance, dose limits established by regulatory bodies in most countries essentially adopt the ICRP recommendations. Other international organizations, such as the International Atomic Energy Agency (IAEA), incorporate much of the ICRP recommendations in their Basic Safety Standards relating to radiation safety. The application of legally binding dose limits by individual countries may be the responsibility of regulatory agencies or recognized standard setting organizations of a given country.

The US Nuclear Regulatory Commission promulgates exposure limits for workers exposed to ionizing radiation in Title 10, Part 20 of the Code of Federal Regulations (10CFR20). The current dose limits established in 10CFR20 are based upon previous ICRP recommendations. The occupational dose limits for workers specified in 10CFR20 and the ICRP 103 recommended dose limits are summarized in Table 10.1.

The major difference between these agencies dose limits is associated with the annual whole-body dose values. Whereas both the NRC and ICRP specify an annual dose limit of 50 mSv, the ICRP recommended dose limit of 100 mSv (10 rem) in any 5 year period and the 20 mSv/year 5-year average essentially results in an annual dose limit of 20 mSv (2 rem) per year. The annual limits on intake and associated derived air concentration values provided by the ICRP are predicated on an annual dose limit of 20 mSv. The ICRP ALI and DAC values for the various radionuclides will differ from those currently provided in 10CFR20 since these are based on an annual dose limit of 50 mSv. Improvements in dose reduction efforts have resulted in a significant reduction in exposures received by workers at LWRs. Many nuclear power plant utilities have established annual administrative control dose values of 10–20 mSv (1–2 rem) per year. These administrative control values essentially maintain annual whole-body exposures below both the NRC regulatory dose limits and the ICRP recommended dose limits.

Whole-body exposure or penetrating, deep dose represents the exposure received by the primary body organs and related vital tissues from radiation sources external to the body. Regulatory agencies and the ICRP define deep dose equivalent (DDE) to be that exposure received at a tissue depth of 1 cm (1000 mg/cm$^2$). Effective dose equivalent (EDE) is the sum of the products of the organ or tissue weighting factors and the committed dose equivalent to the organ or tissue irradiated. The sum of the products of the weighting factors for each of the body organs or tissues that are irradiated and the committed dose equivalent to these organs or tissues is the committed effective dose equivalent (CEDE). Tissue and organ weighting factors are specified by the ICRP and incorporated in 10CFR20. To obtain the total effective dose equivalent you sum these two dose components and therefore:

$$\text{TEDE} = \text{DDE} + \text{CEDE}$$

The basis behind the weighting factors was to address the need for risk-based dose limits with regards to stochastic effects such as cancer. The concept of the organ and tissue weighting factors was basically to equalize the risk associated with whole-body irradiation and that due to localized irradiation of an organ. The weighting factors account for differences in cellular radio-sensitivity and other factors that affect the susceptibility of stochastic effects.

Shallow dose is defined as the dose equivalent due to external exposure to the skin or extremities received at a tissue depth of 0.007 cm (7 mg/cm$^2$) averaged over 1 square centimeter (ICRP) or 10 square centimeters (NRC). Radiation with sufficient energy will penetrate the dead layer of skin, with an assumed thickness of 0.007 cm, reaching the underlying healthy skin tissue. Lens of the eye external exposure is measured at a tissue depth of 0.3 cm (300 mg/cm$^2$).

## 10.3  Personnel Monitoring Devices

Many types of personnel monitoring devices are available for measuring and recording worker exposures. Usually a combination of dosimetry devices are utilized to measure personnel exposures, due to the varying radiological conditions that may be encountered. Various dosimeter designs are available to monitor specific types of radiation dependent upon the energy of the radiation being measured by the device. Personnel dosimetry devices for convenience may be divided into the following categories or groupings: primary and secondary dosimeters or in-direct reading and direct reading dosimeters. Primary dosimeters are those personnel monitoring devices that serve as the "dose of record" and are typically indirect reading type dosimeters that require specialized processing equipment (e.g., TLD readers) or processing techniques to convert the dosimeter response to an integrated dose.

When determining the need for assigning or issuing personnel dosimetry several parameters may be considered. Visitors or one-time only entries to the RCA involving tours or inspection type activities that will result in minimal or essentially no "measurable" dose may not require assignment of a primary dosimeter. Under these circumstances a secondary type dosimeter (e.g., an electronic dosimeter or pocket chamber) may be all that is required or necessary. An additional factor that may be considered under these circumstances is whether or not qualified radiation escorts will be provided. This provision would provide added assurance that tour groups or visitors do not inadvertently enter areas where measurable dose could be received. However, for practical and at least in the USA, for "claims avoidance" or liability reasons a primary dosimeter is often times provided to all individuals entering the RCA regardless of the anticipated exposure. Even though ICRP, IAEA, and 10CFR20 recommendations and regulations all have provisions not requiring the issuance of dosimetry if certain conditions can be met (e.g., if the dose received is less than 10% of the annual exposure limit) it is oftentimes more prudent to provide dosimetry for all entries into the RCA.

### *10.3.1 Primary Dosimeters*

A primary dosimeter is designated to serve as the "dose-of-record." Dose-of-record is the monitored dose utilized to record official personnel exposures and the value that is used to report exposures to individuals to demonstrate compliance with regulatory dose limits. Ideally a primary dosimeter should have a flat response over the radiation energy range of interest, should be stable under the environmental conditions encountered at a LWR, should not be subject to significant fading, and for those primary dosimeters that will be re-used (e.g., TLDs) the residual dose should be low (i.e., after processing the dosimeter should be capable of being "zeroed"). Ideally, the primary dosimeter should have a low threshold of detection (e.g., 10 µSv or 1 mrem) over the wear period. Primary dosimeters should also have characteristics that allow the accurate measurement of personnel exposures over an extended period of time. Primary dosimeters may have a wear period ranging from one to four months or longer. Dosimeters are routinely worn between the neck and waist typically in the area of the chest to measure exposure received by the primary organs. Under various circumstances it may be necessary to relocate the primary dosimeter to another area of the body or to provide multiple dosimeters. This topic is further discussed later in this chapter.

Though several different types of dosimeters are suitable for employment as primary dosimeters the thermoluminescent (or TLD) type dosimeter has gained widespread acceptance over the past few decades. Film badges are also quite common for this purpose. More recently optically stimulated luminescence dosimeters have been introduced and also serve as primary dosimeters. Various regulatory authorities may also specify dosimeter types acceptable for use as the primary dosimeter and for dose-of-record purposes for programs under their jurisdiction. The primary dosimeter must be capable of accurately measuring radiation exposure from the radionuclides of interest at LWRs. The primary objective is to ensure that the dosimeter type has a known response to the types of radiations and their energy spectrum encountered at a given facility. The dosimeter should be rugged and capable of withstanding the environment commonly encountered at a LWR. Dosimeter operational parameters should specify temperature range, humidity and moisture limitations and other factors that could affect the accuracy of dosimeter readings.

A brief summary of the characteristics of primary dosimeters used in the industry is provided below. The reader may consult one of the many texts that are available for a more detailed discussion concerning the principles of operation, calibration and analysis of these dosimetry devices. Technical literature from dosimetry providers should also be consulted.

## 10.3.2 Thermoluminescent Dosimeters

Thermoluminescent dosimeters (TLDs) have gained widespread acceptance in the LWR industry and probably are the most frequently encountered form of dosimeter serving as the dose of record. The benefits of TLDs include the ability to utilize various materials with known energy response characteristics that are essentially tissue equivalent. TLDs have a good energy response over the range of interest for both gamma and beta radiations. The precision of TLDs is excellent over the dose ranges likely to be present at a LWR. The dosimeter response is essentially independent over the range of dose rates to which workers may be exposed.

When a thermoluminescent (TL) material is exposed to ionizing radiation electrons are excited to higher energy bands. To serve as a practical TL material the substance must be able to maintain or "store" this energy for a period of time. The electrons are trapped in lattice imperfections within the TL crystal. An ideal TL material will maintain a stable configuration for an extended period after being excited by exposure to radiation. When TL materials are heated at a later time the electrons are released from the trap sites releasing the excitation energy in the form of light photons. This process is called thermoluminescence. By calibrating a given TL material to known radiation exposures and measuring the amount of light that is given off during the heating cycle, TLDs can be "calibrated" to provide a measurement of the amount of radiation exposure received by the badge.

TLD dosimeters are typically incorporated into a badge holder and may consist of one or more TLD "elements." These elements comprise the thermoluminescent material positioned behind a "shield" or absorber material incorporated into the TLD holder. A common TLD badge used by the LWR industry is the UD-802 badge supplied by Panasonic Industrial Company. This badge holder consists of four TL elements as depicted in Fig. 10.1. By incorporating a specific TL material behind an absorber with unique properties the TLD badge may be designed to measure the dose received at tissue-equivalent depths of 7, 300 and 1,000 mg/cm$^2$. These tissue depths correspond to the depth at which shallow dose, lens of the eye dose and the deep dose equivalent are required to be measured. The various elements in the TLD badge can be used to measure beta dose, dose due to low energy gamma rays, high energy gamma rays and neutron radiation over a wide range of energies. Each element is designed to respond to a given energy range. The combination of all four elements provides a relatively flat response over the energy range from 10 keV to 10 MeV. Table 10.2 summarizes the TL and shielding materials commonly used in the UD-802A dosimeter.

ANSI Standard N13.11, Personnel Dosimetry Performance—Criteria for Testing, establishes the test conditions and various performance criteria for personnel dosimetry systems. The accreditation and quality control measures associated with a personnel dosimetry program are discussed later in this chapter.

To analyze TLDs the individual phosphor or the entire holder containing the TL elements is placed into a reader. A reader is simply a device that heats the TL material to a selected temperature to produce a light output. The associated electronics measures the light output given off by the phosphor during the read cycle. The light output from the reader produces a characteristic glow curve. A glow curve depicts the relative light intensity emitted by the phosphor during the heating cycle. The total amount of light output or the size of the glow curve is proportional to the number of electrons that were excited and trapped in a higher valence state. The number of excited electrons in turn is proportional to the amount of energy absorbed from exposure to radiation.

The area under the glow curve is directly proportional to the radiation absorbed by the phosphor and, hence, to the radiation dose received. The glow curve may consist of several characteristic peaks as electrons are released from various energy-level trap sites. Phosphors are typically heated to about 190° C to release the majority of the trapped electrons. Once a TLD phosphor is processed the TLD is heated to a higher temperature on the order of 300° C to release all remaining trapped electrons. This process is referred to as annealing and is typically performed during the processing cycle. Annealing "zeroes" the TLD for subsequent reuse.

TLD readers may be manual or automatic. Automatic readers are configured to accept several trays in sequence and are capable of processing as many as 500 TLD badges. Utilizing specific calibration factors for each element and incorporating the absorber characteristics associated with each TLD element the amount

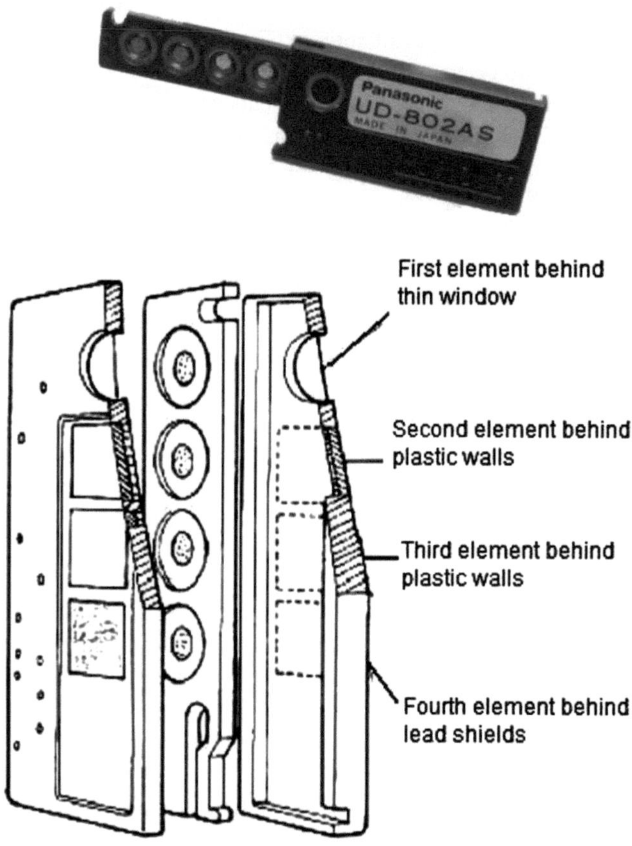



**Fig. 10.1** A Panasonic 802 series TLD showing the four TLD elements and the construction of the TLD badge holder (Courtesy of Panasonic Industrial Company, www.panasonic.com)

**Table 10.2** Composition of the Panasonic UD-802A dosimeter

| UD-802 | Element 1 | Element 2 | Element 3 | Element 4 |
| --- | --- | --- | --- | --- |
| Phosphor | $Li_2B_4O_7$ | $Li_2B_4O_7$ | $CaSO_4$ | $CaSO_4$ |
| Front filtration | Plastic—14 mg/cm$^2$ | Plastic—160 mg/cm$^2$ | Plastic—160 mg/cm$^2$ | Lead—0.7 mm |
| Rear filtration | Plastic—14 mg/cm$^2$ | Plastic—160 mg/cm$^2$ | Plastic—160 mg/cm$^2$ | Lead—0.7 mm |

of radiation exposure from different types of radiations may be determined. By incorporating various absorbers in the badge holder located in front of individual TLD elements limited energy discrimination may also be achieved to distinguish

**Fig. 10.2** The Panasonic UD-7900 model automatic TLD reader (Courtesy of Panasonic Industrial Company, www.panasonic.com)

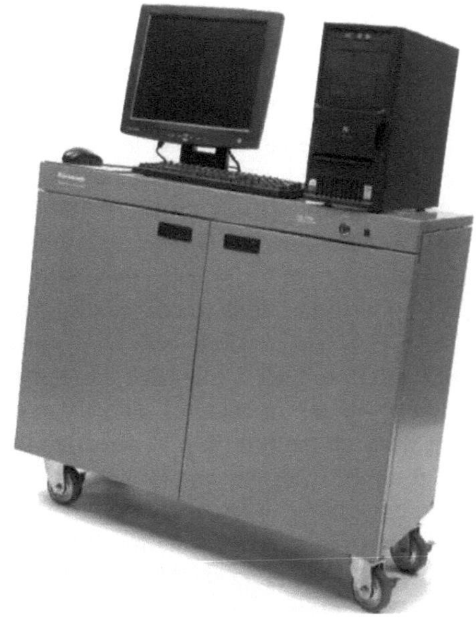

between exposures from low and high energy gammas and discrimination of beta exposure from soft X-rays. Algorithms incorporate various calibration factors, TLD reader parameters and individual phosphor element correction factors to determine exposures measured by the TLD badge due to gamma radiation, low and high-energy beta particles and neutron radiation.

The accuracy of TLD measurements depends heavily on determining calibration factors and ensuring the quality of individual TLD elements. Therefore, a comprehensive program to periodically verify calibration factors and the integrity of TLD elements is an essential ingredient of the dosimetry program. Quality control measures associated with the processing of TLDs may include the use of reference and control TLDs, and background TLDS whenever TLDs are read. Quality control checks are also routinely performed on the reader to ensure that the heating cycle and associated electronics are working properly and are within established operating limits. The Panasonic UD-7900 reader (depicted in Fig. 10.2) is equipped with self diagnostic functions for the optical circuit, heating stability, and operability of the photomultiplier tube and associated electronic circuitry required for the proper processing and analysis of TLD readings. The reader also performs such operations as annealing of the TLDs, optical reading of the TLD identification code and adjustments to the built-in reference light source.

For LWR facilities the number of personnel that are typically monitored for exposure may be in the range of a few hundred to a couple of thousand, depending on how many units are at a given site. The advantage of TLDs, as previously noted, is that hundreds of TLDs may be placed into trays and automatically fed to a reader. This greatly facilitates the time and effort required to obtain dose results for large numbers of dosimeters.

The characteristics of TLDs that contribute to their wide acceptance include the ability to obtain dosimetry results in a short period of time if the need arises and the ability to reuse the dosimeter. The dose measurement threshold for TLDs is extremely low. Depending upon the length of the wear period and fade characteristics of a given TL material, doses as low as 10 µSv (one mrem) may be measured. The combination of the four TL elements in the UD-802 dosimeter has a measurement range of 10 µSv–10 Sv (1 mrem–1,000 rem). TLDs can be fabricated into finger rings or placed in chip holders and used for extremity monitoring. These and other characteristics of TLDs make them ideally suited for LWR personnel dosimetry.

Various other holder configurations may be utilized to measure neutron dose. These TLD badges often work on the principle of measuring the neutrons that are reflected from body tissue back into the TL material in the badge holder. The term "albedo" is the measure of reflectivity of a surface or body. Neutron dosimeters that work on the principle of detecting the "reflected" neutrons are often referred to as albedo dosimeters. Utilizing this technique, TLD holders incorporate a cadmium shield in front of one or more of the TL elements to shield the TL material from neutron radiation. While one or more elements in the same TLD holder have cadmium shields located behind the TLD elements. Since cadmium is an effective neutron absorber the TL material is shielded from direct neutron exposure so the dose measured by the TLD is due to the neutrons reflected from the worker's body. By analyzing the readings on the various elements and using appropriate algorithms exposure due to neutron radiation may be determined.

Neutron TLDs, as well as any other primary personnel dosimeter device used for measuring neutron exposure, should be calibrated to the degree possible for the environment in which they will be used. The design of the TLD holder must be such as to provide reasonable assurance that the neutron phosphor will yield a "calibrated" response to neutrons of a given energy. The energy spectrum of the neutron calibration source should closely match that of the in-plant neutron energy spectrum. Essentially for LWR environments this means that the neutron energy spectrum in those plant areas where monitoring of neutron radiation exposures is required must be characterized. This often entails performing a neutron energy spectrum study within the containment building and drywell areas while at power. Obviously the neutron energy spectrum in these areas will be influenced by the location. Such parameters as whether there is a direct line-of-sight with the reactor vessel and how many "deflections" neutrons have undergone before reaching the measurement location will influence the energy spectrum. Consequently, the neutron energy spectrum at such locations as in the vicinity of reactor coolant pumps, accumulator, rooms, hot legs or on the refuel floor area may be different. Plant locations that are typically accessible while at reactor power have neutron energy spectrums in the range of 250–350 keV. Based upon plant designs, consideration of those plant areas that may need to be accessed during power operations, the neutron dosimeter design, and other parameters a detailed neutron energy spectrum determination may be warranted. Once the neutron energy spectrum is known, or at least approximated, then a neutron TLD dosimeter may be selected that provides an optimal response to the anticipated neutron energy spectrum. For these reasons neutron TLD badges often contain multiple chips with various cadmium absorber configurations to determine personnel exposures from various energy neutrons.

### 10.3.3 Film Badges

Historically the film badge was one of the first primary dosimeters to gain wide spread use in the LWR industry. Film badges work on the principle that radiation exposure will change the density of a photographic emulsion. This change in density is reflected as varying degrees of darkening in the film once the film is developed. Early film badges consisted of packets of x-ray film that were worn by radiologists and other radiation workers. These film packets were periodically developed to determine the degree of darkening of the film. The degree of darkening would be representative of a given exposure to radiation.

Film badges are convenient to use, are small in size and stand up well to environmental conditions encountered at LWRs. The film badge consists of one or more photographic films encased in a light-tight holder. By utilizing a combination of filters and providing an "open window" area in the badge holder, film badges can be used to distinguish between various energy gamma rays and beta radiation. The proper design of the film badge holder is important in order to provide reasonable accuracy for measuring skin dose (shallow dose) and penetrating or deep dose. Various filter materials are also incorporated into film badge holders to provide a more linear energy response of the film. Film badges are calibrated by measuring the film response to known radiation exposures. Various film emulsions are incorporated into the film package to provide an acceptable dose range. Film badges can provide a response over a large dose range. Combinations of film badge holder designs and the use of various emulsions allow film badges to measure exposures in the range of 0.1 mSv–10 Sv (10 mrem–1,000 rem).

The film area under the open-window measures exposure due to gamma and beta radiation. By determining the exposure received on parts of the film badge located behind various absorbers the dose attributable to just the beta component in the open window area of the film badge may be determined. Film badges can also be designed with appropriate filters and absorbers to measure exposures due to neutron radiation. Usually the neutron film badge would be a separate badge. Perhaps the major disadvantage of the film badge that may have contributed to its decrease in popularity is the relatively long length of time required to process a film badge to obtain exposure results. Additionally the analysis of film badges requires skilled individuals trained in interpreting the data. Common practice is to send film badges to an offsite processor for reading and analysis which further extends the time period required to obtain results.

One of the earlier advantages, which may have lost some of its importance over the years, was that the film badge could be read without losing the recorded dose and the film could be saved, thus, providing a permanent record of exposure.

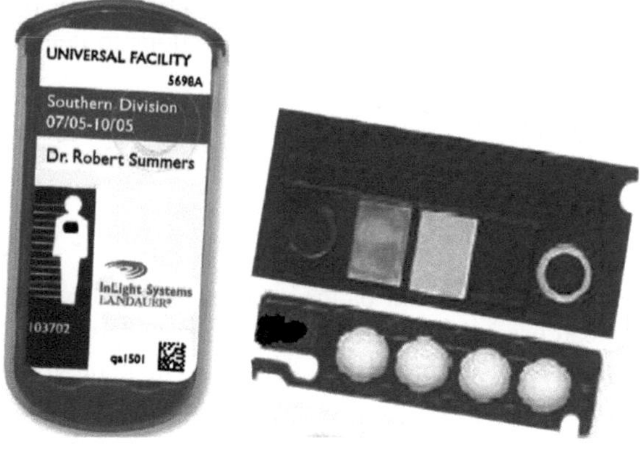

**Fig. 10.3** The Landaurer InLight® dosimeter holder and detector slide (Courtesy of Landauer, www.landauer.com)



Therefore film badge results could be recertified if necessary. The ability to maintain a permanent record was considered an important advantage in the early days as a means to save personal dose records associated with over exposures. Experience has shown that exposures exceeding regulatory limits were isolated events and the need for a "permanent" record lessened. The introduction of TLDs and other dosimetry devices that are more conveniently analyzed and that had equal or better sensitivity characteristics as compared to film diminished the need for primary dosimeters capable of maintaining a permanent dose record. Whereas a high quality TLD may have a detection threshold on the order of tens of μSv (or a few mrem) over a given wear period, the threshold of detection for film badges is approximately 100 μSv (10 mrem).

### 10.3.4 Optically Stimulated Luminescence Dosimeters

A relatively new type of dosimeter that is gaining widespread acceptance is the "optically stimulated luminescence" (OSL) dosimeter. The dosimeter is offered by Landauer, Inc., under its proprietary trade name of "Luxel and Inlight." InLight is a common dosimeter used in the nuclear power industry. The dosimeter works on the principle that when certain crystals are subsequently stimulated by laser or LEDs (light emitting diodes) after being exposed to radiation the material will luminesce. The degree of luminescence is proportional to the amount of radiation received by the material. The most common OSL material consists of crystals of aluminum oxide. The crystals are grown under laboratory conditions with strict specifications along with proprietary dopant material. The OSL dosimeter is capable of measuring exposures due to low and high energy gamma rays, beta radiation along with albedo neutrons. The OSL badge holder can also be equipped with CR39, another neutron sensitive material. Figure 10.3 depicts the InLight® dosimeter holder along with four aluminum oxide detectors.

The OSL dosimeter has good energy response over the range of interest for gamma, beta and albedo neutron radiations. The response of OSL dosimeters over the dose range can be accurately assessed over a wide range of dose rates typically encountered at LWRs. The OSL dosimeter has a detection sensitivity range of approximately 10 μSv–10 Sv (1 mrem–1,000 rem) for gamma radiation and 100 μSv–10 Sv (10 mrem–1,000 rem) for beta radiation. As with TLD processing systems, OSL dosimeters are also capable of being read by automatic readers.

The OSL dosimeter characteristics are similar to those noted above for TLD dosimeters namely, the ability to process dosimeters automatically, capable of onsite processing and the availability of various OSL crystals to measure exposures for the types and energies of radiations encountered at a LWR. Other benefits of the OSL dosimeter include the following: (1) the ability to analyze the dosimeter numerous times. Unlike TLD, the OSL technology is not a destructive readout. This is due to the material engineering enabling little signal depletion in the read area; (2) a 2D engraved barcode contains the element correction factor for all positions—element correction factors do not have to be established each time before use; (3) dosimeters arrive ready to wear with no annealing required. If a dosimeter is to be reused, annealing is available; (4) dosimeters may be archived for subsequent analysis or evaluation; (5) good environmental stability for conditions encountered at LWRs: and, (6) not subject to fading that allow for longer monitoring periods.

An automatic reader for the OSL dosimeters is depicted in Fig. 10.4. This reader is used in conjunction with Landaurer InLight® dosimeters and is capable of processing as many as 500 dosimeters at a time. Dose algorithms are incorporated into the reader along with software to control reader setup and operating parameters, quality control checks, identification of dosimeter serial numbers and recording and retention of individual dosimeter dose results.

## 10.4  Accreditation of Personnel Dosimetry Measurement Programs

The importance of maintaining accurate and reliable personnel exposure records are vital as noted previously. Regulations of the Nuclear Regulatory Commission require that the processing of dosimeters meet certain standards. Dosimeters utilized to monitor personnel exposures to show compliance with regulatory exposure limits must be processed under an accreditation program. In the United States dosimetry processing programs for LWRs are accredited by the National Voluntary Accreditation Program (NVLAP) of the National Institutes of Standards and Technology (NIST). This program was established in 1984 in response to an NRC initiative. Successful accreditation signifies that the dosimetry processing facility or laboratory meets the NVLAP proficiency requirements for processing dosimeters. The proficiency requirements must be satisfied for the specific dosimeter model or type and for the radiations the processor intends to record as the official dose of record for individuals.

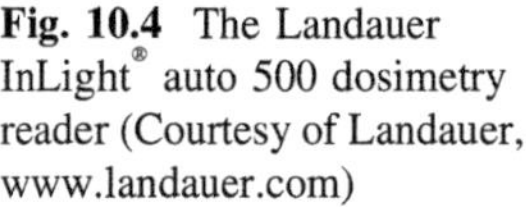

**Fig. 10.4** The Landauer
InLight® auto 500 dosimetry
reader (Courtesy of Landauer,
www.landauer.com)

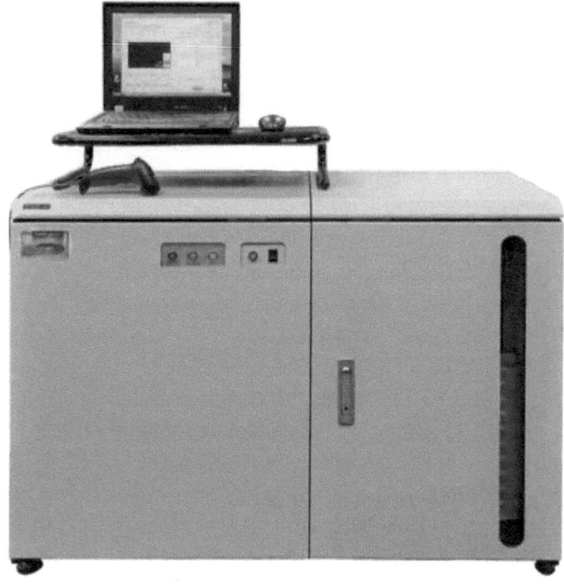

A key element of the NVLAP accreditation is the periodic evaluation of each dosimetry processing laboratory once accreditation is achieved. These comprehensive evaluations ensure that the facility maintains a minimum level of quality in the processing of dosimeters. Adherence to various ANSI personnel dosimetry performance standards as well as the associated NIST requirements is evaluated during these assessments. The qualification and training of personnel responsible for the processing of dosimeters are also reviewed as part of the assessment process. This ensures that the processing facility is staffed by personnel suitably trained and qualified in the processing and analysis of personnel dosimeters. Accredited processing laboratories are also required to participate in a periodic proficiency testing program. This program requires processors to periodically forward dosimeters for which they have received NVLAP accreditation to a proficiency testing laboratory. The proficiency testing laboratories are qualified by NVLAP. Dosimeters are exposed to various types of radiation covering a range of known doses and returned to the processor. The processor evaluates the dose received for each category they have been accredited. Results of this independent proficiency test are reviewed by NVLAP and the proficiency testing laboratory. To maintain NVLAP accreditation, dosimetry processors must demonstrate satisfactory performance during the proficiency test.

Depending upon the circumstances, individual plants or utilities may choose to maintain their own NVLAP accredited dosimetry processing facility. These may include dosimetry processing facilities at each site for those nuclear utilities that have more than one nuclear site. Due to the expense associated with the maintenance of multiple dosimetry laboratories, the necessary staff, dosimeter readers and administrative functions it may be advantageous to maintain one centralized facility. Alternatively many dosimeter providers and distributors now offer commercial processing services. These dosimetry processors maintain NVLAP accreditation, thus, eliminating the need for a given LWR facility to staff and maintain an onsite processing facility. Obviously, the decision to process personnel dosimetry onsite or at a centralized facility or to secure the services of a commercial provider will be based primarily upon the availability of qualified staff, cost and convenience. Regardless of whether or not dosimeter processing is performed onsite or by another entity, provisions should be established to ensure that dosimeters are processed in accordance with a quality control program that continuously assesses the performance of the processing laboratory.

## 10.5  Electronic Dosimeters

Electronic alarming dosimeters (EADs), electronic personnel dosimeters, or simply electronic dosimeters (EDs), have gained widespread acceptance within the LWR industry. The term electronic dosimeter will be used here to represent any electronic device equipped with alarm capabilities utilized for personnel dose monitoring. Electronic dosimeters are versatile devices that have facilitated live-time recording of personnel exposures. Electronic dosimeters come equipped with several attractive features. Their ease of use, their light-weight, their convenient displays, and their ease of reading are some of the features that have contributed to their wide-spread acceptance. Most notably is the ability to pre-establish dose and dose-rate alarms for a specific task or activity that provides early warning to the user that radiological conditions (i.e., radiation levels) in the work area may have changed that could result in unplanned or unanticipated exposures. The ability to interface EDs with the RCA access control process and the ability to communicate with computer data bases are attractive features. The activation of EDs can be configured in such a way as to administratively restrict entry to the RCA until prerequisite requirements such as radiation worker training status, RWP verification, current year-to-date dose, current dose margin, respirator fit status, and other individual data are verified prior to entry.

Existing ED models are far superior in quality, durability, accuracy and ease of use compared to the first generation of electronic dosimeters introduced in the 1980s. Additionally, coupled with automatic readers and interfaced with computerized dosimetry data bases, the Health Physicist has the capability to track and trend daily personnel exposures to specific tasks and to specific plant areas. Various exposure summary reports may be generated to provide current exposure information for work control or dose management purposes. For these and other reasons the ED has essentially replaced the pocket ion chamber (PIC) as the secondary dosimeter of choice in the LWR industry. Advances in technology are constantly resulting in additional applications applied to electronic dosimeters. While improvements in design have resulted in smaller, light-weight electronic dosimeters, the state of technology is such that any physical limitations due to the physical size or weight of the various models available no longer pose an over-riding concern. The continued improvements in the features of electronic dosimeters that increase their versatility as a dosimetric tool (e.g., teledosimeters discussed below) should further expand the beneficial uses of these devices.

Electronic dosimeters may serve as a secondary dosimeter providing real-time personnel exposure results. Typically, EDs are assigned to individuals upon entry into the RCA or assigned at a work area requiring the use of dosimetry. Electronic dosimeters may be pre-assigned to an individual or alternatively a supply of EDs may be made available for general issuance. To fully utilize the features offered by an ED, namely the ability to establish task-specific dose and dose-rate alarm settings, it is beneficial to have a process that automatically assigns these alarm settings upon activation and issuance of the ED. RCA access control systems (as described in Chap. 6) may be equipped with ED readers or computer interface modules, programmed to access a data base of ED set points. The database could be maintained by radiation protection personnel who establish task-specific alarm settings based upon a review of a given task. This function could be assigned to the RP work control representative. Upon entry to the RCA and activation of the ED, the set points are programmed into the worker's ED based upon the RWP or work package number or any convention that links the RCA sign-in process to the worker's task. This mechanism allows the individual's ED alarm settings to be automatically assigned for the task when the ED is activated upon entry into the RCA. If these set points have been properly established then full advantage can be taken of the alarm function. A balance has to be achieved between establishing dose rate alarm settings that serve a meaningful purpose and those that result in frequent "nuisance" alarms. If alarm settings are too close to anticipated radiation levels, resulting in frequent alarms, workers may be conditioned to ignore these alarms or worst, workers may regard these alarms as meaningless. Conversely, alarm settings should not be set so high that they do not provide timely warning of unforeseen dose rates that could result in unnecessary radiation exposures.

Several types of EDs are commonly used in the LWR industry. Figure 10.5 depicts an electronic dosimeter model offered by Mirion Technologies together with an ED reader. Standard features of EDs typically include a visual display (that may be read by the user) with the ability to display dose rate and integrated exposure readings. Separate and distinct audible alarms are provided to warn the individual in the event that either the dose rate or integrated dose alarm set point values have been reached or exceeded. A primary benefit of an ED is its ability to provide instantaneous indication of the presence of unanticipated or unknown radiation fields. To fully utilize this feature it is imperative that workers be trained in the proper response to ED alarms. Notwithstanding, EDs should not be looked upon as a replacement for effective pre-job planning or as a means to forego evaluations of possible work area radiological conditions that could be encountered under various scenarios. EDs are merely a tool in the arsenal of radiological safety equipment and control measures that may be employed for a given task. The proper use and application of EDs provides an effective ALARA tool that may help minimize unnecessary radiation exposures and allow for a timely response to mitigate the consequences of a radiological incident.

**Fig. 10.5** The DMC 2000S electronic dosimeter and the LDM 250 desktop dosimeter reader (Courtesy of Mirion Technologies, www.mirion.com)

Oftentimes the dose rate alarm is designed to silence when the wearer moves away from the source of radiation that actuated the alarm. When the integrated dose alarm is activated the ED may have to be placed in a reader to acknowledge and reset the alarm, depending on the ED model type. This feature preventing a manual reset of the dose alarm prevents the user from inadvertently clearing an integrated dose alarm. Additionally, once the integrated dose alarm is reached the alarm may not silence until the unit has been reset. The need to manually reset the dose alarm could be seen as a nuisance under certain circumstances; however, the importance of ensuring that the individual did not receive an unnecessarily high exposure outweighs any inconvenience of an "annoying" alarm that cannot be readily silenced. In actual practice workers should be instructed to leave the area immediately upon activation of the integrated dose alarm and report to radiation protection. Unless specific instructions to the contrary are provided or if full-time RP job coverage is being provided then immediate evacuation of the area should be the standard practice.

Procedures should specify the requirement to record and evaluate incidents in which ED alarms are encountered for unknown reasons or were otherwise unanticipated. Figure 10.6 depicts the type of information that may be useful in the event of an ED alarm that requires investigation. In addition to the obvious information, such as the identification of the person (or persons) involved, plant location where the alarm was activated, and whether or not the alarm was generated by a true radiation source, it is important to record information that is pertinent to the radiological conditions existing at the time. By capturing this information vital lessons-learned may be identified that could help prevent future radiological incidents. Electronic dosimeter models in use today have been designed to eliminate, or at least minimize, "false" alarms that may be due to radiofrequency interference, welding machines, or other electronic interference that could produce false alarms.

Unanticipated ED alarms should be thoroughly investigated, including the performance of follow-up radiation surveys in the suspect plant location(s) to confirm existing radiation fields. Alarms should not be dismissed unless there is tangible evidence as to the cause of the unanticipated alarm. If unknown radiological conditions caused the alarm, unnecessary additional personnel exposures could occur if timely confirmation surveys are not performed and plant areas posted accordingly. The ED alarm evaluation investigation should address such items as providing an estimated exposure for the incident, the need to process the individuals' primary dosimeter, an evaluation of plant conditions or system

Wearers' Name: ________________ Dosimeter or Badge Number: ________

Department: ______________ ED Serial Number: _______ RWP: _______

Description of the event:

______________________________________________________________

______________________________________________________________

Was ED dropped or bumped? □ Yes □ No     If yes provide details _____

______________________________________________________________

Moisture damage? □ Yes □ No     Welding involved? □ Yes □ No

Radio transmitter in use at the time? □ Yes □ No

Problems with ED display indicators?  □ Yes □ No

### Following Items to be Completed by the Evaluator

1. Investigation survey performed? □ Yes □ No     Survey No.:______
2. Did individual observe last ED reading before failure?
   □ Yes □ No    Dose reading if known: ______
3. Response to RF □ SAT □ UNSAT Comments: _______________________
4. Dose alarm functional? □ SAT □ UNSAT  Comments: _______________
5. Dose rate alarm functional?  □ SAT □ UNSAT  Comments: ________
6. Battery condition: □ SAT □ UNSAT
7. Display condition: □ SAT □ UNSAT
8. Team member ED reading (if appropriate):  □ Yes □ No Dose: ____
9. Individuals dose record updated: □ Yes □ No  Dose credited: ____
10. Investigation report entered into tracking data base? □ Yes □ No

Note:  ED incidents should be entered into a centralized tracking
system to monitor any adverse trends concerning either reliability
issues associated with a given design or model or to identify
potential issues relating to the use of EDs on the part of workers.

Brief explanation of the suspected cause of the failure: __________

______________________________________________________________

Brief description of the methods or basis for assigning the
estimated dose: _________________________________________________

Evaluator's name and date that evaluation was completed.

Note: SAT = satisfactory; UNSAT = unsatisfactory; RF =
radiofrequency

______________________________________________________________

**Fig. 10.6** Electronic dosimeter—alarm evaluation

evolutions that may have contributed to the unanticipated radiation levels, and
whether other individuals were potentially exposed to unknown radiation levels.
Many ED models maintain a history of the dose profile and dose rate profile as a

function of time. Depending upon the ED model, these histories may cover an extended period of time. The ED dose profile may be downloaded to a computer or a reader for analysis. Utilizing the ED time line of the dose rate and integrated dose readings and knowledge of the individual's location as a function of time after entering the RCA it may be possible to identify the location or proximity of the source of radiation that caused the ED alarm. If this feature is available it could provide valuable information when evaluating unknown causes of an ED alarm.

The primary consideration when responding to an ED alarm is to ensure that a process is in place to address these events. Responses to ED alarms should be deliberate and thorough. A complacent attitude towards unanticipated ED alarms may result in missed opportunities to identify unknown situations or plant conditions that impact radiation levels in a given plant area. In extreme cases failure to perform an investigation could subsequently result in radiological incidents and high radiation exposures to personnel due to the presence of unknown radiation sources.

By recording and trending ED alarm events generic issues may be identified that could be associated with a given ED model design. Perhaps the dependability of a given model lessens over time? Battery life-time may be shorter than anticipated or different from that specified by the manufacturer. As a minimum, ED alarm investigations should result in a proper evaluation of the exposure received by the individual, accurate dose assignment, and the identification or confirmation of radiation levels in the affected plant areas.

The calibration and functional checks associated with the use of EDs may depend somewhat on the ED application and use in monitoring daily personnel exposures. An electronic dosimeter oftentimes may serve a vital role in providing early warning of unforeseen radiation levels or may serve as the primary device to warn workers that they are approaching pre-established dose limits or have exceeded limits for a task or activity. Under these circumstances EDs should be included in a comprehensive calibration program to ensure their operability and response characteristics. Operational controls should incorporate a functional check process that confirms the ED is operable and properly charged, and that alarm functions are working prior to issuance.

Many ED systems are equipped with readers that interface directly with the electronic dosimeters. These readers may have self-diagnostic features that verify, as a minimum, such items as alarm functions, remaining battery life and the calibration status of the unit (among others). A common reader for the DMC 2000S EDs is shown in Fig. 10.5.

## 10.6  Teledosimeters

Teledosimetry relates to the ability to provide remote, live-time, monitoring of worker exposures during the performance of work activities. The use of wireless remote monitoring systems for personnel exposure monitoring purposes is referred

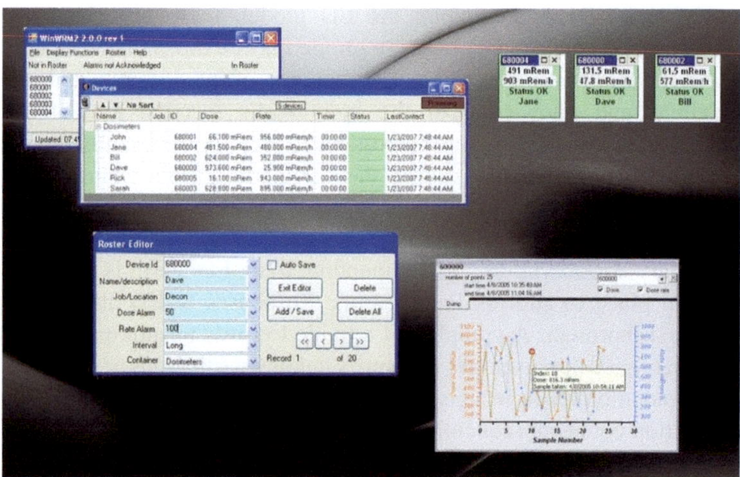

**Fig. 10.7** Computer screen depicting the various parameters displayed by a teledosimetry base station (Courtesy of Mirion Technologies, www.mirion.com)

to as "teledosimetry." A remote monitoring teledosimetry system consists of a base station, a transmitter to receive and send signals, and the electronic teledosimeters. Repeater devices are often employed to extend the useful coverage area and range of the remote monitoring system.

The base station comprises a key component of a teledosimetry system. It provides the interface to a personal computer and software to record and display live-time dose tracking data. The base station may have additional connections or ports to interface with other computer systems. Display features typically provide instantaneous update of each worker's teledosimeter reading including the integrated dose and current dose rate reading of each individual's teledosimeter. Depending upon the software package available other display parameters could include the dose rate and integrated dose alarm set points for each individual's teledosimeter, the time in the area and the status of the communication signal between the base station and the teledosimeter unit. Figure 10.7 depicts a typical teledosimetry base station computer screen. Data capable of being displayed on the **local PC** screen may include such information as the wearer's name, highest dose rate measured by the ED, total dose received, current dose rate where the individual is located, dose and dose rate alarm settings and the radiation work permit or work document associated with the task.

The base station should also provide an alarm in the event of a malfunction, loss of signal, or a fault indication for a specific teledosimeter. In the event of a fault indication (e.g., loss of communication with the base station or a teledosimeter operational issue) the individual should be instructed to immediately leave the work area. The strategic use of teledosimeters makes them an essential tool in minimizing

**Fig. 10.8** The Mirion Technologies wireless remote monitoring system consisting of, from left to right, the WRM2 transmitter unit, the WRM2 repeater unit, and the WRM2 base station transceiver (Courtesy of Mirion Technologies, www.mirion.com)

personnel radiation exposures. Teledosimeters can play a vital role in high dose rate areas and for those worker tasks that involve significant collective exposure.

Several vendors offer teledosimetry systems to the nuclear industry. These systems come complete with base stations, repeater units, teledosimeters and various auxiliary devices to provide additional features and capabilities. The major components of the Mirion Technologies wireless remote monitoring (WRM) system are displayed in Fig. 10.8. The transmitter is coupled with a compatible ED, such as the DMC 2000S model shown in Fig. 10.5. Repeater units are available to extend the range of the wireless remote monitoring system. The base transceiver is capable of supporting as many as 200 connections per base station.

The ability to remotely monitor live-time readout of accumulated dose and changes in dose rates affords unique dose savings opportunities. Assuming that remote communication capability exists (e.g., headsets or radios and video capability) an individual may be stationed in a low-dose area, perhaps even outside of the RCA to monitor dose received by workers from this remote low-dose location. The individual at the remote monitoring base station location could provide directions to the workers to move to a lower dose-rate area, inform workers when they are approaching established dose control values for the task, and in general provide guidance to allow workers to minimize their exposures during the performance of the task. Remote dose monitoring may also reduce the amount of time that RP job coverage personnel, responsible supervisors, and other support personnel spend in the immediate work area. These individuals could monitor progress of the job at the remote monitoring base station location entering the work area only when their services are required, thus reducing exposures to these individuals. A member of the work group could be stationed at the remote monitoring location to record progress of a task or for example to maintain associated data sheets. This reduces the time spent for individuals dressed-out in protective clothing that are making entries on data sheets and work documents while in radiation areas.

Teledosimetry systems may be connected directly to the plant's local area network. This greatly expands both the numbers and locations at which live-time data may be displayed. The capability to access this information at the radiation protection RCA access control office could be beneficial. Additionally, support staff located at remote offices or work areas could access the information at multiple remote locations eliminating the need for support personnel to enter RCA areas unless absolutely necessary.

Wireless remote monitoring technology has resulted in the development and expansion of remote monitoring radiation surveillance systems. Another application has been the use of wireless technology for remote monitoring of radiation levels in such areas as process filter cubicles, demineralizer vessels (resin beds), pipe chases, liquid waste hold-up tanks and retention tanks, and other plant areas that are subject to either high radiation fields or fluctuating radiation fields. The strategic placement of remote wireless units to monitor radiation levels in areas of the RCA can result in significant dose savings related to the performance of radiological surveillance activities.

## 10.7  Extremity and Multi-Badges

The use of multi-badges and extremity badges may be necessary when working in areas where significant dose rate gradients may be encountered. Some tasks may involve work in tight work locations or areas where the highest dose rates may be in close proximity to the hands, head, upper legs or a body location other than where the primary dosimeter is normally worn (e.g., the upper torso or chest area). Under these circumstances consideration may have to be given to monitoring of extremity dose or exposure received to other parts of the body which may be significantly higher than that recorded by the whole-body dosimeter. Certain situations may necessitate the use of extremity badges or the use of multiple dosimeters, or so-called "multi-badges" to assess worker exposures. American National Standard, HPS N13.41, provides criteria and guidance for evaluating the use and placement of multiple dosimeters. The standard serves as a useful guide in ensuring that key program elements are properly addressed when establishing and implementing a multiple dosimetry program.

Under certain circumstances a multi-badge dosimetry pack may include several dosimeters to monitor exposures received by various body locations. These locations could include the head, front torso, back torso, hands, and feet. Tasks involving work in confined areas with large dose rate gradients may require the use of such dosimetry multi-packs. Entries into steam generators often require dose to be monitored at several body locations. For instance a worker's head area in close proximity to the steam generator tube sheet may receive significantly more

**Fig. 10.9** A common multi-badge vest design; note the pockets for the placement of dosimeters in several locations (Courtesy of G/O Corporation, www.gocorp.com)

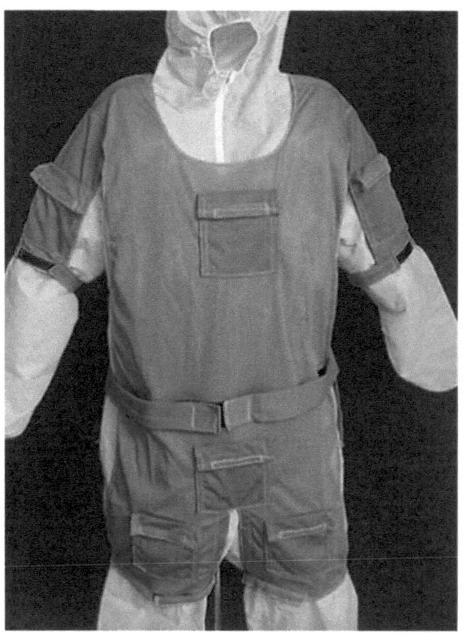

exposure than the main torso positioned in the center of the steam generator bowl. Under these conditions separate dosimeters to monitor head dose (lens of the eye) and the whole-body dose may be required. The main purpose is to ensure that exposures to individuals working in areas with significant dose gradients are properly monitored to ensure that the portion of the body that may receive the highest exposure is measured. Situations whereby individuals working on a platform or a location above the source of radiation may encounter higher radiation levels in the vicinity of their feet or lower extremities when compared to general area dose rates on the platform for instance. Under these circumstances (assuming that shielding cannot be provided) additional dosimeters may be required to monitor dose received by the lower extremities.

Figure 10.9 depicts a multi-badge vest that may be worn by individuals when multi-badges are required. A multi-badge vest is designed to facilitate the placement of dosimeters and provide a convenient mechanism to ensure that a given dosimeter's is secured at the proper location. The vest containing the multiple dosimeters is worn under protective clothing. Typically the entire multi-badge vest is returned to a point of issuance or designated drop-off location to ensure proper tracking to the individual who wore the vest. Measures should be established to ensure that the dosimeter's location within the vest (e.g., right forearm, head, right thigh, or front torso) is recorded and tracked with the wearer. These type vests may be more convenient than taping or strapping dosimeters to several different body locations. Additionally, vests are often more effective in ensuring that dosimeters are not displaced during work performance.

Recording dose to individuals equipped with multiple dosimeters may entail assigning the highest whole-body dosimeter reading as the dose of record. The use of multi dosimeters can significantly overestimate the effective dose equivalent. This is especially the case when the highest dosimeter reading is used regardless of its location on the body. This is the most conservative approach and for purposes of demonstrating compliance with regulatory exposure limits may suffice for the vast majority of situations encountered in LWRs. However, it should be recognized that this practice will overestimate the total effective dose equivalent.

Considering a LWR environment in which internal exposures seldom result in any significant measureable dose the EDE is essentially the TEDE. Externally worn whole-body dosimeters are designed to measure deep dose equivalent (DDE) at a tissue depth of 1 cm. The dose received to internal organs located at a depth greater than 1 cm will be lower than that measured by the external dosimeter. This is particularly true for exposures involving Co-60, the predominant component of personnel exposures at LWRs. Consequently externally worn dosimeters may not provide the "true" dose value for EDE. Since TEDE is a risk-based dose limit, utilizing the highest dosimeter reading not indicative of the dose received to deeper-seated organs and tissues will oftentimes result in overestimating the stochastic risk associated with an exposure and hence, overestimating the EDE. The use of two dosimeters worn on the front and back will oftentimes yield a more accurate value for EDE and a value that is lower than that obtained by using multiple dosimeters. ERPI Reports TR-0101909 and TR-109446 describe an EDE methodology that results in a less conservative assessment of EDE and eliminates the practice of using multi-dosimeters for certain conditions while providing a more accurate assessment of worker exposures. These concepts should be evaluated for use at a particular facility, especially those that rely heavily on the use of multi-badging.

However, under unique circumstances the need to perform a more elaborate assessment of the exposure received by various body locations may be necessary. ANSI Standard HPS N13.41 provides a methodology for assigning dose to specific body locations (or compartments) based on multiple dosimeter readings (that could be utilized as necessary). Regardless of how exposure is assigned when utilizing multiple dosimeters, it is essential that dosimetry results be recorded and retained for each dosimeter even if the dosimeter reading was not used to assign the dose of record.

Oftentimes situations may be encountered whereby the direct handling or close contact with components and equipment comprising the source of the radiation may be necessary. Under these circumstances if contact dose rates are significantly different from those to which the whole-body will be exposed then the use of extremity dosimeters should be considered. In the vast majority of cases this may result in the use of "finger rings" to monitor the extremity dose to the hands. In other situations the feet may be located in a higher dose rate area than that of the whole-body, requiring an extremity dosimeter to be placed on the ankle or lower calf area. The use of a finger ring is often required when working on highly contaminated equipment that has a significant beta dose component. Finger rings may consist of a single TLD chip or OSL dosimeter enclosed in a ring holder or embedded in a piece of flexible

**Fig. 10.10** Finger ring
offered by Landauer
(Courtesy of Landauer,
www.landauer.com)

material that may be wrapped around a finger or a wrist. Figure 10.10 depicts a
finger ring design used to monitor exposure to the hands.


## 10.8  Secondary Dosimeters

It is advantageous to track worker exposures on a daily basis. Additionally, it is
beneficial to determine how much exposure an individual receives for a specific
task or over a given time period. For instance if daily or weekly administrative
exposure controls are established, a method to confirm that workers do not exceed
administrative dose values must be available. Under these circumstances personnel
dose monitoring devices should have characteristics that facilitate reading and
should allow for direct readout of dose that has been received. These type
dosimeters are often referred to as direct-reading, secondary dosimeters. These
dosimeters are classified as secondary dosimeters since they typically do not serve
as the dose of record. The main function of these dosimeters is to provide live-time
measurement of the dose received by workers. The design of these dosimeters
often does not provide the capability to distinguish between exposures due to beta
or gamma radiation. Secondary dosimeters simply respond to radiation over a
given energy range.

The electronic dosimeters discussed previously are often used as secondary
dosimeters since they provide live-time readout of dose received. Some electronic
dosimeter models respond to both gamma and beta radiation and if properly calibrated could also serve as the dose-of-record. However, as noted above, TLDs,
film badge, and OSL dosimeters are the primary dosimeters utilized to record
official exposures.

Historically, pocket ionization chamber (PIC) dosimeters were commonly used
as a secondary dosimeter providing live-time dose measurements. The availability
of electronic dosimeters has essentially supplanted the use of PICs in recent years.
The pocket ionization chamber is a direct reading ionization chamber detector.
The ionization chamber is connected to a quartz fiber electroscope equipped with a

display that depicts the amount of radiation exposure received. The display is in units or subunits of sieverts or rem. The PIC chamber is charged prior to use to "zero" the unit. Exposure to ionizing radiation decreases the charge on the chamber causing the quartz fiber to move up the scale. These devices are designed and calibrated so that a decrease in charge can be equated to a given radiation exposure. A PIC is exposed to a light source to illuminate the scale. Viewing the scale by means of a built-in lens allows the user to read the exposure value of the PIC. The PIC allows workers to obtain their current exposure at any time by simply viewing the reading on the scale. A PIC responds primarily only to gamma radiation.

Pocket ionization chambers are rugged and easy to use and maintain. They are, however, sensitive to shocks due to dropping or bumping. When dropped a PIC may go upscale or even off-scale. Thus, a bumped or dropped PIC could be interpreted as an actual radiation exposure when, in fact, no radiation exposure was received. In order to prevent the unintentional recording of dose under these circumstances, a PIC should be read immediately after being dropped or bumped. If the reading has moved upscale individuals should be instructed to report to radiation protection to have the PIC re-zeroed. An off-scale PIC reading should be treated similar to that of an ED dose alarm. An investigation to evaluate the reason for the off-scale reading should be performed. Investigation results and the dose assigned to the individual should be retained. Additionally, PICs are prone to charge leakage at a slow rate across the insulator. This leakage appears as an upscale movement of the quartz fiber. This results in a reading somewhat higher than the actual exposure received by the wearer. Under most situations this effect is negligible. However, in the event that PICs are assigned to individuals for prolong periods, procedures should be established to require the periodic charging of the PIC to re-zero the unit.

## 10.9 Internal Dosimetry and Bioassay Programs

The ICRP also issues periodic recommendations that provide guidance concerning the use of biokinetic models used in monitoring the intakes of radionuclides by workers. The biokinetic models provide a basis for calculating exposures resulting from internal uptakes of radioactive material. The two primary guidance reports are ICRP 30 and ICRP-68 published in 1979 and 1994, respectively. The guidance provided in ICRP-30 was superseded by ICRP-68. A revised respiratory tract model was provided in ICRP-68. The model provides transport rates between the various compartments of the respiratory tract and gastrointestinal tract in addition to gastrointestinal absorption fractions. The revised modeling resulted in changes to the gastrointestinal absorption fractions for many radionulcides. The gastrointestinal absorption fractions are represented by the symbol "$f_1$" in ICRP publications. Once a radionuclide is absorbed into the blood or extracellular fluids the contamination is referred to as systemic contamination. The radionuclide will then undergo various interactions and transfer processes within the body that ultimately determines its distribution within the body and its eventual elimination. The purpose of the respiratory model is to assign values to these complex mechanisms based on the latest biokinetic information.

The key aspect of these models is that values are assigned to how quickly a given radionuclide is cleared from the body or absorbed into extracellular fluids. This is the key parameter in determining the dose commitment resulting from an uptake of a quantity of radioactive material. These rates are primarily influenced by the physicochemical characteristics of the inhaled substance and, specifically, those factors that impact the solubility of a given radionuclide. Particle size is important when considering deposition in the respiratory tract. Another significant change introduced in the ICRP-68 recommendations involved a revision to the default particle size for inhaled particulates from an AMAD of 1 to 5 μm. The change in the default activity medium aerodynamic diameter (AMAD)[1] value impacts dose assessments. Particle size impacts the regional deposition characteristics (e.g., clearance rates) of particles in the respiratory tract. The half-life and the types and energies of the emitted radiations of a particular radionuclide in turn impact the dose assessment resulting from changes in the retention characteristics of 5 μm-sized particles versus that of 1 μm-sized particles. The significance of any change in dose assessment depends upon the resulting differences in the regional deposition characteristics stemming from the change in the AMAD default value.

Previous ICRP guidance provided three general clearance classes for materials. These clearance classes were referred to as "D" (days), "W" (weeks) and "Y" (years). The general clearance classification provided an estimate of the retention timeframe of a particular radionuclide within the body. These designations were revised in the ICRP-68 report to reflect more closely the dissolution rates and absorption rates into blood. The dissolution rates and absorption rates have a direct influence on how quickly a substance will make it to extracellular fluids and once there how long the substance may reside within the body. The revised clearance classes or absorption types are now designated as Type "F", "M" or "S" in ICRP-68. These types represent fast, intermediate, and slow dissolution and absorption rates, respectively.

Current NRC regulations are based on ICRP-30 and therefore reflect the previous clearance classes and absorption and retention values for specific radionuclides. Therefore, reference to superseded terms may be encountered in various publications.

It is important to distinguish between the terms intake and uptake when dealing with internal exposures. The term "intake" refers to the amount of radioactive material taken into the body. Activity may enter the body via the respiratory tract (i.e., inhalation), the gastrointestinal tract (i.e., ingestion) or via the skin (i.e.,

---

[1] Activity medium aerodynamic diameter simply means that 50% of the particles composing the airborne activity are smaller than the stipulated AMAD value and 50% of the particles are larger than the stipulated AMAD.

absorption). The intact skin usually serves as an effective barrier against absorption of radionulcides, with the exception of tritium, into the body. Cuts, wounds, or skin abrasions should be covered and protected prior to allowing individuals to enter or work in contaminated areas. The primary intake pathway for internal exposures at LWRs is via inhalation. However, ingestion could pose a significant pathway under certain circumstances, such as a result of when poor radiological work practices result in an intake. The term "uptake" refers to the portion of the radioactive material from the intake that passes into body fluids and is retained in body organs or the systemic system. Bioassay programs are designed to determine the amount of uptake.

Once the quantity of an uptake is known then the appropriate dose conversion factors based on the absorption type may be used to assign an exposure to the individual. Bioassay programs supplement the external dosimetry monitoring program and are an integral part of the overall personnel dosimetry program. Whereas personnel dosimeters measure the amount of external exposure received bioassay programs are designed to evaluate exposures resulting from the internal deposition of radioactive material. Bioassay measurements include the direct measurement of internally deposited radionuclides. These measurement techniques are referred to as 'in vivo' measurements. The amount of a radionuclide present in the body may also be determined by indirect methods or "in vitro" analysis, the most common being urine analysis. Fecal sampling and analysis may be necessary (under certain circumstances) when insoluble radionuclides are involved, in order to adequately assess internal exposures. The most likely route of entry into the body for insoluble radionuclides would be as a result of an incident involving ingestion.

Inhalation is the most significant route of accidental entry of radionuclides into the body at a LWR facility. Deposition and subsequent removal of inhaled radionuclides in the lungs are greatly dependent on the physical sizes and chemical properties of the material. Factors to be considered are particle size, shape, density, chemical composition, solubility and others. Radionuclides deposited in the lungs will irradiate sensitive tissues. Larger sized particles will be filtered out in the upper regions of the respiratory tract while smaller particles may reach the alveoli in deeper regions of the respiratory track. The ultimate clearance time or residency time from the respiratory track depends largely on the place of deposition. In general, soluble radionuclides are cleared rapidly (i.e., within a few hours or less) while less soluble radionuclides may reside within the lungs for up to a few days. Insoluble radionuclides may take months or years to be cleared from the lungs. These general clearance patterns are the basis for the "D", "W" and "Y" ALI values provided in 10 CFR 20 and as discussed above.

Airborne particulates or aerosols consist of dust, lint, condensation nuclei, and similar type matter and usually encompass a wide range of particle sizes. Airborne radionuclides may be produced during such activities as grinding, milling, drilling or welding on contaminated components. Radionuclides existing as aerosols can exist in a wide and complex range of chemical states with a wide range of biological half-lives. These radioactive aerosols may be deposited within the lungs and absorbed and concentrated in body tissues.

Since the gaseous radionuclides are inert they pose an external whole-body exposure hazard versus an internal hazard. Half-lives for gaseous radionuclides of interest range from seconds to a few days with the exception of krypton-85 which has a 10-year half-life. The inert gaseous are highly volatile and, thus, easily escape from system components. Tritium gas, that may be present, is readily changed to tritiated water vapor. Tritiated water readily mixes with body fluids and, therefore, represents a whole-body exposure hazard.

The presence of iodine, notably I-131, is of special interest. Iodine is absorbed into the body and preferentially concentrates in the thyroid gland. Radio-iodine is one of the more significant airborne exposure hazards from an exposure perspective if present in significant concentrations. Airborne iodine typically consists as a gas but may also exist as a particulate attached to condensation nuclei.

Ideally, engineering controls, effective contamination control measures, and the use of respiratory protective equipment in conjunction with good radiological work practices would prevent intakes of radioactive material and negate the need for an internal dosimetry program. However, in the event of a radiological incident or unforeseen radiological situation, the need to perform internal dose assessments may be encountered. Consequently, an internal dosimetry program or bioassay program must be established and implemented in conjunction with the external radiation monitoring program to afford effective evaluation of worker exposures.

Bioassay programs must instill a high-level of confidence in the work force. (That is not to say that other elements of the radiation protection program should not aspire to the same level of excellence and quality). Individuals typically have a heightened sense of concern when it comes to internal uptakes of radioactive material into their bodies. The interest shown on the part of individuals concerning internal uptakes is often not proportional to that with regards to an external radiation exposure. Health Physicists may spend many hours assuring an individual of the safety aspects of a "20 $\mu$Sv" (2 mrem) uptake versus explaining why a person received an unnecessary 500 $\mu$Sv (50 mrem) external exposure. The concern shown on the part of the individual may be completely disproportional to the relative health risks. Consequently, it is important to implement an effective bioassay program. Key elements of the bioassay program should be presented in general orientation training and radiation worker training programs to ensure understanding of the purpose and function of the station's bioassay program. Radiation workers need to be assured that evaluation of any such uptake is based on a sound bioassay measurement program.

### 10.9.1 Whole-Body Counting

The assessment of internal exposure is an integral component of the bioassay program. Facilities and equipment should be available in order to assess potential uptakes in a timely manner. Since the primary radionuclides of interest (e.g., Co-58 and Co-60) are gamma emitters the assessment of internal uptakes may be accomplished by means of in vivo measurement techniques. The most common method is often referred to as whole-body counting. Whole-body counters consist of a detector or detector array coupled with a multichannel analyzer system. The primary advantages of whole-body counting include the ability to directly measure the amount of radioactive material in the body, results can be obtained faster when compared to in vitro methods, and does not require the use of indirect analysis methods. Internal contamination involving gamma-emitting radionuclides, whether soluble or insoluble, is capable of being detected by a WBC system. Under certain situations high energy beta-emitting radionuclides may also be detected by whole-body counting techniques provided that the system is properly calibrated. Under these conditions the WBC system would utilize the bremsstrahlung radiation as a means of detecting beta emitters. However, since the vast majority of radionulcides of interest at a LWR that pose an internal contamination concern are either gamma or beta-gamma emitters this discussion will assume that WBC systems are utilized to detect gamma radiation. The measurement of high-energy beta emitters by whole-body counting poses additional challenges such as obtaining an acceptable minimum detectable activity level and the need for longer duration count times. American National Standards Institute, Standard HPS N13.42-1997, Internal Dosimetry for Mixed Fission and Activation Products, addresses those radionuclides most likely to pose internal exposure concerns at LWRs.

It is highly unlikely that a significant uptake of a predominant beta-emitter (e.g., Sr-90) could occur singly in a LWR environment without involving more significant, simultaneous uptakes of gamma-emitting radionuclides (e.g., Cs-137, Co-60 or Co-58) capable of being detected by a WBC system. If such an event were to occur, urine analysis would probably be necessary in order to perform a more complete and accurate dose assessment associated with the uptake and clearance of the beta-emitter from the body. Intakes of beta emitters could also be evaluated indirectly by determining scaling factors or the ratios of certain radionulcides present in contamination or airborne radioactivity. For instance, the ratio of Co-60 to Sr-90 present in airborne radioactivity could be determined and any possible uptake of Sr-90 could be based on the level of Co-60 present in an individual based on WBC data. Obviously such parameters as relative solubility, clearance rates, time after exposure and other biokinetic factors would have to be considered when employing this method, especially in the event of an uptake approaching action levels or regulatory exposure limits.

### 10.9.2 Whole Body Counting Systems

Any detector type capable of detecting gamma rays may be utilized in a whole-body counting system. The most common detectors that have gained wide spread acceptance in whole-body counter designs include sodium iodide (NaI) scintillation detectors and high purity germanium (Ge) semiconductor detectors. Thallium is added to NaI detectors as an impurity or "doping" agent. These detectors are referred to as thallium-activated sodium iodide NaI(Tl) detectors. The thallium activator increases the overall effectiveness of the scintillation process in the NaI crystal. Sodium iodide detectors are ideally suited for performing a "whole-body" scan utilizing large sized NaI(Tl) crystals. Lower limits of detection, in the range of 5–10% of an ALI, or lower, for those radionuclides of interest may be detected utilizing large area NaI detectors. These detection levels are often achieved with count times of a few minutes or less. NaI(Tl) detectors 40 cm in length are commonly incorporated into stand-up type WBCs. Whole-body counting systems that utilize NaI detectors often employ a fixed geometry system in which the detectors remain stationary and the subject is either sitting or standing.

Germanium detectors offer higher resolution than NaI detectors and are more suited for monitoring for the presence of internal contamination located in specific body locations such as the thyroid or lungs. The recent introduction of high-purity germanium (HPGe) detectors has facilitated the use of these detectors for use in whole-body counting systems. Prior to the availability of HPGe detectors the use of lithium-drifted germanium (GeLi) detectors for WBC was limited due to the need to constantly maintain these detectors at liquid nitrogen temperatures. When in use HPGe detectors must be cooled with liquid nitrogen; however, these detectors can be allowed to warm up to room temperature when not in use without damaging the detector. On the other hand if GeLi detectors are allowed to warm up the lithium would diffuse out of the detector rendering the detector useless for gamma spectroscopy purposes. Since WBC systems are typically used on an intermittent basis and may be subject to extended periods of non-use (where the system is basically in a stand-by mode), the need to maintain liquid nitrogen supply to WBC detectors could prove inconvenient. The vast majority of whole-body counts are performed during periods of time leading up to outages and the subsequent outage periods. During this time initial baseline whole-body counts are performed on supplemental outage employees, and the need to perform investigative or incident-related WBCs often reach a peak during outage periods.

Even though relatively large sized HPGe detectors are available their high cost may make them prohibitively expensive for LWR whole-body counting purposes. Common practice is to utilize multiple smaller-sized detectors with a moveable mechanism to scan the subject's body. Detector sizes range from six to ten centimeters in diameter. Longer counting periods to obtain comparable detection limits are required due to the lower detection efficiency of germanium detectors when compared to the larger-sized NaI WBC systems. The higher resolution of germanium detectors and the incorporation of these detectors in a scanning mechanism provide the ability to identify localized areas of internal contamination in a subject. This could prove useful in follow-up investigations and when evaluating the elimination rate of a radionuclide from a specific body location (e.g., the lungs or thyroid).

Various manufacturers supply whole-body counting systems, along with diagnostic software, to the LWR industry. One company, Canberra, offers two systems that are in common use in the industry. Another company, ORTEC, as well as

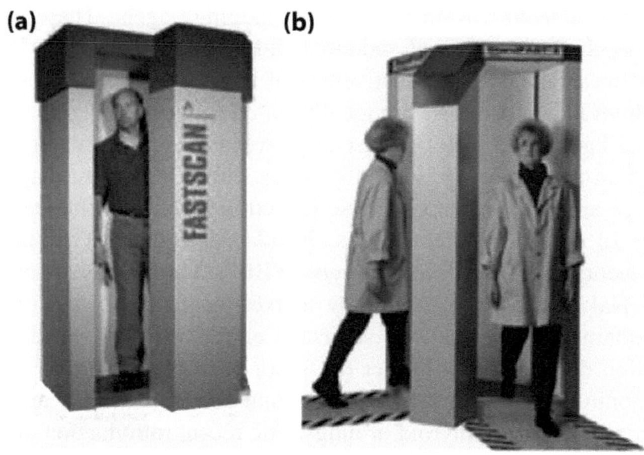


**Fig. 10.11** **a** The Canberra FASTSCAN™ (Courtesy of Canberra www.canberra.com). **b** The ORTEC StandFAST® (Courtesy of ORTEC International, www.Ortec-online.com)stand-up whole-body counters

others, offers systems that are well suited for use in the industry. Canberra whole-body counting systems include the FASTSCAN™ and ACCUSCAN™ systems. The FASTSCAN model utilizes two large sized NaI(Tl) detectors. The detectors are maintained in a fixed position. Individuals remain standing during the analysis period. The detectors are sufficiently sized to monitor major portions of the upper torso. A FASTSCAN™ unit is displayed in Fig. 10.11. The ORTEC StandFAST® whole-body counter also employs large-sized NaI (Tl) detectors and as the name implies also performs the monitoring sequence with individuals in the standing position. A StandFAST® system is also shown in Fig. 10.11. These systems are capable of detecting Co-60 uptakes in the range of 100-200 Bq (several nanocuries) with one-minute count times. These lower limits of detection represent less than one percent of the ALI for Co-60. The Co-60 ALI (ICRP-68) for inhalation is 6.9E5 Bq and 30 μCi (10CFR20). Similar detection limits are achievable for other major gamma-emitting nuclides such as Co-58 and Cs-137.

The "standup" whole-body counting systems are ideal for processing large numbers of individuals during high work load periods such as those encountered during outage periods. The detection limit achievable with short counting periods with standup WBC systems will probably prove adequate for the vast majority of cases. These characteristics of standup whole-body counting systems are advantageous when performing analyzes not involving suspected uptakes. This could include analyzes associated with the performance of routine whole-body counts, initial employment and termination counts and random screening. The standup configuration could also be utilized when assessing positive uptakes by employing longer count times to obtain a more adequate assessment of the uptake.

The germanium based whole-body counting systems include the ACCU-SCAN™ offered by Canberra and a range of models offered by ORTEC and other

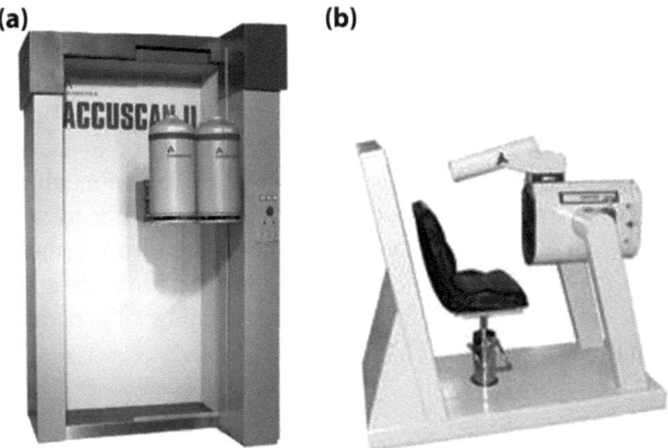


**Fig. 10.12**  **a** The Canberra ACCUSCAN$^{TM}$ (Courtesy of Canberra www.canberra.com). **b** The ORTEC combination HPGe (lung and GI tract) and NaI (thyroid) (Courtesy of ORTEC International, www.Ortec-online.com) whole-body counting systems

firms. Due to the lower detection efficiency of HPGe detectors compared to that of NaI(Tl) detectors, germanium-based WBC systems require longer count times to achieve acceptable lower limits of detection. Figure 10.12a depicts Canberra's germanium-based ACCUSCAN$^{TM}$ whole-body counter. Analysis times of several minutes or longer may be required for Ge-based whole-body counting systems. These systems often employ a chair or bed configuration to allow individuals to remain reasonably stationary while the analysis is performed. Figure 10.12b depicts an ORTEC chair configuration that consists of a combination of an HPGe lung and GI tract and NaI (thyroid) detector configuration. Bed systems usually employ a scanning arrangement whereby the detector array moves along the length of the subject's body. These systems may also be configured to include dual detector systems consisting of a combination of HPGe and NaI(Tl) detectors.

Whole body counts may be performed for a variety of reasons. A good practice is to require an initial "baseline" WBC for persons entering the RCA to perform work activities. This practice may be especially important in identifying internal contamination present in individuals who previously worked at another LWR or a facility involving the handing and exposure to radioactive material. A baseline WBC provides a means to document previous uptakes that did not take place at a particular plant prior to allowing an individual access to RCA areas. Other measures that are commonly encountered at LWR facilities include the use of walk-through or portal type personnel contamination monitors located at the entrance to the restricted or controlled areas. Individuals entering the restricted area are monitored for the potential presence of radioactive contamination. These type monitors are typically gamma sensitive and set to alarm when contamination levels "statistically" greater than background are detected. A secondary benefit accrues from the ability to identify individuals who may have undergone medical procedures involving the use of radioactive materials. Oftentimes the quantities of radioactive material used in medical diagnosis and therapy procedures are sufficient to trip alarm settings on personnel contamination monitors several days following the medical administration of the radionuclide. The ability to detect individuals undergoing nuclear medicine procedures prior to entering the RCA prevents the need to respond to "nuisance" RCA exit point contamination monitor alarms not associated with plant operations. If the radionuclide used in the medical procedure (e.g., an isotope of iodine) cannot be distinguished from those encountered in RCA areas of the plant, then difficulties may arise in determining the source or cause of the contamination and assigning an accurate dose assessment to the individual.

### 10.9.3 Interpretation and Evaluation of Whole-Body Counting Results

Whole body counts serve an important function in evaluating potential internal exposures due to contamination events or exposure to unknown airborne radioactivity concentrations. The need to perform a WBC may be triggered for the following reasons:

- Event involving detectable nasal contamination
- Exposure greater than 10 DAC-hours (or other appropriate value) or an administrative control value based on a percentage of the ALI (e.g., an exposure greater than 1 or 5% of the applicable ALI)
- A personnel contamination monitor alarm that does not involve external contamination of the individual
- Respirator equipment failure that may have resulted in an internal exposure
- Event involving exposure to unknown levels of airborne contamination
- As a result of an operational event
- Situations dictated by procedure guidelines

When WBC results serve as the basis for assigning dose then procedural guidelines should specify details associated with performing follow-up WBCs. These details should include the frequency at which follow-up WBCs are to be performed and at what value (perhaps radionuclide-specific) no further actions are required. A good practice may include the requirement to perform follow-up WBCs until measurements fall below the established minimum detection activities for the radionuclide(s) involved. Events that result in measurable uptakes are often associated with a radiological incident that may have involved personnel contamination. Under these circumstances it is important to ensure that whole-body counting results are not misinterpreted to include external contamination. A good practice when external contamination is suspect is to require the individual to shower prior to the whole-body count. If the individuals' personal clothing is suspected of being contaminated then any suspect clothing articles should also be removed. A supply of modesty garments or other suitable supply of "clean" or one-time use clothing should be available to minimize the chances of inadvertently performing WBCs on individuals who may have external skin or clothing contamination. The removal of jewelry, watches and metal objects prior to performing a WBC is a common practice to minimize possible natural and manmade radionuclide interference.

Even if WBC results do not detect internal contamination the results serve to confirm that any uptakes resulting from a radiological incident are less than a certain level not posing a significant radiological safety concern. During those periods involving significant radiological work activities random WBCs performed on selected individuals provide an indication of the effectiveness of radiological control measures and radiological work practices. For instance during a refueling outage individuals working extended periods of time in the containment building or in the drywell, not necessarily working in airborne radioactivity areas or areas requiring the use of RPE, could be selected for a random WBC. A random whole-body counting program during intense radiological work periods could serve a vital function in the early detection of unplanned uptakes. Early detection allows timely assessment of the situation and affords a higher chance of identifying the time of the uptake as well as the plant conditions that may have caused the uptake. Early detection of an uptake is important in order to determine the initial amount of an uptake. The timely identification of unplanned uptakes also affords an opportunity to investigate the cause or source of the contamination and implement corrective measures to prevent further radiological incidents. Even though work place air monitoring and radiological surveillance activities are designed to detect and prevent unplanned internal exposures, random whole-body counting serves as another means to confirm the effectiveness of radiological control measures implemented in the field.

Based on the biokinetic properties of many radionuclides of interest at LWRs, clearance times on the order of days or weeks are common (i.e., clearance Type F and M). Consequently, the timing of a WBC is critical to detect unplanned uptakes. This concept is often referred to as "missed dose." Ideally incidents that could result in measurable internal exposures will not go undetected, thus, affording the opportunity to perform WBCs prior to clearance of the material from the body. Since unplanned exposures, by definition, may go undetected, a length of time sufficient to allow clearance of the material could elapse from the time of the exposure to the individual's next scheduled in vivo bioassay. Obviously a balance must be achieved between how frequently WBCs are performed and the possibility that an exposure event could go undetected between scheduled WBCs. Work place monitoring, continuous airborne monitor alarms, personnel contamination monitor alarms, and personnel contamination events all serve to provide early indication of a potential internal uptake. These and other radiological monitoring indicators may be cause to perform timely investigative or follow-up WBCs. Controlling airborne contamination at the source, stringent contamination control measures, strategic placement of monitoring equipment and effective job planning and preparation are key components in minimizing the chance of an unplanned uptake.

Action levels and procedural requirements should be established to provide guidelines to determine when a dose assessment and formal follow-up investigation are required based on WBC results. These actions should be triggered whenever a WBC indicates the potential that an uptake may have occurred. ANSI Standard N13.42 establishes the term "investigation level." Investigation level is defined as an intake of a single radionuclide equal to 0.02 of the stochastic ALI, or for a mixture of radionuclides a value of 0.02 for the sum of the individual intakes relative to the stochastic ALI values. These values correspond to a committed effective dose equivalent of 1 mSv (100 mrem).

Software programs provided with WBC systems calculate exposures based on WBC results. Assuming that the quantity of the uptake is a small fraction of the ALI and no further assessment is required to determine clearance time or other factors then the most conservative dose conversion factor (DCF) may be utilized. Alternatively if the clearance type is known for a given radionuclide based on previous experience or individual studies conducted at a plant then the appropriate DCF should be utilized. Dose conversion factors for inhalation are assumed to be the default values used in dose calculations. Incidents for which ingestion is the primary route of entry should utilize the DCFs for the ingestion pathway.

The DCFs recommended by the ICRP may be used to determine exposures once the amount of an uptake has been determined. The DCFs provided in ICRP-68, for both inhalation and ingestion exposure pathways, are listed in Appendix B for selected radionuclides. The ALIs are also listed for the ingestion and inhalation pathways. The ICRP values are predicated on an annual exposure limit of 20 mSv; consequently, the ALI values are not the same as those provided in Appendix B to 10CFR20. Exposure received by an individual as a result of an uptake may be calculated by multiplying the uptake (in Bq) by the applicable DCF (Sv/Bq) for a given radionuclide.

Another factor required to perform a more accurate dose assessment based on WBC results concerns assigning a time at which the uptake occurred. Obviously WBC data provides an estimate of the quantity of radioactive material present in the subject at the time of the analysis. Depending upon when the exposure took place a considerable fraction of the initial uptake for radionuclides with a short biological half-life may have been eliminated prior to the WBC. Consequently consideration should be given when determining the time of exposure, particularly when uptakes exceed established action levels or are a significant fraction of the ALI. Fortunately in a LWR environment chronic exposures are not likely to occur. Furthermore routine in-plant radiological surveillance activities, the use of whole-body personnel contamination monitors at RCA exit points and other radiological control measures limit the chance of any significant chronic exposures going undetected. The role of PCMs serving as "passive" whole-body counters is discussed later in this chapter. Significant uptakes most likely will be the result of a radiological incident for which the time of exposure should be known with reasonable accuracy. The time at which the exposure occurred should be used whenever positive WBCs are encountered or at a minimum whenever established action levels are exceeded.

In addition to the above parameters associated with uncertainties in dose assessments based on WBC data for uptakes at or above regulatory limits a more accurate assessment may be necessary. Such parameters as how the distribution of the activity within the subject compares to the activity distribution in the phantom used to "calibrate" the WBC may have to be considered when determining the quantity of the uptake. Particle size determinations may also be performed along with chemical analyzes to determine solubility characteristics of the material.

Notwithstanding, the accuracy achievable with a properly calibrated and maintained whole-body counting system is sufficient for the vast majority of in vivo assessments. Positive whole-body counts that are less than one percent of an ALI and not the result of a radiological incident may simply be recorded with no further action necessary. A significant radiological event would have to take place at a LWR with concurrent breakdown in radiological control measures to cause an uptake approaching significant fractions of an ALI. In the event of an exposure situation approaching regulatory limits or in the event of an over-exposure, extraordinary measures could be implemented. These measures could include performing confirmatory whole-body counts at an offsite location, such as a research center or medical facility if available. Additionally, WBC data could be supplemented with urine analysis, for example, to quantify uptake amounts and the rate of elimination. Fecal samples may also be required for significant uptakes involving insoluble radionulcides.

Consideration should also be given to restricting individuals from the RCA with a positive WBC that exceeds administrative control values or that are a significant fraction of an ALI until the assessment is completed. It may be necessary to prevent additional potential exposure to internal contamination until the biokinetic characteristics have been sufficiently determined to allow for an accurate dose assessment. Depending upon the individual's job classification (e.g., an auxiliary plant operator or a contract welder involved with critical outage tasks) restricting individuals from the RCA could pose a significant burden. However, necessary measures should be implemented to ensure a proper dose assessment is obtained.

Insoluble radionuclide deposition in the lungs may be indicative of an ingestion intake versus inhalation due to airborne contamination. Under these circumstances any follow-up investigation into the cause of the intake should evaluate whether or not poor radiological work practices or unforeseen work-related issues may have contributed to the intake. This possibility should be considered especially when airborne radioactivity measurements indicate that airborne contamination was not the likely source of the intake or radiological survey data are otherwise inconclusive.

### 10.9.4 Calibration of Whole-Body Counters

To quantify the amount of a radioactive material uptake whole-body counting systems must be properly calibrated. Since the "subjects" analyzed by a

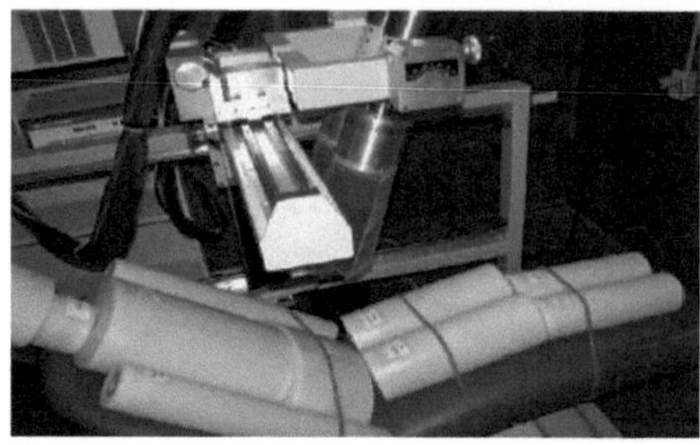

**Fig. 10.13** A phantom utilized to perform WBC system calibrations (Courtesy of ORTEC International, www.Ortec-online.com)

whole-body counting system do not represent a repeatable geometry allowances must be made to adequately assess or quantify measurement results. It is not practical to calibrate a whole-body counting system for the entire range of heights, weights, and body sizes and shapes represented by the work force. ANSI Standard, ANSI/HPS N13.35-2009, Specifications for the Bottle Manikin Absorption Phantom, provides specifications for phantoms that may be utilized for the calibration of in vivo measurement systems. The phantom is composed of cylinders of various sizes and shapes to approximate the major parts of the human body. The phantom is referred to as the Bottle Manikin Absorption Phantom (BOMAB). The standard provides the necessary specifications to construct the phantom. Details associated with the material composition and physical specifications of the phantom along with average height and weight values for different age groups are provided. Figure 10.13 depicts a BOMAB phantom. Use of such a phantom provides a means to quantify internal uptakes to a reasonable level of accuracy for the wide variety of body sizes and shapes that may be encountered in the work force.

Whole-body counting systems must be routinely calibrated. At a minimum whole-body counting systems should be calibrated on an annual basis or based on the operating history of a given system. Calibrations could be performed utilizing an elaborate phantom as described above or a specific calibration fixture uniquely designed for a given whole-body counting system. The calibration is performed utilizing known activity amounts of multiple radionulcides. The calibration source(s) should include gamma emitters with energies to cover the range of interest. In other words the response of the whole-body counting system for key radionulcides at the low-end of the energy range (e.g., I-131) and at the high end of the energy range (e.g., Co-60) should be encompassed during the calibration. Mixed radionuclide calibration sources are available from various suppliers that are ideally suited for use in performing these calibrations. Due to the unique nature of these calibrations the services of a specialized firm are often used to perform calibrations. These firms can often provide more elaborate calibration phantoms resulting in a more accurate overall calibration of the system. The supplier of the whole-body counting system may also provide routine calibration services utilizing a phantom specifically designed for a given system.

An operational check source or a suitable reference source with a known and repeatable geometry should be available to serve as a calibration check between calibration periods. The WBC system response to this operational check source should be determined at the time of the most recent calibration. The response of the "calibrated" WBC to the operational check source is then determined. This response as a minimum should include a range of control values to confirm that the activity of the reference source and energy channel values fall within acceptable limits. Emphasis should be placed on ensuring that a daily operational check is performed to verify that system operability and performance parameters are maintained within acceptable ranges between calibration periods. The daily operational check should verify that the energy channels of the gamma emitters have not drifted outside acceptance values. The operational check source should also consist of a known activity level. The daily operational check should confirm that the system detector response falls within acceptable ranges indicating an accurate measurement of the operational check source activity. In addition to these daily operational checks other parameters such as system background checks, detector voltage and other operational parameters important to ensuring that the system is properly operating should be verified. The use of daily control charts or automated programs to warn operators when system performance falls outside operational parameters should be established. Verification that whole-body counting systems are within established operating parameters prior to use is essential in order to ensure the proper assessment of internal uptakes.

### 10.9.5  Passive Whole-Body Counting

Whole-body contamination monitors are prevalent at LWRs for monitoring personnel for the presence of external contamination. These devices are often found at the entrance and exit points of the restricted area and at exits from RCAs. Additional units may be located within RCA areas at convenient locations to allow workers to be monitored upon exit from contaminated areas. An indirect advantage of these units is that they essentially perform a "passive" whole-body count on individuals when they are monitored for external contamination. Depending upon the design, model type, operational configuration and ambient background radiation levels, modern, state-of-the-art personnel contamination monitors (PCM) may serve as a passive WBC.

For instance assume that Co-58, Co-60, and Cs-137 comprise the majority of the contamination encountered at a given plant. If a PCM can detect 5–10% of an ALI for these radionuclides then a PCM could serve as a mechanism for initiation of a formal follow-up whole-body count for an individual. Obviously this situation assumes that the presence of external contamination has been shown not to be present or is otherwise not a contributor to the PCM alarm. This is not to suggest that PCMs be utilized to serve as the primary method for determining the presence of an internal uptake. Rather, lacking other radiological survey information the fact that an individual successfully "cleared" a PCM could be used as supplemental (or retroactive) information to verify that any suspected uptake was less than a certain quantity. The response of PCM units to known activity levels of various radiation sources (e.g., Co-60) could be determined utilizing appropriate phantoms or source geometries to reflect the configuration of a person when monitored by a PCM. The count time of the PCM would have to be taken into consideration since the length of the PCM counting sequence would significantly impact the lower limit of detection achievable by the specific PCM model. This data could be used to determine an effective lower limit of detection for PCMs. The calibration (or source response check) would not have to be as formal or rigorous as that described above for a whole-body counter. Under these conditions the PCM would simply provide an indication for the potential presence of internal contamination and would not be used to quantify an uptake. Such factors as determining the detection level as a function of the PCM count time and the ability to utilize a suitable phantom or calibration device to determine the detection capability of PCMs compared to the respective ALIs would be required. The primary objective of this discussion is for RP personnel to realize that in the event of an unplanned uptake or other radiological incident the health physicist should utilize various tools and techniques that may already be available but not necessarily serving that purpose.

### 10.9.6  Urine Analysis

The vast majority of bioassay measurements at a LWR consist of whole-body counts. Whole-body count measurements are convenient and their analysis relatively straightforward. Since the radionuclides most likely to comprise significant percentages of airborne radioactivity concentrations are beta-gamma emitters, WBC is a suitable bioassay technique. However, there may be occasions when WBC data may have to be supplemented with in vitro measurements. In vitro analyzes may be required when the predominant nuclide comprising the uptake is a pure beta-emitter, such as Sr-90. The need to perform dose assessments due to tritium exposures may also require an in vitro measurement. The most common in vitro measurement technique involves urine analysis and perhaps, on rare occasion, supplemented with fecal sampling. Fortunately most radionuclides of interest are eventually excreted in the urine, most notably tritium.

A urine analysis program should be established as part of the overall personnel dosimetry program. The analysis of urine samples may be performed onsite or at an offsite location. Oftentimes the analysis of urine samples may be performed by a suitably qualified contract laboratory. Though the need to obtain urine samples for dose assessment purposes may be infrequent, procedures and administrative aspects associated with the collection, analysis and reporting of urine analysis results should be available. Urine samples obtained to perform dose assessments associated with uptakes approaching an ALI may be crucial in determining an accurate dose to an individual. Applicable program procedures, with regards to tritium urine analysis, should address the following areas, as a minimum.

- Threshold for initiating the collection of urine samples based on air sample data or DAC-hour exposure estimates
- Collection and drop-off location for samples
- Guidelines specifying the length of the collection period (e.g., a 24-hour collection period or a one-time voiding)
- Instructions provided to individuals concerning the collection, handling, and submittal of urine samples
- Process associated with the packaging and mailing of samples to an offsite contractor laboratory, as applicable
- Guidelines detailing the need for follow-up or additional samples

Tritium concentrations in urine are most often determined by liquid scintillation counting techniques. Since bioassay samples involve low concentrations of radioactive substances, precautions should be taken to ensure that samples are not cross-contaminated. Consequently, the processing and analysis of urine samples should be performed in a clean laboratory separate from the primary radio-analytical laboratory. Based on the number of samples typically processed over a given period it may be more economical to secure the services of a contract firm.

Even though the likelihood of experiencing a significant airborne exposure to tritium is relatively low at a LWR, program elements should be in place to ensure the accurate and timely assessment of urine samples. Additionally, in the event of a significant uptake, especially one approaching or exceeding dose limits, the collection and analysis of urine samples may be vitally important in the accurate assessment of the exposure received by the individual.

## 10.10  Summary

The monitoring, analysis and recording of personnel exposures is a vital function of LWR radiation protection programs. Personnel dosimetry devices should be subject to strict quality control measures and covered by a comprehensive calibration program. Personnel responsible for maintaining and analyzing primary dosimeters must be suitably trained and qualified in their areas of responsibility. The personnel radiation monitoring and bioassay program should maintain equipment and supplies necessary to support all facets of a LWR personnel dosimetry program. Radiological incidents requiring a dose assessment should be thoroughly investigated in a timely manner. Incident investigation results should be entered into the stations' corrective action program. Lessons-learned and contributing causes should be identified and corrective actions implemented to prevent recurrence of events. The whole-body counting program in conjunction with the radiological surveillance program is a crucial element in detecting unplanned uptakes of radioactive material. The personnel dosimetry program should be conducted and implemented in such a fashion as to provide a high level of confidence to radiation workers that their radiation exposures are accurately assessed, recorded and maintained.

## Bibliography


1. American National Standard ANSI/HPS N13.32, Performance Testing of Extremity Dosimeters, 1995
2. American National Standard ANSI/HPS N13.39, Design of Internal Dosimetry Programs, 2001
3. American National Standard ANSI/HPS N13.41, Criteria for Performing Multiple Dosimetry, 1997
4. Auman L., et al. An Intercomparison of Neutron Dosimeters and Detectors for In-Containment Dosimetry, Health Physics, 62(2):190–193, 1992
5. Clarke R., and Valentin J., A History of the International Commission on Radiological Protection, Health Physics, 88(5):407–422, 2005
6. Electric Power Research Institute, Assessment of the Effective Dose Equivalent for External Photon Radiation Volume 1: Calculational Results for Beam and Point Source Geometries, TR-101909, College Station, TX, 1993
7. Electric Power Research Institute, Assessment of the Effective Dose Equivalent for External Photon Radiation Volume 2: Calculational Techniques for Estimating External Effective Dose Equivalent from Dosimeter Readings, TR-101909, College Station, TX, 1995
8. Electric Power Research Institute, Criteria and Methods for Estimating External Effective Dose Equivalent from Personnel Monitoring Results, EDE implementation Guide, TR-109446. Palo Alto, Ac, 1998
9. Glickstein S., Analytical Modeling of Thermoluminescent Albedo Detectors for Neutron Dosimetry, Health Physics, 44(2):103–114, 1983
10. International Atomic Energy Agency, Safety Guide No. RS-G-1.2, Assessment of Occupational Exposure Due to Intakes of Radionuclides, Vienna, 1999
11. International Commission of Radiological Protection, 1994, Dose Coefficients for Intakes of Radionulcides by Workers, ICRP Publication 68, Pergamon Press, Oxford
12. International Commission of Radiological Protection, 2007 Recommendations of the International Commission on Radiological Protection, ICRP Publication 103, Pergamon Press, Oxford
13. International Commission on Radiological Protection, 1979, Limits for Intakes by Workers, ICRP Publication 30, Part 1, Pergamon Press, Oxford
14. International Commission on Radiological Protection, 1990 Recommendations of the International Commission on Radiological Protection, ICRP Publication 60, Pergamon Press, Oxford
15. International Commission on Radiological Protection, Annals of the ICRP 1977, ICRP Publication 26, Pergamon Press, Oxford
16. International Commission on Radiological Protection, Individual Monitoring for Intakes of Radionuclides by Workers: Design and Interpretation, ICRP Publication 54, Pergamon Press, New York, 1987
17. Leggett R., and Eckerman K., Dosimetric Significance of the ICRP's Updated Guidance and Models, 1989-2003, and Implications for U.S. Federal Guidance, Oak Ridge National Laboratory Report ORNL/TM-2003/207, Oak Ridge, Tennessee, 2003
18. National Council on Radiation Protection and Measurements, Deposition, Retention and Dosimetry of Inhaled Radioactive Substances, NCRP Report No. 125, Bethesda, MD, 1997
19. National Council on Radiation Protection and Measurements, Dose Control at Nuclear Power Plants, NCRP Report No. 120, Bethesda, MD, 1994
20. National Council on Radiation Protection and Measurements, Limitation of Exposure to Ionizing Radiation, NCRP Report No. 116, Bethesda, MD, 1993
21. National Council on Radiation Protection and Measurements, Use of Personnel Monitors to Estimate Effective Dose Equivalent and Effective Dose to Workers for External Exposure to Low-LET Radiation, NCRP Report No. 122, Bethesda, MD, 1995

22. National Institute of Standards and Testing, National Voluntary Laboratory Accreditation Program, NIST Handbook 150-2D, Calibration Laboratories, Technical Guide for Ionizing Radiation Measurements, 2004
23. National Institute of Standards and Testing, National Voluntary Laboratory Accreditation Program, NIST Handbook 150-4, Ionizing Radiation Dosimetry, 2005
24. Swaja R., and Sims C., Neutron Personnel Dosimetry Intercomparison Studies at the Oak Ridge National Laboratory, Health Physics, 55(3):549–564, 1988
25. U.S. Nuclear Regulatory Commission, Regulatory Guide 8.26, Applications of Bioassay for Fission and Activation Products, September 1980
26. U.S. Nuclear Regulatory Commission, Regulatory Guide 8.32, Criteria for Establishing a Tritium Bioassay Program, July 1988
27. U.S. Nuclear Regulatory Commission, Regulatory Guide 8.34, Monitoring Criteria and Methods to Calculate Occupational Radiation Doses, July 1992
28. U.S. Nuclear Regulatory Commission, Regulatory Guide 8.40, Methods for Measuring Effective Dose Equivalent from External Exposure, July 2010
29. U.S. Nuclear Regulatory Commission, Regulatory Guide 8.7, Instructions for Recording and Reporting Occupational Radiation Dose Data, Revision 2, November 2005
30. U.S. Nuclear Regulatory Commission, Regulatory Guide 8.9, Acceptable Concepts, Models, Equations, and Assumptions for a Bioassay Program, Revision 1, July 1993
31. U.S. Nuclear Regulatory Commission, NRC Regulatory Issue Summary 2003-04, Use of the Effective Dose Equivalent in Place of the Deep Dose Equivalent in Dose Assessments, February 2003
32. U.S. Nuclear Regulatory Commission, NRC Regulatory Issue Summary 2004-01, Method for Estimating Effective Dose Equivalent from External Radiation Sources Using Two Dosimeters, February 2004
33. Yigal S. Horowitz, Thermoluminescence and Thermoluminescent Dosimetry, Volume 1, CRC Press, Boca Raton, Florida, 1984

# Chapter 11
# Radiological Survey and Monitoring Instrumentation

## 11.1 Overview

The measurement of in plant radiation fields is one of the most important aspects of a LWR radiation protection program. Radiation areas must be identified in a timely manner, with both the extent and magnitude of radiation areas properly characterized. Accurate dose rate information must be obtained to assess pre-job dose estimates and to support dose mitigation efforts (e.g., the placement of temporary shielding). All these aspects are crucial in ensuring the radiological safety of employees and in minimizing unplanned radiation exposure events. To achieve these objectives an inventory of properly calibrated dose rate and contamination measurement survey instruments must be maintained. The inventory of dose rate instruments must be sufficient to allow the measurement of beta-gamma and neutron radiation levels over a measurement range that covers plant operational phases and potential plant excursions. Various detector types are incorporated into a wide range of portable and fixed survey instrumentation. Common types of detectors include gas filled ionization and Geiger-Mueller detectors, and proportional and scintillation detectors among others.

A wide range of radiological survey instrumentation is available to measure and evaluate radiation fields and contamination levels encountered at LWR facilities.[1] Survey instruments must be designed to function reliably in this environment and be properly maintained and calibrated to ensure the accuracy of radiological survey data. Radiological survey equipment required to support radiation protection program functions are presented in this chapter. A calibration and instrument functional performance check program is an essential element in ensuring the

---

[1] Specific instrument models discussed in this chapter and throughout this text are representative of instrumentation available to the industry and should not be considered as an endorsement. There are many excellent models of survey meters offered by well-established firms in addition to those described here. The description of specific models serves to present the range of features and the capabilities associated with radiological survey instruments commonly used in the LWR industry.

R. Prince, *Radiation Protection at Light Water Reactors*,                                        289
DOI: 10.1007/978-3-642-28388-8_11, © Springer-Verlag Berlin Heidelberg 2012

accuracy and reliability of survey instrumentation. The elements of a portable instrument calibration program are presented in general detail only. Description of the specific calibration methods and procedures associated with a given instrument model are beyond the scope of this text. Emphasis is placed on the applied operational aspects associated with the use of survey and monitoring equipment to ensure the operability of radiological survey instrumentation prior to use. Theory associated with the principles of detection for various detector types is provided in sufficient detail to describe the operational characteristics of a given instrument design. The purpose and operational capabilities of common survey instrumentation is presented. More detailed information relating to the technical aspects associated with the detection principles of a given detector type may be found in one of the many excellent references available, some of which are provided in the reference listing for this chapter.


## 11.2  Ionization Detector Radiation Survey Instruments


Numerous studies and reports have shown that the vast majority of the radiation source term derives from the presence of Co-60 and Co-58. This is a direct consequence of the fact that these two isotopes comprise the majority of the ex-core activation corrosion products. Therefore dose rate survey meters should have a well characterized response to gamma ray energies bounding these two radionuclides. Properly designed ionization type detectors, have a relatively flat energy response over a wide energy range. Ionization survey meters with accuracy of 10–20% over an energy range of 20–40 keV to 2 MeV are common. For the vast majority of purposes this accuracy is acceptable when performing in-plant radiation surveys. Depending on the detector volume and specific design features of the ionization detector some models may have a gamma response over a larger energy range.

Special interest should be given when performing surveys in areas of the plant where N-16 may be a significant contributor to ambient radiation fields. Due to the high-energy, 6-MeV gamma, emitted by N-16 an ionization chamber survey meter with a known energy response in this range should be used when performing surveys in areas where N-16 may be encountered. Areas inside the biological shield wall of PWR containment buildings, or within the loop rooms, and inside the drywell at BWR units while at power may have a significant N-16 component. Additionally, steam lines and associated piping and components at BWRs will contain N-16 due to the minimal decay time since transit from the reactor core. Instrument response to N-16 should be well characterized to ensure dose rates are not underestimated when N-16 is present in plant areas.

Portable hand-held ionization survey meters are typically air-filled detectors vented to atmosphere. Survey meters may be equipped with a cover or slide mechanism (so-called beta shield) that may be opened to expose the detector window. When the window is exposed the ionization detector is capable of detecting beta particles in addition to gamma photons. When the window is closed the detector allows measurement of the penetrating or "deep" dose comprised essentially of the gamma component. The housing surrounding the sensitive area of the detector is often constructed with 1,000 mg/cm$^2$ equivalent wall thickness to allow measurement of the deep dose. When the beta window is open an ionization detector with a window covering of 7 mg/cm$^2$ equivalent thickness allows for the measurement of shallow, or skin, dose. Window and housing thickness specifications are provided by the manufacturer and radiation protection personnel should have knowledge of these values when using a specific survey instrument to ensure proper interpretation of survey results.

Hand-held portable ionization survey meters should have the following capabilities and characteristics:

- Measure gamma and X-ray exposure rates
- Equipped with a beta window to measure beta dose rates
- Auto range switching or multiple measurement ranges
- Effective wall thickness equivalent to 1,000 mg/cm$^2$
- Effective beta window thickness of 7 mg/cm$^2$
- Flat energy response no greater than $\pm10$–30% over entire energy range
- Temperature compensated measurements
- Rugged and dependable when used under the environmental conditions encountered at LWRs
- Backlit display
- Long battery life

Ionization survey meters are often the meter of choice when setting dose rates due to the flat energy response over the range of interest at LWRs. Numerous ionization-type survey meters are commonly encountered in use at LWRs. Some of the available models are described below. This is by no means an exhaustive listing of ionization survey meters used in the industry. Additionally improvements and enhancements to the design of survey instruments are constantly being introduced by manufacturers and suppliers of these instruments. RP personnel should keep abreast of innovations relating to the design and capabilities of newly introduced survey instrumentation.

The RO-20 ionization survey meter (displayed in Fig. 11.1), offered by the Thermo Fisher Scientific Corporation, utilizes an air-filled ionization chamber detector. The instrument is capable of measuring beta-gamma dose rates to 500 mSv/h (50 R/h). As is common with most hand-held ionization survey meters the Mylar window thickness is approximately 7 mg/cm$^2$ while the detector housing and beta window cover have a thickness of approximately 1,000 mg/cm$^2$. The material comprising the beta window with an equivalent thickness of 7 mg/cm$^2$ results in a nominal thickness in the range of 20–30 μm. When obtaining survey measurements RP personnel should take precautions to protect the thin beta window from damage or punctures.

The Ludlum model 9-3 ion chamber survey meter (depicted in Fig. 11.2) is designed to measure beta–gamma radiation over a dose rate range up to 500 mSv/h

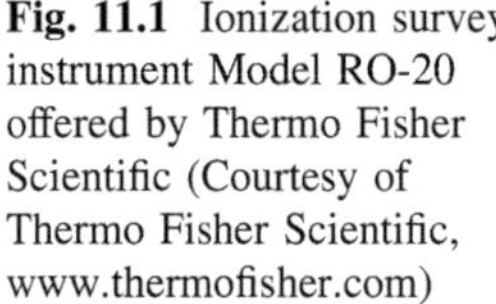

**Fig. 11.1** Ionization survey instrument Model RO-20 offered by Thermo Fisher Scientific (Courtesy of Thermo Fisher Scientific, www.thermofisher.com)

(50 R/h). The detector wall and beta window have equivalent thicknesses of 1,000 and 7 mg/cm$^2$ respectively. A five-decade selector switch provides convenient scale ranges that cover dose rate ranges commonly encountered at LWRs. Again this is an example of the typical ionization survey instrument offered by various vendors.

The available ionization survey meter models, in general, provide an energy response that is acceptable over the energy range of interest at LWRs, they are capable of measuring dose rates over the range that represents the vast majority of situations normally encountered (that may be appropriately surveyed with a hand-held instrument), and they are equipped with various operability status indicators, and other features.

Many instrument manufacturers now offer models equipped with digital readout displays versus analog scales. Other features may include auto-ranging displays, capability to display integrated dose in addition to dose rate, and various data logging functions. Survey meters equipped with computer interface capabilities facilitate the recording and acquisition of field data. Survey meters may be equipped with a USB port to store data on a thumb drive. Subsequent down loading of survey data to spread sheets or a trending program could be automated

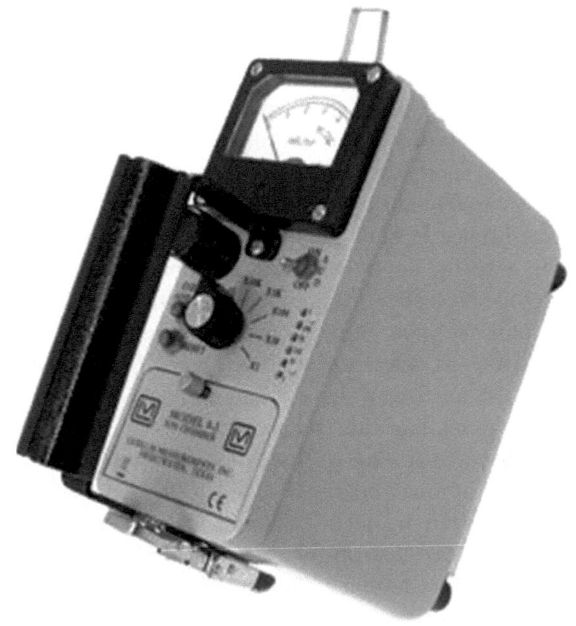

**Fig. 11.2** Ludlum Model 9-3 ionization survey meter (Courtesy of Ludlum Measurements, Inc. www.ludlums.com)

with these type survey meters. Such models often have self diagnostic features to verify operability and to ensure that appropriate acceptance criteria are achieved during the calibration process.

## 11.3 Geiger–Mueller Detector Survey Instruments

Geiger–Mueller (G–M) detectors are also gas filled detectors that provide a proven and rugged detector design that has been incorporated into many types of portable survey meters used in the LWR industry. A distinct benefit of the G–M detector is the large size of the output pulse generated by the detector. The large output pulse greatly simplifies the electronics required to process and analyze the output signal from G–M detectors. This feature makes G–M detectors highly suited for use in hand-held survey meters and portable radiation detection equipment.

Regardless of the initiating ionization event within the detector a G–M detector will produce the same size output pulse. Consequently G–M detectors are not capable of providing energy discrimination capability. Certain detector designs are available that provide limited capability to distinguish between gamma and beta radiation levels. This feature is achieved by placing the G–M detector within a shielded probe that incorporates a design that allows the shield to be rotated or opened to expose the detector to the radiation field. These designs offer a gross estimate of the beta component that may be present in the radiation field. The "closed" window reading measures the gamma radiation dose rate while the "open" window reading provides an estimate of the total beta-gamma radiation level. By subtracting the closed reading (gamma only) from the open reading (gamma plus beta) an approximation of the beta component may be obtained. A similar process is used utilizing the open window and closed window readings for ionization survey meters. Even though the beta dose rate component may not be accurately quantified, this design feature allows RP personnel to determine the significance of beta radiation levels. Based on survey results precautions such as the need for finger rings or additional protective clothing may be prescribed to provide protection from the beta radiation. The main advantages of G–M detectors include the high output signal, their rugged design, and relatively low cost that makes G–M detectors the ideal choice for many survey instrument applications.

G–M hand-held survey meters may be utilized for both dose rate and radio-active contamination measurements. G–M detector response is more energy dependent than that of ionization detectors. G–M detectors will inherently over respond at lower energy levels. Unshielded or uncompensated G–M detectors are characterized by significant changes in detection sensitivity for photons with energies less than 100 keV. However, many G–M survey meters commonly available are designed to provide a relatively flat energy response over a large energy range. To flatten out the energy response of a G–M detector various shields may be used to enclose the detector to minimize over response of G–M detectors at the low energy range. This feature essentially causes the G–M detector to under respond to low energy gammas thus compensating for the inherent over response at low energies. Additionally, unique design features can provide a relatively flat energy response to photons in the range of a 100–200 keV up to about 2 MeV. These design features result in what is commonly referred to as energy "compensated" G–M detectors. When selecting a G–M survey meter the manu-facturer's energy response characteristics and specifications should be consulted to ensure that a given model is suitable for its intended use.

The various designs of G–M survey instruments include dose rate meters with internal G–M detectors, G–M detectors coupled with a rate meter for contamination monitoring and instruments with both internal detectors and the option to connect with various sizes and designs of external G–M probes. Figure 11.3a depicts the Ludlum Model 14C G–M survey meter that utilizes an internal G–M detector and allows the user to switch to an externally mounted probe. The Mirion Technologies RDS-30 hand-held dose rate meter shown in Fig. 11.3b contains an energy com-pensated G–M detector. A typical G–M external probe, Ludlum Model 44-6, is depicted in Fig. 11.3c. Note the window area in the 44-6 probe, which may be rotated to expose the G–M detector to measure beta-gamma radiation. This feature is common in many of the external G–M probes available to the industry. These are just a few examples depicting the variety of G–M survey meters and external probe designs available, and an indication of the versatility of uses for G–M detectors.

In order to provide a large range of dose rate measurements with a specific meter, survey meters may include two or more energy-compensated G–M detectors. The incorporation of two or more G–M detectors in a survey meter extends the useful

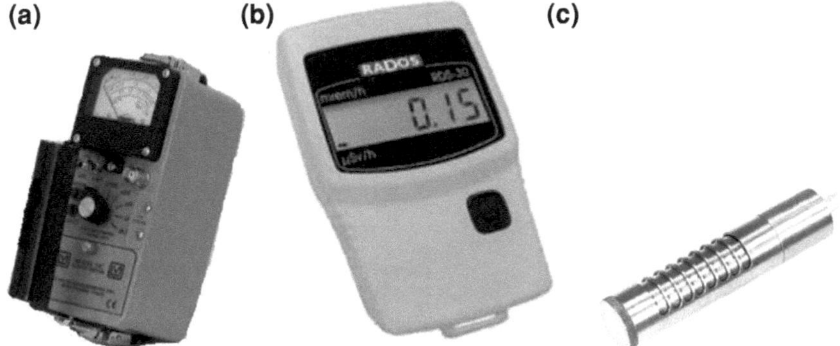



**Fig. 11.3**  **a** Ludlum Model 14C G–M survey meter. **c** Ludlum Model 44-6 G–M probe (Courtesy of Ludlum Measurements, Inc. www.ludlums.com). **b** Mirion Technologies RDS-30 G–M dose rate meter (Courtesy of Mirion Technologies www.mirion.com)

range of the dose rate levels that can be accurately measured. The larger of the GM tubes is designed to respond to lower dose rates while the smaller GM tube (with a smaller active detector volume) measures higher dose rates.

Due to their relatively simple design and rugged features G–M detectors have been incorporated into survey meters equipped with a telescoping mechanism. The telescoping extensions range up to approximately 4 meters on some models. These designs are capable of measuring gamma dose rates up to 10 Sv/h (1,000 R/h). This feature allows the individual to remain at a distance while obtaining survey measurements. These detector designs are ideal when performing surveys in high radiation areas to take advantage of the "distance" principle to lower the exposure received by the individual while performing the survey. Additionally the telescoping feature allows surveys to be taken in overhead areas, through wall penetrations and at other difficult to reach locations.

Two popular designs include the model 6112 M "Teletector" offered by Automess of Germany and the Model 78 "Stretch Scope" offered by Ludlum Measurements of the USA. Both of these models have an upper measurement range of 10 Sv/h (1,000 rem/h). The models are used for measuring gamma radiation levels and are equipped with two energy-compensated G–M detectors. Figure 11.4 depicts the Teletector models 6150 and 6112 M offered by Automess. These models measure gamma radiation and "detect" beta radiation. Various dose rate meters are used in conjunction with the 6150 unit. The dose rate meters are mounted on the probe housing as depicted in Fig. 11.4. The rate meter allows for the simultaneous measurement of such parameters as dose rate, dose rate mean value and the maximum dose rate. These models are equipped with an auto-ranging feature thus eliminating the need for a selector switch. The dual compensated G–M detectors provide an effective energy range of 65 keV–3 MeV. The liquid crystal display for the 6112 M allows the user to select functions from a

**Fig. 11.4** The Automess Teletector Model 6150 shown with an attached dose rate meter and a close-up of the model 6112 M display (Courtesy of Automess, www.automess.de)

menu. These display features include various operability functions. The unit is also programmed with three languages (English, German, and French). This language program feature is common on various monitoring equipment such as the PCM units described in Chap. 6.

The Ludlum Model 78 Stretch Scope is similar in design and capability as the Teletector and is shown is Fig. 11.5. A liquid crystal display model provides messages related to the operational status of the instrument.


## 11.4 Neutron Radiation Survey Instruments

Neutron radiation fields are encountered in those plant areas in the vicinity of the reactor vessel while the unit is at power. Neutron radiation levels are directly proportional to reactor power, increasing as the number of fissions within the core increase. Intense neutron fields are limited to those areas inside the biological shield wall and at pipe penetrations with a direct line of sight to the reactor vessel. Access to these areas is strictly limited during periods of power operation. However, access to certain areas of the containment building and reactor building may result in exposure to neutron radiation fields. Consequently the need will arise to perform neutron radiation surveys to support plant operations.

Since neutrons cause ionization indirectly the measurement of neutron radiation dose rates requires an intermediate neutron interaction to produce an ionization event to yield an electrical pulse that can be analyzed and processed. A common type of neutron survey meter consists of a hydrogenous moderator material to slow down neutrons to thermal energies to take advantage of slow neutron interactions. By enclosing a radiation detector within a uniquely designed spherical moderator, neutron radiation survey meters may be designed to be tissue-equivalent to allow neutron dose rates to be measured in Sv/h. Unfortunately this results in neutron survey meters weighing up to several kilograms.

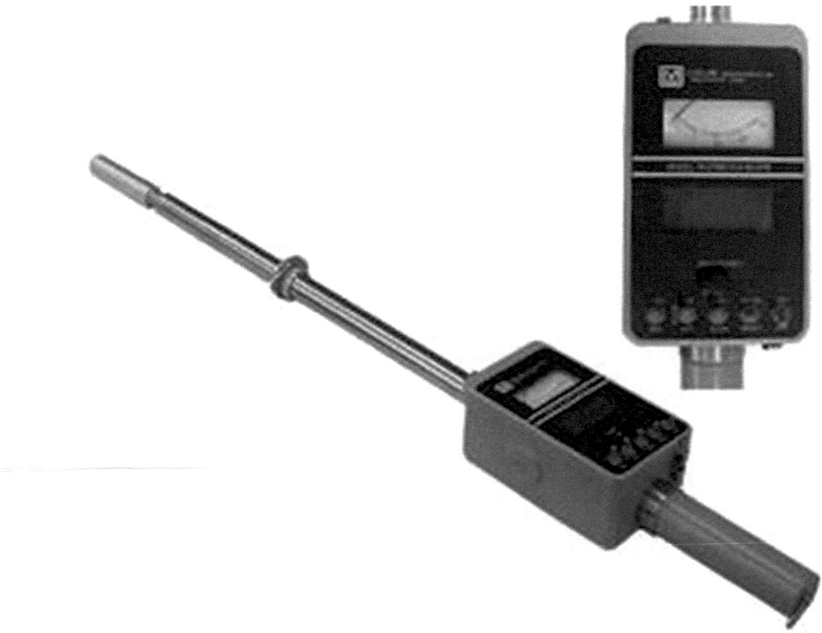

**Fig. 11.5** Ludlum stretch scope dose rate survey instrument (Courtesy of Ludlum Measurements, Inc. www.ludlums.com)

A popular detector type used to measure the ionization produced via various neutron interactions is the proportional detector. Portable neutron dose rate survey instruments are usually equipped with boron trifluoride ($BF_3$) or helium-3 ($^3He$) gas-filled proportional detectors. The boron n-$\alpha$ reaction results in the production of a highly ionizing particle. Absorption of a neutron with a helium-3 nucleus causes the prompt emission of a proton. The slow or thermal neutron reactions with these filling gasses are as follows:

$$^{10}B + {}^1n \rightarrow {}^7Li + {}^4\alpha$$
$$^{3}He + {}^1n \rightarrow {}^3H + {}^1p$$

The resulting charged particles produce ionization in the filling gas. The ionized species are collected and the subsequent electrical pulse processed by the electronic circuitry of the survey meter. Thus the interaction characteristics of neutrons are utilized to produce ionizing particles which are then collected and measured, incorporating the same detection principles as those employed with directly ionizing radiations.

Significant neutron radiation fields are encountered in conjunction with gamma radiation in LWR environments. Oftentimes the gamma radiation dose rates may be orders of magnitude greater than those associated with neutron dose rates. Therefore it may be assumed that whenever neutron surveys are conducted the

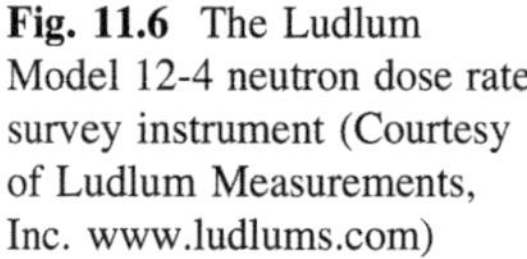

**Fig. 11.6** The Ludlum
Model 12-4 neutron dose rate
survey instrument (Courtesy
of Ludlum Measurements,
Inc. www.ludlums.com)

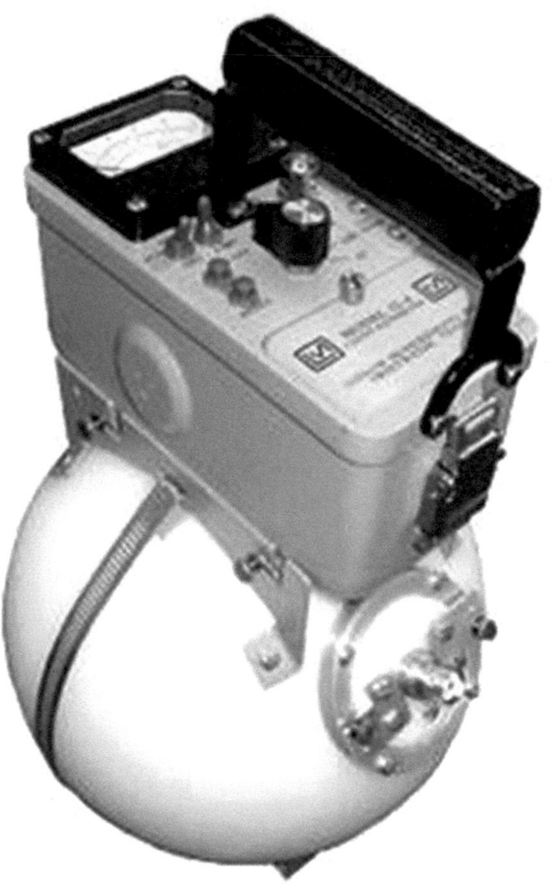

neutron dose rate survey meter will also be exposed to gamma radiation fields. The amount of secondary charge produced by a proportional detector is predicated on the amount of primary ions initially produced. Consequently the electronic circuitry of neutron survey meters is designed to discriminate the lower pulse-size signals produced by the gamma component from the alpha or proton-produced signals. Due to the significant difference between these pulse sizes neutron dose rate measurements may be obtained even in the presence of high gamma radiation fields.

The Ludlum model 12-4 neutron survey meter (depicted in Fig. 11.6) has a measurement range of 0–100 mSv/h (0–10 rem/h). This model utilizes a $^3$He detector enclosed within a cadmium loaded polyethylene sphere approximately 23 cm in diameter. The design offers gamma background rejection to allow use in high gamma radiation fields typically encountered in those plant areas with significant neutron radiation fields.

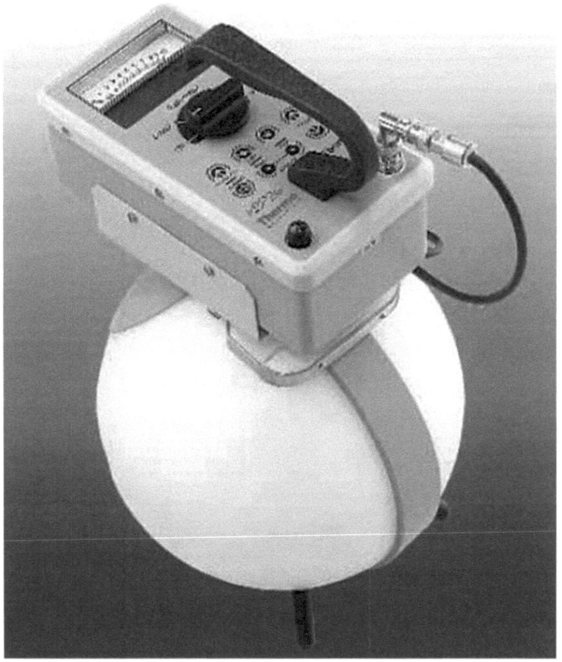

**Fig. 11.7** The Thermo Fisher Scientific neutron survey meter Model ASP-2e/NRD (Courtesy of Thermo Fisher Scientific, www.thermofisher.com)

The Thermo Fisher Scientific ASP-2e/NRD neutron survey meter (depicted in Fig. 11.7) utilizes a $BF_3$ detector and is similar to the Ludlum Model 12-4 with regards to neutron energy range, gamma rejection, and design of the cadmium loaded polyethylene sphere to moderate fast neutrons.

Canberra also offers a portable neutron survey meter with a distinct weight advantage. The Dineutron survey meter is equipped with two $^3$He detectors that are placed within two different diameter moderating spheres. The two spheres are approximately 6.3 and 10.7 cm in diameter respectively, considerably smaller than the 23 cm diameter spheres associated with the common neutron "rem" ball meters. This results in a weight for the Dineutron survey instrument of 3.5 kg. This model is capable of measuring neutron dose rates as high as 100 mSv/h (10 rem/h) and the two moderating spheres allows the quality factor associated with the detected neutron energies to be determined and displayed. The energy response covers a useful range from thermal to 15 MeV neutrons. Figure 11.8 depicts the Dineutron survey meter.

As noted above the need to perform neutron radiation surveys will be primarily limited to those occasions requiring entry inside the biological shield wall of containment buildings, drywell (if assessable) or various areas of the reactor building depending upon the design of the BWR unit. Oftentimes neutron radiation surveys coupled with stay times are utilized to estimate worker exposures. Since neutron dose results are usually not available until the primary or neutron-issued dosimeter is read a method to estimate neutron exposures received in the interim

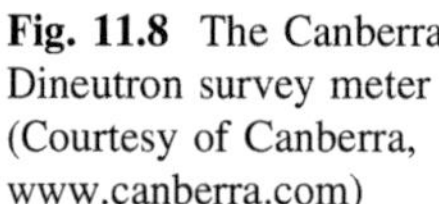

**Fig. 11.8** The Canberra
Dineutron survey meter
(Courtesy of Canberra,
www.canberra.com)

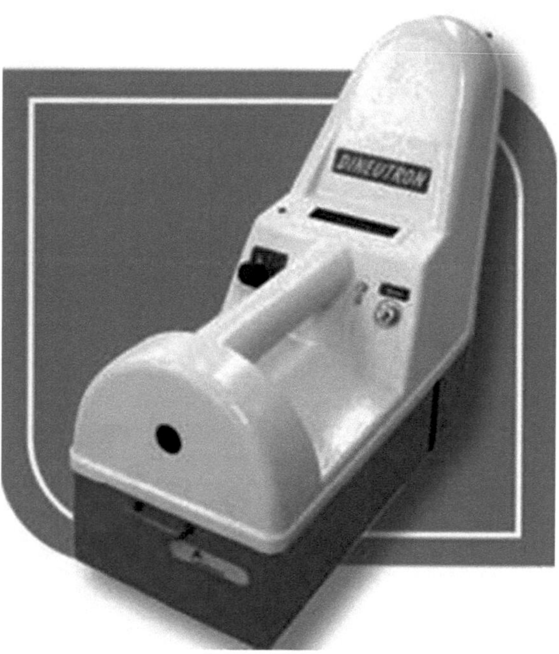


period must be established. If stay times are utilized for this purpose then it is
essential to ensure that neutron radiation surveys are performed during these type
entries. Dose results based on stay times should be entered into the exposure data
base to ensure that an individual's dose is maintained current.

## 11.5 Contamination Survey Instruments

Contamination survey meters are the other major category of survey instrumentation necessary to support radiological survey functions. The measurement of
fixed and transferrable levels of contamination constitutes an important facet of the
LWR radiation protection surveillance program. Contamination survey meters are
utilized to perform such functions as direct surveys on equipment and components,
floor areas, and personnel. Oftentimes the amount of fixed contamination present
on floor areas or components must be evaluated in order to assess the potential for
generating airborne contamination or spreading surface contamination during the
performance of tasks that may dislodge contamination. Welding and grinding on
valve seats with fixed contamination may pose airborne contamination concerns,
for example. Consequently, a suitable inventory of portable, hand-held survey
meters should be available to support radiological survey functions.

**Fig. 11.9** Ludlum Model 177 rate meter and the Ludlum Model 44-9 G–M pancake probe (Courtesy of Ludlum Measurements, Inc. www.ludlums.com)

A typical contamination survey meter consists of a flat detector probe coupled to a rate meter. These units are often referred to as "friskers" since their design facilitates their use as a scanning device used to "frisk" personnel and monitor equipment for the presence of contamination. The use of friskers within the RCA for personnel contamination monitoring purposes was discussed in Chap. 6. These frisker units are often stationed at strategic locations throughout the RCA for use by plant personnel. The availability of frisker stations provides a means to detect the presence of contamination on workers to minimize the inadvertent spread of contamination. Individuals exiting a contaminated work area may proceed to the nearest frisker and monitor themselves for the presence of contamination versus proceeding to the RCA exit point PCMs. The benefit of detecting the presence of contamination on workers at the earliest possible opportunity allows corrective measures to be implemented in a timely manner, minimizes personnel exposures, and prevents the spread of contamination to clean areas of the RCA.

Frisker units are typically equipped with an open window G–M detector. The "pancake" detector design has gained widespread acceptance within the industry. This design provides a relatively large detector surface area and is convenient to use. Figure 11.9 depicts the Ludlum Model 177 rate meter along with the Model 44-9 pancake type G–M detector probe. The active area of the probe is approximately 15 cm$^2$. Thermo Fisher Scientific offers lead or tungsten shielded G–M pancake probes that improve the overall detection capability of friskers when background radiation levels may be of concern. Thermo Fisher Scientifics' Model HP-210 series hand probes are depicted in Fig. 11.10 along with the Model RM-25 rate meter. These rate meters (i.e., the model Thermo Fisher RM-25 and Ludlum 177) coupled with the appropriate G–M probe are ideally suited for detecting the presence of beta-gamma contamination on personnel and equipment. Both rate meters are equipped with an internal rechargeable battery to facilitate their use in the field without the need to be connected to a power supply. Additionally both units may be used with other detector probes thus increasing their versatility.

Personnel may also utilize frisker stations to monitor for the presence of contamination on hand tools, personnel items and whenever the presence of

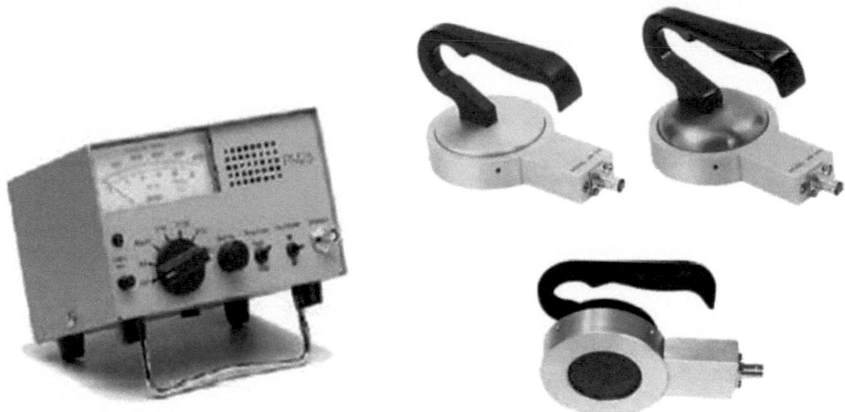

**Fig. 11.10** The Thermo Fischer Scientific RM-25 rate meter and the HP-210 probe and the tungsten shielded HP-210 probe (Courtesy of Thermo Fisher Scientific, www.thermofisher.com)

contamination is suspect while performing duties within the RCA. Depending on ambient radiation levels present at the frisker location, shielded probes (such as the HP-210 pancake design) may be required to provide an acceptable limit of detection. Frisker stations that are utilized primarily for personnel contamination monitoring purposes may require a more elaborate means to shield against ambient background radiation levels. Enclosures erected around frisker stations that allow workers to step inside a shielded area may be necessary in order to reduce background radiation to a level suitable for monitoring personnel for contamination. Examples of shielded frisking enclosures are depicted in Fig. 6.17.

Oftentimes smears and large area swipes may be evaluated by means of a portable contamination survey meter. Frisker units described above are often used for this purpose to "screen" smears in the field. However, quantitative surveys may require the use of more elaborate survey equipment. These counting systems are described later in this chapter.

Instruments used to evaluate alpha contamination typically employ scintillation or proportional detectors. The window thickness associated with G–M detectors prevents the detection of alpha particles or at the very least offers very low detection efficiency. The vast majority of alpha particles will be absorbed by the windows of G–M detectors. Scintillation detectors equipped with a very thin Mylar covering and windowless proportional detectors are more suitable for the detection of alpha particles.

Scintillation detectors are probably the most common detector type utilized in portable alpha survey equipment. The primary disadvantage of utilizing windowless proportional detectors as a portable survey meter is associated with the need to provide a supply of proportional counting gas with the instrument. A common detector used to monitor for the presence of alpha contamination

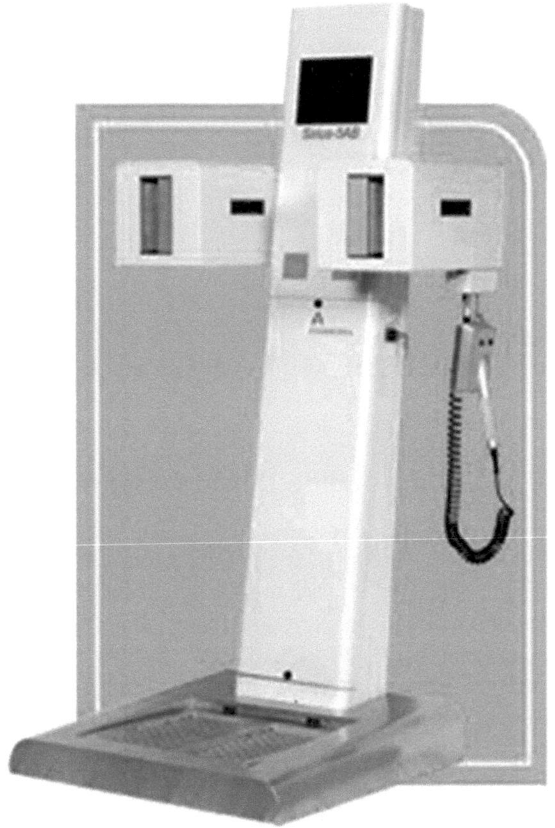

**Fig. 11.11** The Canberra
Sirius-5™ hand-and-foot
monitor with optional frisker
unit (Courtesy of Canberra,
www.canberra.com)

incorporates a ZnS(Ag) scintillation probe. These detectors are reasonably rugged and can be used with a variety of rate meters.

Alpha contamination survey meters consist of an appropriate probe coupled with a rate meter. Alpha probes of various sizes are available with the most common being those probes with a surface area of 100 cm$^2$. Some PCM and hand-and-foot monitors are equipped with an attached hand-held probe that is used to monitor for the presence of contamination on workers. These probes may be dual purpose alpha and beta detectors and may have an active detector window as large as 500 cm$^2$. Figure 11.11 depicts a hand-and-foot monitor equipped with an attached frisker. This model is available with either gas flow proportional detectors or plastic scintillation detectors.

A common alpha probe design is shown is Fig. 11.12. The detector is a ZnS(Ag) plastic scintillator with a surface area of 100 cm$^2$. The protective screen covering results in an active surface area of approximately 89 cm$^2$. A precaution when performing surveys with alpha scintillation detectors is to protect the thin Mylar window from damage. Damage to the Mylar covering could expose the

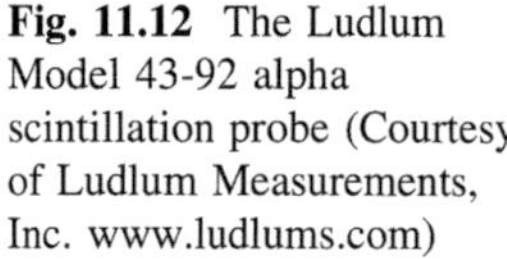

**Fig. 11.12** The Ludlum
Model 43-92 alpha
scintillation probe (Courtesy
of Ludlum Measurements,
Inc. www.ludlums.com)

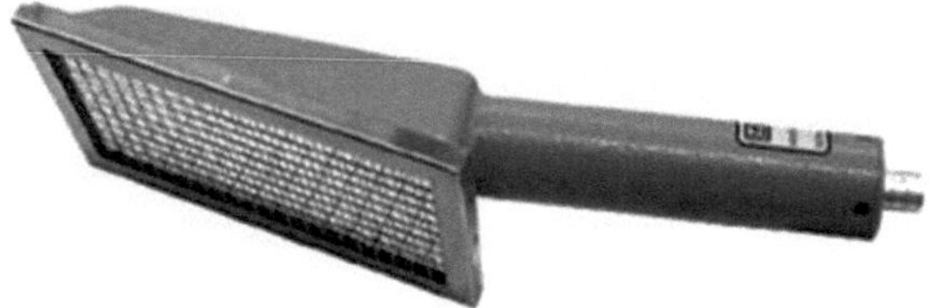

scintillation detector to light resulting in high readings. This detector and similar designs are equipped with "overload" protection features that provide indication of punctures in the Mylar detector covering.

## 11.6 Sodium Iodide Detector Survey Instruments

Sodium iodide (NaI) detectors are often used for portable survey meters. These solid crystal detectors are more efficient for detecting gamma rays versus gas-filled ionization and G–M survey meters. A range of crystal sizes are available and NaI is reasonably rugged to withstand the environmental conditions encountered at LWRs. Radiation levels of interest within the RCA area are easily measured by G–M and ionization survey meters. NaI-equipped survey meters are ideally suited for performing gamma radiation surveys in low ambient dose rate areas, taking advantage of the better detection sensitivity offered by these detectors. Hand-held NaI detector survey meters are available that measure essentially background radiation levels and small differences in background radiation fields. Survey meters capable of measuring gamma radiation levels in the nSv/h (or µR/h) range are common. This detection sensitivity makes NaI survey meters the preferred choice when performing radiation surveys in those plant locations where radioactive material is not expected to be present. Routine surveys may be performed in such locations as material and equipment storage warehouses, clean-side trash collection locations, and general area surveys conducted outside of the RCA but within the plant protected area, for example. The purpose of these type surveys is to provide added assurance that radioactive material has not been inadvertently released from the RCA. The ability to detect elevated radiation levels a few nSv/h above normal background radiation levels could indicate the presence of radioactive material in unwarranted locations.

The Ludlum Model 19 (depicted in Fig. 11.13) referred to as a µR meter, indicating the radiation levels that the survey instrument is capable of measuring, is equipped with a 2.5 cm diameter NaI crystal that is 2.5 cm thick. The survey meter has a range of 0–50 µSv/h (0–5,000 µR/h). This and similar model µR meters are ideal for performing surveys in plant areas where the presence of radioactive material is not suspected.

**Fig. 11.13** The Ludlum
Model 19 μR survey meter
(Courtesy of Ludlum
Measurements, Inc.
www.ludlums.com)

## 11.7  Instrument Source Response Checks

Survey meters should be verified to be operational prior to performing surveys in the field. Operational source response checks consist of verifying that a detector responds to radiation. A more comprehensive check may be performed on a routine basis, such as weekly. These checks may utilize a uniquely designed source holder or configuration to reproduce a known response range for a specific survey instrument model. Upon completion of calibration an instrument response should be verified against the check source that will be utilized to confirm that a given instrument is within an acceptable response range. These routine instrument response checks verify that an instruments' response remains within acceptable limits during a calibration cycle. An instrument that fails a response check should be taken out of service and recalibrated. The need to perform elaborate routine instrument response checks or operability performance checks may be determined by the instrument design and available features. Many survey instruments available today incorporate self diagnostic capabilities and other features that verify operability status and provide warning in the event that the instrument performance falls outside established operability limits.

In the event that an instrument fails to pass a response check certain actions may be required beyond just removing the instrument from service. Procedures should require an evaluation concerning the recent use of the survey meter. For instance, if the survey meter was recently used for a pre-job survey in order to estimate preliminary dose estimates it may be necessary to re-confirm dose rate data. Routine surveys performed with the instrument to confirm radiological postings within the plant may need to be validated. Obviously the rigor and depth of such a review would be based on the extent of the instrument response check failure. If the meter readings fell outside the acceptance range by a small percentage on only one scale or the meter failed in a conservative direction (i.e., readings were higher than the acceptance range) then no further action may be necessary. On the other hand if an instrument failed the performance test by a wide margin on multiple scales then immediate corrective measures may be warranted. The key point is to ensure that instrument response check failures are evaluated in a timely manner to ensure that no radiological safety issues have gone unnoticed. Survey meters with self-diagnostic features that provide fault indicators when an instrument performance parameter is out-of-tolerance minimize the chances of unknowingly using a faulty instrument in-the-field.

Personnel contamination monitors and in-plant frisker stations should also be incorporated into the instrument response check program. Due to the important function of PCMs located at the RCA exit point these units should undergo an extensive routine source response check program. Daily source checks or bi-weekly source checks may be appropriate. The frequency of source checks may be predicated on the self-diagnostic features available with various PCM models. Units that perform self-diagnostic checks continuously and go into a "fault" mode upon detection of a problem may require less frequent response checks. Frisker stations and PCMs located within the RCA may be subject to less frequent, and perhaps not as elaborate source response checks, as those performed on the RCA exit PCM units.

## 11.8 Laboratory Counting Equipment

A wide range of radio-analytical counting equipment is required to support the radiological monitoring and radiation protection surveillance program. Though this text has focused on the operational aspects of a LWR radiation protection program; sampling activities associated with the stations' environmental and effluent monitoring program, and radiochemistry and radioactive waste processing programs require specific radionuclide analyzes on a routine basis. These program requirements often determine the type and range of laboratory analysis equipment required to support daily plant operations. Several approaches may be taken regarding the establishment and daily operation of the radio-analytical laboratory. The specific approach often hinges on the site-specific organization structure. Environmental and effluent monitoring may be a stand-alone organization with a significant portion of the environmental sample analyzes performed offsite or contracted to a firm that provides these services. Radiochemistry analyzes require the use of specialized laboratory equipment such as gamma spectroscopy systems, liquid scintillation counters and low-background alpha–beta counting systems.

The costs associated with the acquisition and maintenance of these and other specialized radio-analytical systems may preclude the establishment of multiple laboratories for chemistry and radiation protection. It may be more efficient to establish one centralized laboratory equipped with all the necessary counting systems and equipment to support RP and other program areas.

For purposes of this text it will be assumed that a centralized laboratory equipped with gamma spectroscopy systems and liquid scintillation counters has been established that supports RP sample analysis requirements. A common practice, at least for plants in the USA, is for the chemistry department to have primary responsibility for the radio-analytical laboratory. There are some practical advantages to this approach. The chemistry sampling and analysis work load is typically highest while the unit is operating while the RP sampling and analysis workload reaches a peak during plant outages and other periods when the unit is off-line. Consequently, the workload of one centralized laboratory would be complimentary to the needs of these two program areas. Therefore RP samples' requiring specific radionuclide analyzes such as air samples will be processed by the centralized laboratory, including tritium analyzes. It will be assumed that the maintenance of multi-channel analyzer systems (e.g., GeLi, HPGe, and NaI), liquid scintillation counting systems, and other specialized radio-analytical counting equipment is maintained by a separate organization that provides support to the operational RP program.

Notwithstanding the above considerations, specific laboratory counting equipment will be required to be maintained by the RP department to support radiological surveillance activities. This equipment will consists of automated smear counters for the analysis of gross activity on air samples and for contamination swipe surveys. Additionally, multiple counting stations to analyze single air samples and smears should also be available.

Routine contamination surveys that are performed daily may result in the collection of a large number of smears requiring analysis. It would be tedious to analyze a large number of smears individually. Depending upon the location and nature of a routine contamination survey some of these surveys may require a more quantitative analysis then what can be obtained by a frisker unit, for example. Routine contamination surveys in clean areas of the RCA should be analyzed to ensure that contamination levels are below the established values used to demarcate contaminated areas. Additionally, numerous general area air monitoring stations equipped with a fixed-head air sample may be established throughout the RCA. The routine collection and analysis of these air samples require a counting system capable of measuring gross airborne radioactivity concentrations to levels not typically achieved by a simple scaler-detector counting system. An automated system capable of handling numerous samples with an automatic sample changer would be useful for these purposes.

Automatic sample changer systems utilized to count smears and air samples often use thin window gas-flow proportional detectors. The use of the thin window and specific design of these systems also allows for the simultaneous detection and analysis of both alpha and beta contamination levels. Even though alpha contamination may

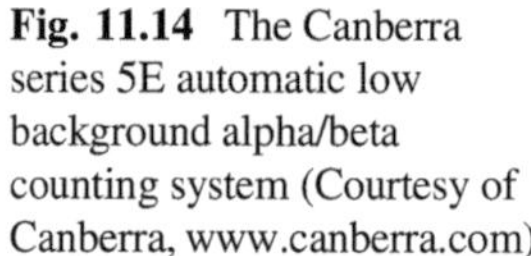

**Fig. 11.14** The Canberra series 5E automatic low background alpha/beta counting system (Courtesy of Canberra, www.canberra.com)

not be present alpha/beta counting systems can be useful in the early identification of the presence of alpha contamination.

The Canberra Series 5E automatic counting system (displayed in Fig. 11.14) is one such system capable of automatically counting smear samples. The system has a capacity of 100 samples. Smears (or air sample filters) are placed onto planchets which are loaded into the automatic sample chamber. Unique serial numbers are transcribed on each planchet holder in order to relate results to a given sample number. A sample run could be loaded with a calibration source and blanks to perform QC checks for each batch of samples. Self diagnostic software monitors such parameters as gas pressure and flow, detector voltage and other critical parameters. In the event that a parameter falls outside established acceptance criteria a fault indication is provided.

These automatic counting systems allow samples to be analyzed for extended periods (e.g., air samples) without the need for an attendant. The ability to load a

**Fig. 11.15** The Ludlum Model 2000 scaler commonly used as a bench-top counting station (Courtesy of Ludlum Measurements, Inc. www.ludlums.com)

large number of air samples, for example, and count each sample for an extended period without the need to manually change each sample is a convenient feature. Software and various other features provide the operator with the ability to customize printed reports. Reports are printed automatically and may be formatted to meet specific needs of the user. Sample data is retained in the system memory until cleared by the operator. Again as with many of the various instrument models and analytical equipment described in this chapter, the incorporation and availability of computer interface features provided with radiation monitoring and analysis equipment greatly facilitates the handling and processing of data.

The primary RP count room may also have several bench-top counting stations. These stations may be used to analyze smears taken in support of free-release surveys. Items or equipment that is requested to be free-released must be surveyed to verify that both direct and transferable contamination levels are below free release limits. These surveys usually require a more quantitative analysis to confirm transferable contamination levels. A simple scaler coupled to a suitable detector is ideally suited for these type surveys. The scaler should have adjustable controls to set a pre-set count time and a high voltage adjustment to allow different detector types to be used. Figure 11.15 depicts the Ludlum Model 2000 scaler.

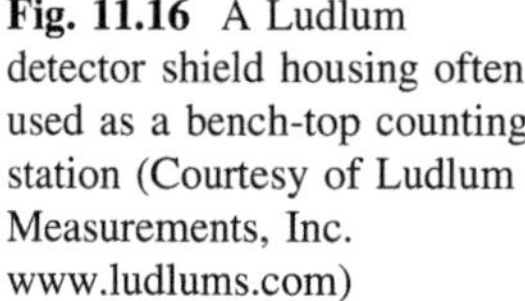

**Fig. 11.16** A Ludlum detector shield housing often used as a bench-top counting station (Courtesy of Ludlum Measurements, Inc. www.ludlums.com)

The scaler may be used with G–M, proportional or scintillation detectors. The scaler may be connected to a personal computer or a printer for direct output of counting results.

Depending on the background in the RP count room scalers may be used in conjunction with a smear holder or a shielded housing as necessary to reduce background radiation levels.

The use of a NaI detector with a scaler may be used as a stand-alone counting system. Figure 11.16 depicts a detector shield housing a NaI detector. These detector shield combinations are often used with a bench top scaler station. Note that the NaI detector is enclosed within the shield to reduce background radiation levels. Detector shield arrangements may consist of a simple arrangement of lead bricks surrounding a detector or the use of pre-fabricated shields such as the one depicted in Fig. 11.16. The shield in this figure weighs approximately 118 kgs. The Ludlum model 2000 scaler (Fig. 11.15) in combination with various detectors is often used as a general purpose counting system.

Temporary counting facilities are often required to support RP field activities. During outages and maintenance periods work stations, more conveniently located near job sites are often utilized. The capability to screen air samples and smears may be required to support rapid turnaround of field radiological survey data to support work activities. The availability of local counting stations eliminates the time required to take samples back to the primary radio-analytical laboratory or RP lab for analysis. Field surveys may require nothing more than a quick confirmation of the magnitude of contamination levels or to verify the effectiveness of contamination control measures or the presence of contamination, for example. While air samples may be screened to verify that gross beta-gamma airborne activity concentrations are less than established administrative control levels. Results may be more qualitative then quantitative under these circumstances.

Local counting stations may consist simply of a rate meter or a scaler coupled to a suitable radiation detector. These systems may be similar to the bench-top counting stations utilized in the primary RP count room. Depending upon the location of the local counting station, ambient radiation levels may require the use of shielded detector housings. These temporary counting stations are commonly established in conjunction with the secondary RP control points discussed in Chap. 6. Detectors ideally suited for use with a scaler for in-plant counting stations include GM and NaI scintillation detectors. These detectors provide good performance in plant areas that may be subject to high temperatures and less than ideal field conditions encountered inside containment and reactor buildings during outage periods.

## 11.9  Area Radiation Monitors

For purposes of this text area radiation monitors (ARMs) consist of two distinct systems; the plant installed or permanent area radiation monitors and portable units utilized to provide monitoring coverage on a short-term basis. Though the operability and maintenance of the plant-installed ARM system may not be the responsibility of the radiation protection group, RP personnel will most likely be required to respond to ARM alarms and therefore should be intimately aware of the location and overall purpose and function of the various ARM channels available at their facility. Since the plant-installed ARM system is not intended to provide total coverage of all plant areas the need may arise from time-to-time to provide temporary local ARM units in support of maintenance or outage-related activities. This section provides a general overview of the purpose and function of the plant-installed ARM system and describes those plant locations that are typically included within the scope of such a system. The use of portable ARMs for job-coverage purposes is also presented.

### *11.9.1  Installed Area Radiation Monitors*

The in-plant area radiation monitoring system is an integral component of the overall protection systems associated with the operation of a nuclear power plant. A properly calibrated and functioning radiation monitoring system can provide the initial warning or indication of elevated radiation levels within various areas of the plant. The installed radiation monitoring system consists of general area, gamma radiation detectors referred to as ARMs. These monitors play a vital role in monitoring ambient radiation levels in plant areas. The extent and location of installed ARM channels differs among plants. These differences, in addition to the obvious difference pertaining to PWRs and BWRs, may be attributable to the size and age of the plant, plant layout and unique design features, manufacturer or supplier of the original monitoring system, and managements' approach or philosophy pertaining to the use and purpose of an in-plant radiation monitoring system among others. It is essential that RP personnel have in-depth knowledge of the monitoring channels provided, including their alarm functions, pathway or location monitored, readout capabilities and sensor or detector locations.

Regardless of the number or types of ARM channels available at a particular LWR facility; the system serves essentially the same purpose and functions. The installed radiation monitoring system serves to protect plant personnel from unnecessary radiation exposure and to initiate plant actions in a timely manner to minimize the radiological consequences stemming from malfunctioning or damaged equipment or components. The plant radiation monitoring system also serves to minimize radiological impacts to the general public and the environment. Monitors are normally equipped with adjustable alarm thresholds which, when exceeded activate an audible and perhaps a visual alarm. In the event of an alarm personnel should be instructed to evacuate the affected areas to prevent or minimize unnecessary personnel exposures.

Due to the complexity of a nuclear power station, it is not possible or even feasible to continuously monitor all plant locations. However, the simplicity and advances in remote monitoring and telemetry systems has greatly expanded the number of areas that may be reasonably monitored with modest expenditures. The use of alarming dosimeter telemetry systems is a convenient way to expand the coverage of plant areas that may be monitored. Those areas where significant radiation sources may be present or dose rates may fluctuate or be subject to sudden changes should be provided area radiation monitor coverage. The ARM channels may consist of either permanently installed or portable (or temporary) monitors with usually some combination of the two encountered in actual practice. Typically a number of fixed ARMs are permanently installed in LWRs. These ARMs provide local readout and alarm functions while some channels may also display and alarm in the main control room.

Permanent ARMs may be provided in such locations as those described below.

(1) Exposure rates emanating from various filter housings or their shielded compartments are usually monitored on a continuous basis. This could include the CVCS letdown or purification line filters, RCWC, and reactor cavity and spent fuel pool purification systems filters. ARMs in these areas provide indication that the activity collected on the filters has reached a level requiring the filters to be replaced to prevent unnecessarily high personnel exposures during change-out. Additionally, these ARMs help to ensure that spent filters may be safely transported as radioactive waste in designated transport casks or shipping containers based on radiation levels.

(2) Usually one or more ARMs are provided at the entrance to the containment or reactor building or the drywell airlocks to warn personnel of high exposure rates inside these buildings. This prevents personnel from entering these areas when high radiation fields are present. The detectors may be located just inside the entrance to these areas while the display is located at a convenient location outside the entrance.

(3) Typically low and high range ARMs are installed inside the containment or drywell that provide readout in the main control room. The high range ARM detector is located directly over the reactor vessel on or above the refuel floor. The high range channel serves to provide information concerning dose rates inside these areas in the event of a severe accident involving fuel damage and should be capable of measuring dose rates in the range of tens of thousands of Sv/h (millions of rem/h). These ARMs are usually referred to as post-accident detectors.

(4) Radiation levels are measured at the reactor cavity and spent fuel pool water surfaces to detect elevated dose rates. This function is especially important during outages when spent fuel is being handled. High radiation levels may result from lifting an activated core component or spent fuel assembly to near the water surface, or the presence of damaged fuel elements, or a decrease in the water level over the core or spent fuel pool storage racks. The actuation of alarms on these ARMs may initiate automatic actions such as isolation of the reactor building or fuel building ventilation systems or route ventilation flow to an emergency filtration system.

(5) As noted earlier ARMs are usually placed in those plant areas where high radiation levels are expected or where dose rates may change rapidly without warning, This could include such areas as RHR pump and heat exchanger rooms, charging pumps, liquid waste collection and holdup tanks, gas storage tanks, primary system sample room, spent resin sluice lines or pipe chases, and radioactive waste evaporator room (if present).

(6) ARMs may be provided in certain plant areas to verify that dose rates near shield walls are within acceptable limits.

There may be other plant locations in addition to those noted above equipped with a permanent ARM. Radiation protection personnel should be knowledgeable of the ARM channels and trained in the necessary response actions to take in the event of an alarm. Each ARM channel may have specific response actions. These response actions could include requirements to perform confirmatory radiation surveys in locations impacted by the ARM alarm. For example if an ARM located in a liquid waste hold-up tank alarms then response actions could include performing surveys on those process lines that were feeding the hold-up tank at the time of the alarm actuation.

## 11.10 Portable Area Radiation Monitors

During maintenance and outage activities the need may arise to provide local area radiation monitoring coverage for various reasons. The use of portable ARMs serve to provide timely indication of elevated radiation levels in plant areas where the temporary ARMs are located. The strategic use and placement of portable ARMs during outage periods serves to minimize unplanned exposure events and to provide early warning in the event of higher than expected radiation levels in plants areas or during the performance of work activities. Consider the situation whereby a maintenance work crew will require several days to repair a valve located in a pipe chase, valve alley or other plant location in close proximity to other systems containing radioactive fluids. The operational status of these systems may be subject to change over the extended work period or may be called upon to support system configuration changes to support ongoing outage activities. The placement of a portable ARM in the work location could provide early warning of elevated radiation levels to the work crew. Even if workers are provided with alarming EDs these portable ARMs still provide a vital function. In the event that radiation levels fluctuate when no workers are in the immediate vicinity of the area, the actuation of the portable ARM alarm would provide warning to individuals in adjacent plant areas of a potential radiological issue. Early notification of operations and RP under these conditions would allow for a more timely response to the situation and initiation of any required corrective actions.

Portable area radiation monitors consist of a gamma-sensitive detector and a display panel or unit. Portable ARM units may also be equipped with neutron or beta-gamma detection capabilities. For purposes of this discussion the primary concern is related to identifying unanticipated changes in gamma radiation fields due to operational or process changes during outage periods. The detector is placed in a designated location to monitor radiation levels. Thus the detector may often be located many meters away from the display panel. The display unit is positioned in a strategic location that is visible to individuals working in the vicinity or at the entrance to the work area. The display unit may be remotely located from the detectors' monitoring position. The key aspect is to ensure that the warning display is located where workers will be likely to see or hear the alarm signals. Most ARM units are equipped with both audible and visual warning indicators.

Various firms offer portable area radiation monitoring units. Many of these units may be networked with multiple channels in the field. Additionally multiple channels may be connected to a central display panel. The central display unit may be located at an RP control point or the primary RCA access control area or other suitable location. Figure 11.17a depicts an ARM offered by Mirion Technologies. Figure 11.17b shows the same unit with the detector connected to a cable allowing the detector to be placed in a remote location to be monitored and the display module to be located some distance away from the detector.

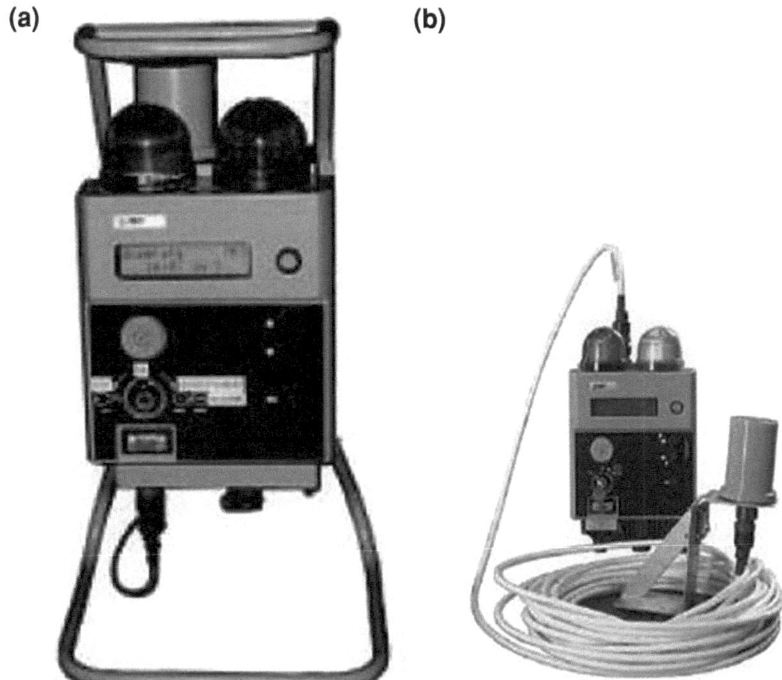

**Fig. 11.17** The Mirion Technologies area radiation monitor Model GIM 204 K (Courtesy of Mirion Technologies www.mirion.com). **a** Depicts an ARM offered by Mirion Technologies. **b** Shows the same unit with the detector connected to a cable

The number, location and positioning of portable ARMs may be based on previous experience. During outage periods the complexity of maintenance activities, changing plant conditions, the movement of radioactive components and materials, and processing and transfer of large volumes of radioactive liquids all pose situations whereby plant areas could be subject to rapid and perhaps unanticipated changes in radiological conditions. If incidents were encountered during previous outages in this regard, then these lessons-learned should be utilized to identify plant locations suitable for placement of temporary ARM units during outage periods.

## 11.11  Air Sampling Equipment

A wide range of portable air sampling equipment is available that is utilized to draw air through various filter media as described in Chap. 6. Various vacuum pumps are used in conjunction with filter holders that are designed to hold the filter media. Filter holders provide a physical housing to enclose the filter media while

**(a)**                                    **(b)**

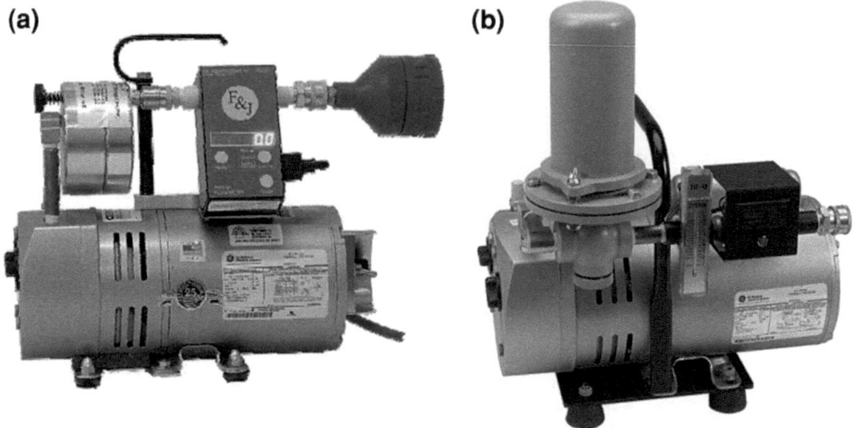

**Fig. 11.18** Examples of low volume air samplers. Figure **a** depicts the F&J Model DF-1 series low volume air sampler with a digital flow meter (Courtesy of F&J Specialty Products www.fjspecialty.com). Figure **b** depicts the HI-Q Model series VS-23 low volume air sampler (Courtesy of HI-Q Environmental Products Company www.hi-q.net)

in service and the necessary support to ensure the integrity of the filter. Particulate filters are relatively fragile and brittle and are supported by a screen or mesh backing material within the filter holder. The in-line filter holders must be properly secured to ensure the proper flow of air through the filter and to prevent any bypass flow around the filter while in service. Figure 6.9 depicts a common filter holder for use with a charcoal and particulate filter. Figure 6.12 displays common particulate filter holders.

The specific air sampler used in a given situation often is determined by the requirement to sample enough air volume in order to achieve a minimum level of detection. For example, obtaining a 2-minute air sample with a low volume air sampler operating at a flow rate of a few liters per minute may not provide sufficient volume to obtain the required lower limit of detection. Assuming Co-60 is the primary constituent of the airborne activity the sample time and flow rate must be sufficient to ensure that the counting system can detect a reasonable fraction of the Co-60 DAC-value of 2E3 Bq/m$^3$ (1 $\times$ 10$^{-8}$ μCi/ml).[2] The type of air sampler used is often determined by the projected time over which the air sample is to be obtained in addition to the type of airborne contaminant to be collected.

Air samplers used for sampling of particulates include both low and high volume air samplers. Low-volume air samplers are ideally suited for long-duration sampling periods. Many models are available that are designed to run continuously for extended periods of time. These units are equipped with vacuum pumps capable of providing flow rates in the range of a few liters per minute to a few

---

[2] Both the ICRP-68 and 10CFR20 DAC values are provided for reference purposes only.

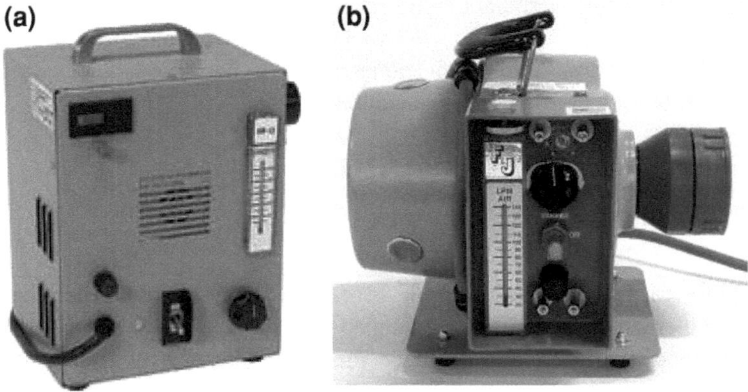

**Fig. 11.19** Examples of portable high volume air samplers. Figure **a** depicts the HI-Q Model CF-1000BRL high volume air sampler (Courtesy of HI-Q Environmental Products Company www.hi-q.net). Figure **b** depicts the F&J Model HV-1 series high volume air sampler (Courtesy of F&J Specialty Products www.fjspecialty.com)

hundred liters per minute. Figure 11.18 depicts two examples of low volume, continuous duty, air samplers commonly used for the collection of long-duration air samples.

Iodine collection efficiency of charcoal filters is a function of flow rate as noted in Chap. 6. Iodine collection efficiency starts to decrease significantly at flow rates approaching 10–15 l/min (or at approximately 0.5 cfm). The application of iodine collection efficiencies can often be ignored at flow rates less than several liters per minute. However, when performing dose assessments or incident evaluations where a higher degree of accuracy may be necessary then appropriate correction factors should be applied. Since the analysis of charcoal samples is performed by gamma spectroscopy methods, appropriate iodine collection efficiency factors as a function of flow rate, could simply be incorporated into the algorithm of the laboratory analysis system.

High volume air samplers provide air flows as high as a couple thousand liters per minute (approximately 60–70 cfm). These samplers are ideal for obtaining particulate air samples when a short duration release is anticipated such as during the initial breach of a contaminated or potentially contaminated system. Short-duration air samples or grab samples are frequently collected during maintenance periods. When responding to radiological incidents, such as a system leak, the need to confirm airborne radioactivity concentrations in a timely manner is often accomplished by obtaining a short-duration high-volume air sample.

The high flow rates of high volume air samplers make them unsuitable for the collection of iodine samples. Due to the high flow rates of high volume air samplers' precautions should be taken to ensure that filter media does not inadvertently collect contamination not representative of airborne contaminants. High volume air samplers may be placed on tripods or similar devices to elevate the unit off the floor or to otherwise maintain a sufficient distance from contaminated equipment or components. Figure 11.19 depicts two examples of high volume air samplers.

Most high and low volume air sample models are compatible with a wide variety of filter heads from various manufacturers.

## 11.12  Continuous Air Monitors

The ability to provide early warning of elevated airborne concentrations is an important element in minimizing unplanned exposures. Early detection of elevated airborne concentration levels could be the first indication of a plant problem. The availability of continuous air monitor (CAMs) units can provide warning of system leaks or other operational excursions that result in elevated airborne concentration levels. CAM units strategically placed in plant areas provide early warning to allow plant operators to initiate actions to prevent or mitigate plant upset conditions that pose potential radiological safety concerns. Pump seal failures, valve gasket leaks, flange leaks, overflow of drains or floor sumps or liquid waste collection tanks could all result in localized areas with elevated airborne concentrations. Early detection of these type events allows for the timely establishment of access controls to affected areas of the plant and for early corrective measures, such as switching to an alternate train to isolate a leaking pump seal or valve, for instance.

Continuous air monitors may consist of a single beta-gamma detector to monitor gross airborne radioactivity levels or may be elaborate three-channel units. Three-channel CAM units monitor for the presence of particulate, iodine, and noble gas (i.e., PING CAM units) radioactive species and often provide independent alarm capabilities. As with ARMs, the number and type of CAM units employed at LWR facilities varies from one plant to the next. The inventory of radiological monitoring equipment should include not only the fixed-station units described here but also portable units. Both multi-channel and single-channel portable CAMs should be available. Portable CAM units are utilized for job coverage activities and short-term use to monitor airborne conditions when recovering from an event until airborne concentrations have returned to normal levels or have stabilized.

### *11.12.1  Installed Continuous Air Monitors*

Permanently installed CAM units are typically provided to monitor airborne radioactivity levels in several areas common to most LWRs. Though not strictly the responsibility of RP oftentimes RP personnel may be responsible for the daily performance checks of installed CAM units. Many of the installed CAM units may be more closely associated with effluent pathways versus in-plant monitoring. As noted above for plant-installed ARM channels RP personnel will most likely be required to respond to a CAM alarm and therefore, should be intimately aware of the location and overall purpose and function of the various CAM channels available at their facility. The discussion that follows will focus primarily on those CAM channels that monitor in-plant locations. With the exception of the primary stack CAM those channels monitoring other liquid or gaseous release pathways are beyond the scope of this text.

Airborne contamination levels are routinely monitored inside containment buildings, reactor buildings, and drywells, as applicable. These CAM units are typically three-channel units capable of measuring particulate, iodine, and gaseous airborne radioactivity concentrations. Sample lines draw air samples from several locations within these buildings and route the sample air stream to the CAM unit for analysis. Common designs incorporate either a fixed-head or sequentially running filter to evaluate particulate airborne concentrations. The sample stream is routed through a charcoal cartridge to collect any radio-iodine present. Activity measured on the cartridge, often by a NaI detector, provides a measurement of radio-iodine airborne radioactivity concentrations. The sample air stream is then routed through an ionization chamber type detector for analysis of the gaseous activity. Upon detection of high airborne contamination in the monitored air stream, ventilation systems (e.g., the containment building) may be isolated or re-directed to an emergency or standby filtration system. These standby ventilation systems may be equipped with additional filter banks and charcoal filters to provide additional cleanup capacity. These CAMS serve an essential function since they can provide early warning of system leaks. Any significant primary system leak will result in increased airborne radioactivity levels in these areas.

In the case of PWR units with a separate fuel storage building or multiple unit sites that may have a centralized spent fuel storage building a CAM unit is often located adjacent to the fuel pool area. This CAM provides a similar function as that described above for the containment and reactor buildings.

Several other plant areas, where airborne contamination could be encountered, may also be equipped with plant-installed CAM units. These areas may include various pump and heat exchanger rooms (e.g., CVCS and RHR), general areas within the auxiliary building, radioactive waste processing, handling, and storage areas, and the decontamination work area. These CAM units typically monitor airborne particulate concentrations only and may consist of a fixed filter medium that is changed-out on a routine basis or equipped with a mechanism that periodically advances the filter paper.

One of the most important CAM units is the stack effluent monitor. Stack effluent discharges are continuously monitored in order to account for radioactivity releases to the environment. The stack CAM is a vital component of the installed plant radiation monitoring system. The stack CAM unit measures the activity concentration in the stack release downstream from final treatment and filtration of the effluent stream. The stack monitor is usually a three channel CAM unit capable of independently measuring particulate, iodine, and gaseous activity concentrations. Depending upon the situation and regulatory requirements the stack monitoring system may consist of a gamma spectroscopy detection system capable of providing radionuclide-specific concentration values in the effluent stream. Detection of elevated radioactivity concentrations results in the actuation of alarms and various automatic actions. Typically the stack CAM provides a low-level and a high-level alarm. The low level alarm provides early warning of elevated concentrations that allows for the initiation of corrective actions or measures to address the cause or source of the elevated concentrations. The high-level alarm initiates automatic actions such as terminating a waste gas storage tank release by automatically sending a signal to close isolation valves to a waste gas storage tank or routing the stack effluent stream to a standby treatment system.

Elevated stack readings indicate potential problems, perhaps serious, and should be investigated to identify the source of the activity. The activity levels monitored by this CAM should be closely followed with emphasis placed on the monitoring of any trends that may indicate potential radiological problems. Though a detailed description of the stack effluent monitoring system is beyond the scope of this discussion, RP personnel should have a basic understanding of the function of the stack effluent monitor and the potential radiological implications associated with high or elevated stack discharge radioactivity concentrations.

## 11.12.2 Portable Continuous Air Monitors

Recent innovations in detector designs and technological advancements have allowed portable CAMs to be designed that provide reliable service in relatively harsh environments. Improved electronics and use of digital and user friendly display features has simplified set-up and operational features. Radon compensation has allowed for development of small-sized units that are portable. Microprocessor functions allow for essentially live-time evaluation and immediate alarm indications to provide warning when airborne concentrations exceed anticipated values. These design features have greatly expanded the versatility and use of portable CAM units. Both single and multiple channel portable CAMs are available. Microprocessor based CAMs can be easily transported to the monitoring location and when used strategically can minimize personnel exposures due to unplanned releases of airborne contamination and minimize the use of respiratory protection equipment.

During plant outages the use of portable CAM units can supplement the installed CAM channels. During the early stages of refueling outages short-lived airborne species such as xenon and other radioactive noble gases, along with their short-lived daughter products and other short-lived particulates may be present. The levels of airborne concentrations of radio-iodine isotopes will most likely be highest during the initial stages of an outage, decreasing as these isotopes decay over the first several days of an outage and as process ventilation filtration systems reduce airborne concentrations. For these reasons three-channel units are often located on the refuel floor, in the vicinity of the spent fuel pool, and adjacent to major work areas, especially during the early stages of refueling outages. During these periods radio-iodine and short-lived gaseous activity may be present in significant concentrations. Under these conditions three-channel CAM units may provide an important and necessary monitoring function. Three-channel CAM units often remain at these locations throughout an outage. The iodine monitoring channel provides an important function to monitor for the presence of iodine during fuel handling operations.

Portable single channel particulate monitoring CAM units facilitate ease of movement and set-up. These type CAMs may be utilized for specific job coverage activities. The presence of significant airborne concentrations of radio-iodine or noble gases is assumed not to be present under these circumstances. Both beta-gamma and alpha single channel CAM units are available. Assuming that no significant fuel cladding damage has been experienced, beta-gamma single channel CAM units may be sufficient for task-specific monitoring purposes.

Many CAM units provide a direct readout of the airborne concentration as a percentage of the DAC and an integrated value for the DAC-hour exposure over the sampling period. Again, the ability to take advantage of these features rests heavily on the establishment of a proper DAC value based on knowledge of anticipated constituents of airborne contamination. Whether the particular CAM model allows for the selection of pre-established DAC values or if a specific value can be specified, the basis for selecting a DAC value should be well understood. The inadvertent selection of the wrong DAC value (or airborne radioactivity concentration) for the alarm threshold could result in unplanned exposures. The parameters to consider when establishing CAM alarm thresholds are further discussed below.

Many portable CAM units offered by vendors are suitable for use in LWR environments. The units described here reflect the range of functions and features that are available. Canberra offers the iCAM alpha/beta air monitor. This unit is shown in Fig. 11.20. The unit utilizes a passivated ion-implanted planar (PIP) silicon radiation detector. The use of a PIP radiation detector allows for the simultaneous measurement of both alpha and beta airborne radioactivity concentrations. These units are equipped with either a fixed filter assembly or a moving filter mechanism. Self diagnostic software provides trouble-shooting functions along with automatic calibration status performance checks.

A model offered by Bladewerx™ is the SabreBPM™ beta particulate monitor. This unit is shown in Fig. 11.21. This CAM measures beta airborne concentrations and is equipped with software for radon background subtraction. A beta-sensitive solid state ion-implanted silicon detector is also utilized with this CAM model.

The AMS-4 continuous air monitor offered by Thermo Fisher Scientific is another portable CAM unit utilized in the LWR industry. The CAM is equipped with a sealed gas proportional detector. The AMS-4 has optional features that allow a radial inline module to be attached for stack or ventilation duct monitoring. The inline modular unit utilizes a sealed gas proportional detector. Figure 11.22 depicts the AMS-4 unit with the optional inline module. The unit has real time gamma background subtraction capability. The microprocessor based unit has the

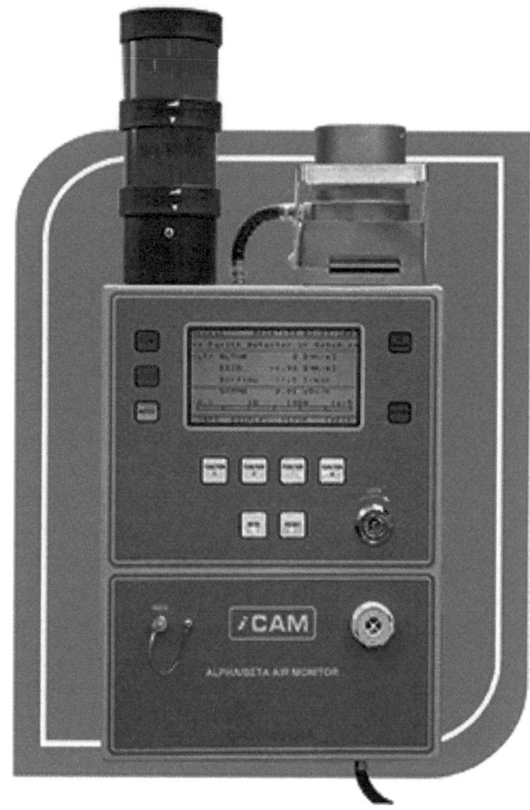

**Fig. 11.20** The Canberra iCAM alpha/beta air monitor (Courtesy of Canberra, www.canberra.com)

capability to display a large range of functions and instrument status indicators including DAC values, DAC-hours and other parameters associated with the measurement of airborne radioactivity concentrations.

All the models discussed above have both audible and visual alarm indicators. The units provide fault indications for such parameters as low flow rate, high background, power faults, detector voltage, and other key parameters. The units have user-friendly display panels that allow the operator to scroll through various screens. These screens depict such features as activity concentrations, DAC-hour values, and air sample volume in addition to instrument status information. Communication interfaces and the storage and transfer of stored data are common features. Network connections to a local computer or other plant computer systems allows for the transfer of monitoring data.

When selecting a particular portable CAM unit for job coverage purposes the manufacturers' specifications should be consulted to ensure that a given model is capable of detecting the desired levels of airborne radioactivity concentrations. The three CAM units described above represent a sampling of the models available and provide an indication of the diverse range of features and functions available with portable CAM units. The features and functions of a specific unit should

**Fig. 11.21** The Bladewerx
SabreBPM$^{TM}$ beta particulate
monitor (Courtesy of
Bladewerx,
www.bladewerx.com)

**Fig. 11.22** Depicts the
AMS-4 unit with the optional
inline module (Courtesy of
Thermo Fisher Scientific,
www.thermofisher.com)

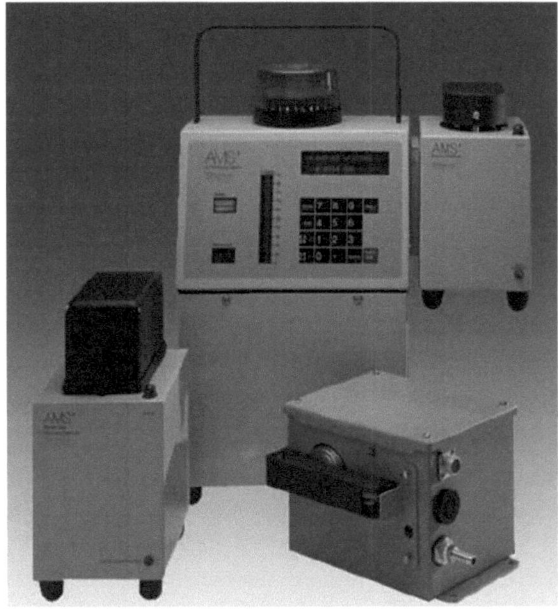

be reviewed prior to placing a unit in service. Radiation protection personnel should verify that the unit is suitable for the intended purpose. Such parameters as the minimum detection level as a function of radiation background, the useful measurement range of the unit, radon compensation features, sampling flow rates, and alarm features should be verified to ensure that the CAM unit satisfies the required monitoring conditions.

To maximize the use of CAM units when providing live-time monitoring of work activities with the potential to generate airborne contamination a meaningful alarm set point should be established. A balance should be achieved between setting alarm thresholds either too high or too low. This is especially important when CAM units are utilized to minimize the use of respiratory protection equipment, when the potential for airborne contamination is low or if potential airborne concentrations are anticipated to be a small fraction of the applicable DAC value. Alarm settings set too low may cause repetitive "nuisance" alarms that do not result in any additional radiological controls being applied. Alarm settings that are too high could result in unplanned exposures or worst result in exposures that go undetected.

Strategic use of reactor coolant system chemistry data leading up to a scheduled outage can be evaluated to provide an indication of the radionuclide mixture most likely to comprise the majority of airborne contamination when systems are opened or when initially breached. When used for job coverage purposes CAM alarm thresholds should be based upon the most restrictive radionuclide known to be present. A conservative approach may be to assume that Sr-90 is the limiting radionuclide when considering potential DAC values. This assumes that long-lived alpha radionuclides can be neglected. However; if supporting radiological analyzes (e.g., primary system chemistry samples or specific Sr-90 analyzes of smears) are available and indicate that Sr-90 is not present then a more appropriate DAC value may be chosen.

Oftentimes a continuous low volume air sampler, equipped with a particulate filter or a combination particulate and charcoal cartridge holder is used to monitor airborne contamination levels as discussed in Sect. 11.10. Unlike CAM units, that provide live-time readout of activity levels, air samplers require that the filter media be removed and analyzed by laboratory counting equipment. This method provides quantitative results after the fact, a limitation which must be taken into consideration when deploying these units for continuous airborne measurement purposes. Notwithstanding, continuous air samplers are utilized in many plant locations for the evaluation of airborne contamination levels. They are inexpensive and easy to operate when compared to a CAM.

## 11.13 Summary

A wide and diverse range of radiological survey equipment is available to the LWR industry. Portable survey equipment must be properly maintained and calibrated to support daily radiological surveillance activities. RP personnel should be properly trained and qualified in the use of radiological surveillance equipment. The limitations, uses, and applications of a given instrument should be understood by the user to ensure the instrument is suitable for the intended purpose. The strategic use of portable ARM and CAM units can minimize the severity and number of radiological incidents or unplanned radiation exposures. Programs should be in place to ensure that the RP group stays abreast of the latest industry developments concerning improvements and introduction of new technology that improves radiological surveillance capabilities.

## Bibliography


1. Cember H., and T. Johnson, Introduction to Health Physics, Fourth Edition, McGraw Hill, 2000
2. International Commission on Radiation Units and Measurements, Radiation Protection Instrumentation and Its Application, ICRU Report 20, Washington, D.C., 1971
3. Knoll G.F., Radiation Detection and Measurement, Fourth Edition, John Wiley & Sons, 2010
4. National Council on Radiation Protection and Measurements, Calibration of Survey Instruments Used in Radiation Protection for the Assessment of Ionizing Radiation Fields and Radioactive Surface Contamination, NCRP Report No. 112, Bethesda, MD, 1991
5. National Council on Radiation Protection and Measurements, Instrumentation and Monitoring Methods for Radiation Protection, NCRP Report No. 57, Bethesda, MD, 1978
6. Price W.J., Nuclear Radiation Detection, McGraw-Hill, New York, 1958
7. Shapiro, J., Radiation Protection, A Guide for Scientists and Physicians, Third Edition, Harvard University Press, 1990

# Appendix A

<br>

**Table A.1** Annual limits on intake (ALI) and derived air concentration values for selected radionuclides (based on 10CFR20 and ICRP-30)

| Radionuclide | ALI-Ingestion ($\mu$Ci) | ALI-Inhalation ($\mu$Ci) | DAC ($\mu$Ci/cm$^3$) |
| --- | --- | --- | --- |
| H-3 | 8E4 | 8E4 | 2E-5 |
| Cr-51 | 4E4 | 2E4 | 8E-6 |
| Mn-54 | 2E3 | 8E2 | 3E-7 |
| Mn-56 | 5E3 | 2E4 | 9E-6 |
| Co-57 | 4E3 | 7E2 | 3E-7 |
| Co-58 | 1E3 | 7E2 | 3E-7 |
| Co-60 | 2E2 | 3E1 | 1E-8 |
| Fe-59 | 8E2 | 3E2 | 1E-7 |
| Ni-65 | 8E3 | 2E4 | 7E-6 |
| Zn-65 | 4E2 | 3E2 | 1E-7 |
| Kr-85[a] | – | – | 1E-4 |
| Kr-85 m | – | – | 2E-5 |
| Rb-88 | 2E4 | 6E4 | 3E-5 |
| Sr-90 | 3E1 | 2E1 | 2E-9 |
| Nb-95 | 2E3 | 1E3 | 5E-7 |
| Zr-95 | 1E3 | 1E2 | 5E-8 |
| Zr-97 | 6E2 | 1E3 | 5E-7 |
| I-131 | 3E1 | 5E1 | 2E-8 |
| I-132 | 4E3 | 8E3 | 3E-6 |
| I-133 | 1E2 | 3E2 | 1E-7 |
| I-134 | 2E4 | 5E4 | 2E-5 |
| I-135 | 8E2 | 2E3 | 7E-7 |
| Xe-131 m[a] | – | – | 4E-4 |
| Xe-133 m[a] | – | – | 1E-4 |
| Xe-133[a] | – | – | 1E-4 |
| Xe-135[a] | – | – | 2E-5 |
| Xe-138[a] | – | – | 4E-6 |
| Cs-134 | 7E1 | 1E2 | 4E-8 |

(continued)

R. Prince, *Radiation Protection at Light Water Reactors*,
DOI: 10.1007/978-3-642-28388-8, © Springer-Verlag Berlin Heidelberg 2012

**Table A.1** (continued)

| Radionuclide | ALI-Ingestion ($\mu$Ci) | ALI-Inhalation ($\mu$Ci) | DAC ($\mu$Ci/cm$^3$) |
|---|---|---|---|
| Cs-137 | 1E2 | 2E2 | 6E-8 |
| Cs-138 | 2E4 | 6E4 | 2E-5 |
| La-140 | 6E2 | 1E3 | 5E-7 |

[a] *Submersion* values given are for submersion in a hemispherical semi-infinite cloud of airborne material

*Note* The ALI values provided in this table are based on a committed effective dose equivalent of 50 mSv (5 rems) or a committed dose equivalent of 0.5 Sv (50 rems) to any individual organ or tissue, whichever ALI is smaller. Appendix B depicts the equivalent ALI and DAC values based on an annual dose equivalent limit of 20 mSv

# Appendix B

**Appendix B** Dose conversion factors (DCF) and Annual limits on intake (ALI) for selected radionulcides (ALIs derived from DCFs; ALI = 20 mSv/DCF) (based on ICRP-68)

| Radionuclide | Inhalation DCF (Sv/Bq) | Inhalation ALI (Bq) | Ingestion DCF (Sv/Bq) | Ingestion ALI (Bq) |
| --- | --- | --- | --- | --- |
| H-3 | 2.0E-11 | 1.0E9 | 2.0E-11 | 1.0E9 |
| Cr-51 | 3.6E-11 | 5.6E8 | 3.8E-11 | 5.3E8 |
| Mn-54 | 1.2E-9 | 1.7E7 | 7.1E-10 | 2.8E7 |
| Mn-56 | 2.0E-10 | 1.0E8 | 2.5E-10 | 8.0E7 |
| Co-57 | 6.0E-10 | 3.3E7 | 2.1E-10 | 9.5E7 |
| Co-58 | 1.7E-9 | 1.2E7 | 7.4E-10 | 2.7E7 |
| Co-60 | 1.7E-8 | 1.2E6 | 3.4E-9 | 5.9E6 |
| Fe-59 | 3.2E-9 | 6.3E6 | 1.8E-9 | 1.1E7 |
| Ni-63 | 5.2E-10 | 3.8E7 | 1.5E-10 | 1.3E8 |
| Zn-65 | 2.8E-9 | 7.1E6 | 3.9E-9 | 5.1E6 |
| Kr-85 | 2.2E-11 | 9.1E8 | – | – |
| Rb-86 | 1.3E-9 | 1.5E7 | 2.8E-9 | 7.1E6 |
| Sr-90 | 7.7E-8 | 2.6E5 | 2.8E-8 | 7.1E5 |
| I-131 | 2.0E-8 | 1.0E6 | 2.2E-8 | 9.1E5 |
| I-132 | 3.1E-10 | 6.5E7 | 2.9E-10 | 6.9E7 |
| Xe-133 | 1.2E-10 | 6.7E5 | – | – |
| Xe-135 | 9.6E-10 | 8.3E4 | – | – |
| Cs-134 | 9.6E-9 | 2.1E6 | 1.9E-8 | 1.1E6 |
| Cs-137 | 6.7E-9 | 3.0E6 | 1.3E-8 | 1.5E6 |

R. Prince, *Radiation Protection at Light Water Reactors*,
DOI: 10.1007/978-3-642-28388-8, © Springer-Verlag Berlin Heidelberg 2012

# Appendix C

# Problems

**Chapters 2 and 3**

1. List the major components of the primary system of a PWR?
2. List the major components located within a BWR reactor vessel and the associated circulation system?
3. For what LWR design (PWR or BWR) does the secondary side (i.e., turbine building) normally pose radiological safety concerns?
4. Describe the flow path of the primary system water (i.e., the coolant) in a PWR?
5. What is the purpose of the pressurizer?
6. How do steam generators transfer heat from the primary side to the secondary side of a PWR? How is steam produced in a BWR? Where is the steam separated and dried in a BWR?
7. Describe the radiological conditions at 100% reactor power encountered inside the bio-shield and loop rooms of a PWR and the drywell of a BWR? How do these conditions change when the reactor is shutdown?
8. What is the purpose of the chemical and volume control system (CVCS) in a PWR?
9. Briefly describe the radiological conditions associated with the residual heat removal (RHR) or decay heat removal system? What parameter primarily impacts RHR system dose rates?

**Chapter 4**

1. How is Co-58 and Co-60 produced? List some of the possible sources of these radionuclides?
2. What type of activities or processes may generate airborne contamination?
3. What factors influence the amount of airborne contamination level of a specific radionuclide?

R. Prince, *Radiation Protection at Light Water Reactors,*
DOI: 10.1007/978-3-642-28388-8, © Springer-Verlag Berlin Heidelberg 2012

4. What are some measures that may be employed to minimize exposure to airborne contamination?
5. What role does coolant chemistry play in minimizing activation product source terms?

**Chapter 5**

1. What conditions must be met to post an area as a high radiation area? A locked HRA? A Very HRA? Describe the administrative and access control requirements associated with a locked-HRA? Where are these requirements specified?
2. When would an area have to be posted as an airborne radioactivity area as defined in 10CFR20?

**Chapter 6**

1. While at power a flange leak on the CVCS system resulted in Xe-133 airborne concentrations of 3.5 E-3 $\mu Ci/cm^3$. (a) Under these conditions how many DAC-hours of exposure would an individual receive if present for 1 h? (b) An entry team will require 3 person-hours to repair the leak. Should respiratory protection be prescribed for members of the repair team? (Hint: see footnotes to Appendix B, Table B.1 for submersion dose in 10CFR20.) (c) What is the total dose received by the entry team? (d) What dose rate is associated with the measured Xe-133 airborne concentration?
2. An individual reports to the radiological control area (RCA) access control station with an ED dose alarm. As the RP on duty what are your immediate actions?
3. A particulate and charcoal low-volume air sample obtained during a pump replacement during an outage revealed the following airborne concentration levels based on gamma spectroscopy analysis of the filters.

$$
\begin{array}{lll}
\text{I-131} &= 3.5\text{E-8 } \mu Ci/cm^3 & (\text{DAC: } 2\text{E-8 } \mu Ci/cm^3) \\
\text{Co}-60 &= 3.0\text{E-8 } \mu Ci/cm^3 & (\text{DAC: } 1\text{E-8 } \mu Ci/cm^3) \\
\text{Co}-58 &= 4.5\text{E-7 } \mu Ci/cm^3 & (\text{DAC: } 3\text{E-7 } \mu Ci/cm^3) \\
\text{Mn}-54 &= 8.5\text{E-8 } \mu Ci/cm^3 & (\text{DAC: } 3\text{E-7 } \mu Ci/cm^3)
\end{array}
$$

(a) Using the DAC values from 10CFR20 Appendix B, Table B.1 determine the DAC-hour exposure for a worker in the area for 2 h.
(b) If respiratory protection equipment (RPE) was not utilized were any exposure limits exceeded?
(c) Should the work area have been posted as an Airborne Radioactivity Area per 10CFR20.1003?

4. A particulate and charcoal low-volume air sample obtained during a pump replacement during an outage revealed the following airborne concentration levels based on gamma spectroscopy analysis of the filters. Determine the fractional DAC (FDAC) for the air sample.

$$\begin{aligned}
\text{I-131} &= 9.3\text{E2 Bq/m}^3 \quad (\text{DAC: } 4\text{E2 Bq/m}^3) \\
\text{Co-60} &= 3.7\text{E3 Bq/m}^3 \quad (\text{DAC: } 2\text{E3 Bq/m}^3) \\
\text{Co-58} &= 2.8\text{E3 Bq/m}^3 \quad (\text{DAC: } 4\text{E3 Bq/m}^3) \\
\text{Mn-54} &= 3.0\text{E3 Bq/m}^3 \quad (\text{DAC: } 6\text{E3 Bq/m}^3)
\end{aligned}$$

5. An air sample has been analyzed by gamma spectroscopy and the following results obtained:

$$\begin{aligned}
&\text{Mn-54: } 3.8\text{E-8 } \mu\text{Ci/cm}^3 \quad \text{DAC} = 3\text{E-7 } \mu\text{Ci/cm}^3 \quad \text{ALI} = 800\ \mu\text{Ci} \\
&\text{Co-60: } 8.0\text{E-7 } \mu\text{Ci/cm}^3 \quad \text{DAC} = 1\text{E-8 } \mu\text{Ci/cm}^3 \quad \text{ALI} = 30\ \mu\text{Ci} \\
&\text{Cs-137: } 3.0\text{E-8 } \mu\text{Ci/cm}^3 \quad \text{DAC} = 6\text{E-8 } \mu\text{Ci/cm}^3 \quad \text{ALI} = 200\ \mu\text{Ci}
\end{aligned}$$

For each radionuclide detected the corresponding DAC and ALI values from 10CFR20 Appendix B, Table B.1, for occupational exposure are provided.

(a) Calculate the FDAC value for each radionuclide present and
(b) The total DAC value for the air sample.
(c) What is the equivalent exposure rate to the whole-body assuming that the ALIs are based on limiting the whole body exposure to less than 5 rem in a year?
(d) Based on the above information what radionuclide is the most limiting?

6. Give at least three reasons why you would perform a radiation survey?
7. When obtaining air samples for internal dose assessment purposes what precautions are necessary in order to obtain a representative air sample?
8. During a maintenance activity to repair an RHR heat exchanger flange leak approximately 500 l of highly contaminated water spilled onto the floor in the heat exchanger room. The vast majority of the spilled water ran into a floor drain located in the room. The floor drain header is routed to a floor drain collection tank located two levels below the RHR heat exchanger room. Describe the immediate actions that should be taken to address the radiological hazards associated with this event.

## Chapter 7

1. A work crew has to perform a seal repair on a leaking CVCS check valve located within a steam generator cubicle. Welding will have to be performed to repair the leaking check valve. Work area survey data is provided below. A summary of the major work steps is also provided.

Radiological Conditions:

Dose rate at contact to valve = 3.50 mSv/h
Dose rate at 50 cm from the valve = 0.80 mSv/h
General area dose rates within 1–2 m of valve = 0.40 mSv/h
Low dose waiting area for the job = 2 μSv/h

Work steps to repair valve:

Time to prep and set up area = 1 h each (2 welders)
Time required to prep valve surface for welding = 0.5 h
Time to perform weld repairs = 3 h (1 welder)
Time to perform dye penetrant testing and weld inspections = 0.25 h
(1 person)
Time required to clean work area and remove equipment and materials = 0.5 h
each (3 workers)
RP job-coverage requires 0.1 h of RPT time in close proximity to the valve

Develop a dose estimate for the job. Provide basis for exposure rates utilized in
the development of pre-planning dose estimates. Assume that manual welding
(i.e., stick welding) will be utilized and will require the welder to be within
30–100 cm of the valve while welding. Dose rates are as specified and assume
that any dose reduction methods (e.g., temporary shielding) have been
incorporated into the dose rate figures. The welding unit and other support
equipment are located at a remote location in a low-dose area. Assume that the
valve has been decontaminated to a level whereby airborne contamination
levels are not a concern.

(a) Develop a dose estimate to complete the repairs.
(b) What measures could be taken to reduce personnel exposures?

2. A valve replacement will be performed for a valve in a system that
   communicates directly with the RCS (reactor coolant system). You are the
   RP performing the ALARA review for the design package to replace the valve.
   What recommendations would you offer to minimize both the short and long
   term radiological issues associated with this valve replacement?
3. Several days after entering a refueling outage a CVCS resin bed change out was
   required. The operation involves sluicing resin from a CVCS resin bed to the
   spent resin storage tank. A 20 m length of the spent resin sluice line runs along
   a pipe chase. This pipe chase also contains piping and associated valves for
   several plant auxiliary systems. Valve work, hanger inspections and general
   maintenance activities are performed in the pipe chase and adjacent areas
   during outages.
   Normally the CVCS resin bed is changed out prior to an outage to ensure that a
   fresh resin bed is available with sufficient resin capacity to remove impurities
   (i.e., activation and corrosion products) from the RCS system to support
   refueling operations. However, the existing bed loaded-up (i.e., reached break-
   through) earlier than originally planned.
   The portion (20 m) of the spent resin sluice line that runs through the pipe chase
   is unshielded. Surveys on the CVCS resin bed housing (teletector readings
   obtained via survey port in the resin bed shield housing) indicate dose rates as
   high as 7 Sv/h (700 R/h) on contact. Wire mesh gates are located at either end
   of the pipe chase and are normally left open to facilitate access by Operators
   and other personnel who routinely enter the area.

Describe the radiological control measures you would employ to support sluicing of the resin to the spent resin storage tank? Discuss measures associated with radiological surveys, access controls, postings, pre-job planning, and any other radiological safety aspect that may be pertinent to this activity.

4. An individual enters the pipe chase two days after the sluicing of the resin bed described in the above problem. The individual's ED dose rate and dose alarms activate (set at 500 µSv/h (50 mrem/h) and 200 µSv (20 mrem), respectively) after being in the area for less than one minute. Surveys prior to the resin transfer indicated general area dose rates of 20–100 µSv/h (2–10 mrem/h) with localized readings on contact to components of 200–400 µSv/h (20–40 mrem/h.) Upon investigation, follow-up surveys in the pipe chase measured general area dose rates as high as 800 µSv/h (80 mrem/h) with contact readings on the spent resin transfer header of several mSv/h (several hundred mrem/h). What are the lessons-learned from this event and what actions would you implement to prevent recurrence?

5. Repairs are required to be performed on a charging pump shaft. Repairs involve machining and welding operations. Decontamination activities were successful in removing all loose transferable contamination. However, direct surveys of the shaft area to be machined (12 cm length, 360° circumferential area) indicate residual fixed contamination levels of 10–20 µSv/h (1–2 mrem/h) on contact to the pump shaft. Repairs will be performed in the main maintenance shop area (i.e., the "cold" machine shop) due to the lack of required machinery in the "hot" shop. The maintenance shop area where the work will be performed is located outside the established RCA. Ongoing work will be performed in the maintenance area due to work activities in support of the ongoing outage. Describe the radiological safety and contamination control measures necessary to perform this task in the cold machine shop area.

## Chapter 8

1. What property does conductivity measure? What impact could out-of-tolerance conductivity levels have on plant radiological conditions?

## Chapter 10

1. What special radiological hazards does iodine pose with regard to airborne contamination and internal dose concerns?
2. Why are the ALI values referenced in 10CFR20 different than those provided in current ICRP recommendations?

## Chapter 11

1. A CAM unit alarms in the auxiliary building, describe the initial actions you would take as an RP.

# Appendix D

# Problem Solutions

### Chapters 2 and 3

1. Reactor vessel, reactor coolant pumps, steam generators, pressurizer and associated primary system piping.
2. Reactor vessel, steam dryers, steam separator, jet pumps, recirculation pumps, and the associated recirculation system piping.
3. BWR
4. The reactor coolant pumps pump water via the cold leg into the reactor vessel, after traveling through the core, the hot primary coolant passes through the tube side of one of the steam generators via the hot leg from the reactor vessel. Suction of the RCP causes water to flow from the outlet of the steam generator, via the crossover leg, through the RCP that pumps the water back to the core via the cold leg.
5. The pressurizer supplies and controls the primary system pressure. The pressurizer is equipped with internal heaters that may be activated to increase system pressure and a set of spay nozzles that allow cold water to be sprayed into the pressurizer to cool the primary system water to reduce pressure.
6. Hot water from the reactor core of a PWR flows through tubes located inside of the SG (each SG may have as many as 3,000 or more tubes). Feed water from the secondary plant side flows through the shell side of the SG, essentially immersing the SG tubes in water. Heat from the hot primary system water is transferred through the SG tubes to the water on the secondary side heating the secondary side water. The secondary side water turns to steam and is routed, via the main steam lines, to the turbine-generator to produce electricity.
   In a BWR water is allowed to turn to steam within the reactor vessel. Directly above the core, inside the reactor vessel sits the steam separator and above that the steam dryer. Dried steam exits the reactor vessel and is routed to the turbine-generator.

R. Prince, *Radiation Protection at Light Water Reactors*,
DOI: 10.1007/978-3-642-28388-8, © Springer-Verlag Berlin Heidelberg 2012

7. While at power neutron activation of air produces various radioactive species, the most important being N-16 with its 6 MeV gamma ray. This in addition to the intense direct neutron and gamma radiation fields emanating from the reactor vessel produces high radiation fields inside the loop rooms and drywell areas. General area dose rates within these areas may be on the order of 100–250 mSv/h (10–25 R/h), and perhaps higher in localized areas when at power. Neutron radiation levels increase as reactor power (i.e., the number of fissions) increases.

8. The CVCS system draws a letdown flow from the RCS to purify the RCS via filtration and demineralization. The CVCS maintains the RCS inventory and provides a means to maintain boron concentration of the RCS. The CVCS charging pumps also serve as part of the emergency core cooling system (ECCS) by providing cooling water in the event of an upset condition.

9. Answer: Radiation levels in the vicinity of RHR pumps and heat exchangers and associated piping may be on the order of a couple of mSv/h (100 mrem/h) or more when in service. The primary parameters that influence RHR system radiation levels are its operating state and whether or not the plant has been operating with fuel failures. Radiation levels in the vicinity of the RHR train in standby mode will be much lower than that of the operating train, on the order of tens of µSv/h (a few mrem/h) to perhaps 100–200 µSv/h (10–20 mrem/h).

## Chapter 4

1. Answer: Co-58 is produced by an n-p reaction with Ni-58: $^{58}\text{Ni}(n,p)^{58}\text{Co}$; while Co-60 is produced by an n-$\gamma$ reaction with Co-59: $^{59}\text{Co}(n,\gamma)^{60}\text{Co}$.
   The primary sources of Co-58 and 60 include corrosion of stainless steel that is used in the construction of various core components and primary system materials. The use of stellite in valve seats to provide a hardened seat that is less susceptible to wear is also a primary contributor.

2. Answer: welding, grinding, machining, cutting or any mechanical type activity performed on a contaminated component and/or in a contaminated area; air currents or movement of air across contaminated surfaces.

3. Answer: factors include such parameters as (1) volatility of the radionuclide; (2) the half-life; (3) solubility of the radionuclide in the reactor coolant water; (4) ventilation removal rate in the affected area; (5) dispersion and diffusion properties of the radionuclide.

4. Answer: (1) eliminate the source; (2) establish controls such as posting and access controls; (3) engineering controls such as temporary filtration units or enclosures; (4) use of respiratory protection equipment—as the last resort.

5. Answer: The maintenance of good plant chemistry minimizes the corrosion rate of plant components. By minimizing corrosion rates less corrosion products will be transported through the core where they may become activated. Good plant chemistry also minimizes the deposition of crud in ex-core locations thus minimizing plant radiation levels. Therefore, the adherence to good plant chemistry will minimize plant radiation levels over the operational life of a plant.

## Chapter 5

1. Answer: (a) HRA ($\geq 0.1$ and $\leq 1$ mSv/h)

   - Each entryway shall be barricaded and posted as a HRA
   - Access to and activities controlled by an RWP
   - If continuous HP escort no RWP required
   - Individual or group entering area shall possess:

     - A radiation monitoring device (i.e., survey meter) or
     - Integrated dose device (i.e., an ED) with alarm function or
     - Teledosimetry system monitored by RP personnel or
     - A self-reading dosimeter (PIC or ED) and

       (a) Be under the surveillance of RP while in the area and RP must have a survey meter or
       (b) Be under closed circuit surveillance by RP and have a means of communicating with individuals while in the area

   Answer: (b) LHRA ($\geq 1.0$ mSv/h)

   - Each entryway shall be posted as a HRA and locked or continuously guarded with the keys controlled by the radiation protection manager (RPM) and shift manager
   - Access to and activities controlled by an RWP
   - Individual or group entering area shall possess:

     - Integrated dose device (i.e., an ED) with alarm function or
     - Teledosimetry system monitored by RP personnel or
     - A self-reading dosimeter (PIC or ED) and

       (c) Be under the surveillance of RP while in the area and RP must have a survey meter or
       (d) Be under closed circuit surveillance by RP and have a means of communicating with individuals while in the area

   In addition the student should also reference the applicable requirements of 10CFR20.1601 (a), (b), and (c).

2. Answer: An area in which airborne radioactive material exists in concentrations in excess of the derived air concentrations specified in Appendix B to 10CFR20 would be required to be posted as an airborne radioactivity area. Those areas with airborne radioactivity concentrations less than a DAC but based on occupancy times in the area, could result in an individual exceeding an exposure of 12 DAC-hours in a week, would also have to be posted as an airborne radioactivity area.

## Chapter 6

1. Answers: (a) 35 DAC-hours. The DAC value for Xe-133 for occupational exposure is 1E-4 $\mu$Ci/cm$^3$. (b) No. The student should recognize that Xe represents a submersion dose not an internal exposure hazard, consequently respiratory protection would have no benefit in reducing worker exposures. (c) 35 DAC-hours x 2.5 mrem/h per DAC-hour x 3 h = 262.5 mrem. (d) 35 DAC-hours x 2.5 mrem/h per DAC-hour = 77.5 mrem/h. The student should use the footnotes to Table 1 to obtain the dose rate value based on ALI's and the fact that xenon is an external exposure hazard with its DAC predicated on an exposure limit of 5 rem/year.

2. Answer: Immediate actions should include evacuating individuals from the affected area where the individual was located when the ED alarmed. Secure access to the area by either posting the area or stationing another RP (or individual) to warn others. Have individuals who were in the area verify their ED dose readings. A concerted effort should be undertaken to identify the source of the elevated radiation levels (assuming that the cause for the ED alarm is not readily identifiable.) Place the ED into a reader or computer interface and obtain a display of the dose and dose rate history recorded by the ED for the time period leading up to the alarm. The ED history could be used to "back track" the person's whereabouts to assist in identifying the plant location where the high dose rate may have been encountered. A formal ED alarm investigation report should be initiated in accordance with plant procedures or polices.

3. Answers:

   (a) DAC-hour exposure:
   The FDAC is calculated by dividing the airborne concentration of each radionuclide by its associated DAC value.

$$I\text{-}131 \quad FDAC = 3.5E\text{-}8/2E\text{-}8 = 1.75$$
$$Co\text{-}58 \quad FDAC = 4.5E\text{-}7/3E\text{-}7 = 1.5$$
$$Co\text{-}60 \quad FDAC = 3.0E\text{-}8/1E\text{-}8 = 3.0$$
$$Mn\text{-}54 \quad FDAC = 8.5E\text{-}8/3E\text{-}7 = 0.12$$

Therefore the total FDAC = 6.37 and the DAC-hour exposure for 2 h = 6.37 x 2 h = 12.74

   (b) Were any exposure limits exceeded? The worker's exposure was 12.74 DAC-hours. An individual can be exposed to 2,000 DAC-hours in a year assuming that the individual receives no other exposure from either internal or external sources. In actual practice this would be a relatively high airborne concentration area and controls to either eliminate or reduce the amount of airborne activity would be applied. If these controls prove unsuccessful then respiratory protection would probably be utilized.

(c) Should the area have been posted as an Airborne Radioactivity area per 10CFR20.1003? Yes.

The area should be posted as an airborne radioactivity area since an individual present in the area without respiratory protective equipment could exceed an exposure of 12 DAC-hours during a 40 h work week. This equates to 0.6% of the annual limit on intake (ALI), which is the posting requirement of 10CFR20. Alternatively, if the area is not routinely accessed by workers and it could be demonstrated that stay-times in the area are less than 1 h over a typical work week then posting may not be required. However administratively this may be difficult to demonstrate and the licensee would have to ensure that access controls are established to ensure that stay times in the area do not exceed a 12 DAC-hour exposure for any individual over a 40 h period. Obviously since the airborne concentration exceeds 12 DAC-hours it would be more convenient to post the area.

4. Answer:

$$I\text{-}131 \quad FDAC = 9.3E2/4E2 = 2.3$$
$$Co\text{-}60 \quad FDAC = 3.7E3/2E3 = 1.8$$
$$Co\text{-}58 \quad FDAC = 2.8E3/4E3 = 0.7$$
$$Mn\text{-}54 \quad FDAC = 3.0E3/6E3 = 0.5$$

The total FDAC = 5. 3

5. Answers:

(a) FDAC's are calculated by dividing the measured airborne concentrations for each nuclide by their respective DAC value. (Notice that the units of $\mu Ci/cm^3$ cancel out).

$$Mn\text{-}54 \quad : 3.8E\text{-}8 \ \mu Ci/cm^3 / 3E\text{-}7 \ \mu Ci/cm^3 = 0.127 \ FDA$$
$$Co\text{-}60 \quad : 8.0E\text{-}8 \ \mu Ci/cm^3 / 1E\text{-}8 \ \mu Ci/cm^3 = 8.0 \ FDAC$$
$$Cs\text{-}137 \quad : 3.0E\text{-}8 \ \mu Ci/cm^3 / 6E\text{-}8 \ \mu Ci/cm^3 = 0.5 \ FDAC$$

(b) Therefore the total FDAC for the air sample is: $0.127 + 8.0 + 0.5 = 8.63$

(c) The equivalent exposure rate is calculated as follows: Since the ALI is based on 5,000 mrem/year and dividing by 2,000 h/year (for a standard work year) = 2.5 mrem/h for one DAC. (Notice that the units for "year" drop out and we are left with mrem/h).

Therefore we have for the equivalent exposure rate: 8.6 DACs x 2.5 mrem/h per DAC = 21.5 mrem/h

(d) Obviously for this sample Co-60 is the most limiting radionuclide since its relative concentration compared to its DAC value is greater than that of the other radionuclides present. Based on biological effectiveness Co-60 is obviously the limiting nuclide since its ALI is lower than that of the other radionulcides. In other words based on a per $\mu Ci$ basis, internally deposited Co-60, will deliver a higher dose to the whole body when compared to the other two nuclides.

6. Answer: (1) confirm radiological conditions in the area; (2) verify postings accurately reflect radiological conditions; (3) to identify unanticipated radiological conditions; (4) obtain dose rate information to project pre-job dose estimates; (4) in response to a radiological incident such as a CAM or ARM alarm, spill or plant upset condition

7. Answer: (1) air sample should be representative of the actual exposure; (2) air sample should be obtained as close as possible to the breathing zone; (3) guard against cross-contamination of the air filter.

8. Answer: Immediate actions should include the need to perform radiation and contamination surveys in the heat exchanger room and ensure that postings are updated as necessary. Once radiological conditions are known and/or stabilized then arrangements to decontaminate the affected area should be initiated. Air samples may be prudent to ensure airborne contamination levels are within acceptable limits to allow access for the decontamination work group.

   The student should also recognize the need to survey the floor drain collection tank room to verify radiation levels in the vicinity of the tank. The volume of highly contaminated water associated with this event may have been sufficient to impact radiation levels of the collection tank resulting in the need to update radiological postings in the vicinity of the tank.

**Chapter 7**

1. Answer:

   (a) Dose estimates for the various tasks:

   Area set-up: 1 h x 2 workers x 400 μSv/h  = **800 μSv**
   Prep valve surface: 0.5 h x 800 μSv/h = **400 μSv**
   Weld repair: 3 h x 800 μSv/h = **2.4 mSv** (the dose rate at 50 cm may be used as the average radiation level that the welder may be exposed to.
   Testing and weld inspection: 0.25 h x 800 μSv/h = **200 μSv**
   Clean job site: 0.5 h x 3 workers x 600 μSv/h = **900 μSv** (the average dose for cleanup should assume some time spent close to the valve area or at least a recognition that cleanup will involve work in various locations anywhere from a few meters away to perhaps several centimeters from the valve. This estimate could be highly variable as long as the student provides a reasonable basis for the selection of the dose rates used in developing the estimate.)
   RP job-coverage: 0.1 h x 200 μSv/h = **20 μSv**
   Total dose estimate for the valve repair = **4.72 mSv or 5 person-mSv**

Note: the actual dose estimate calculated is not the over-riding concern as long as the student provides justification for the dose estimate and demonstrates an understanding of the factors that need to be considered when deciding what dose rate values to use, in developing the pre-job dose estimate. Also the students should realize or at least ask questions concerning possible dose to the welder's hands. Should extremity/finger-rings be considered for the task due to the high contact dose rate (could be a good follow-on question?)

(b) If automatic welding is an option that should be considered. Flushing the line and/or valve to reduce radiation levels may be a possibility. Can the work be performed at a later time when radiological conditions of adjacent systems and components are optimized for dose reduction? Specifically is the secondary side of the steam generator filled with water to reduce general area dose rates? (This is the key parameter that the student should recognize.) Assuming that the dose rates provided in this problem were obtained with the SG drained then filling the SG secondary side could reduce general area dose rates by a factor of 2 or more.

2. Answer: Several key items should be identified by the student. The major items include: (a) the design of the valve should facilitate installation—bolted versus welded flanges for example; (b) specify the use of low cobalt material if possible—no nickel; (c) valve design should minimize hide-out or low spots that could become crud traps over the long term, producing radiation hot spots; (d) design characteristics of the valve should be evaluated to minimize routine maintenance activities and provide long service life.

3. Answer: As a minimum the student should recognize that the task poses significant radiological safety concerns with the potential of significant personnel exposures if the activity is not properly planned and controlled. This type activity should require an ALARA review and some type of formal pre-planning effort. The ALARA review should address controls necessary to prevent access to the pipe chase area while resin is being sluiced. Work control planning efforts and ALARA reviews should ensure that measures are established or identified to preclude any work activity being conducted or entries into those plant areas impacted by the transfer of resin, concurrently while resin is being sluiced. All entrances leading to the pipe chase area should be locked and properly posted as a LHRA and access strictly controlled during the evolution. These controls should be established at a location equipped with a door or gate that can be locked. If the student decides to establish posting and access controls at a location(s) that does not provide a physical barrier and/or is not capable of being locked then ensure that appropriate controls are specified. This may include the requirement to station an RPT at each location, the use of HRA lights (in accordance with plant technical specifications), video coverage of entry locations and similar controls or measures. However, ideally the student should realize that based on the potential for significant dose rates that the preferred method would be to establish positive controls in lieu of passive measures.

Additional measures that could be prescribed could include the use of remote ARM's staged at strategic locations along the spent resin header and perhaps ARMs placed at every entrance leading to a potential HRA. These ARMs should be equipped with a local alarm and posted with a warning to evacuate the area in the event of an alarm. A good practice is to have the control room announce over the in-plant communication system that a resin transfer is eminent and that personnel should vacate the area. In addition a sweep of the area(s) where high dose rates will be encountered should be performed just prior to the start of the job and the area secured.

Radiological planning should also include a requirement to perform a post-job survey of the exposed (unshielded) portions of the spent resin header to ensure that all the resin was sluiced to the spent resin tank. Specific requirements should emphasize the need to survey any low points in the header, pipe bends and any other location with the potential to trap resin during the sluicing operation.

4. Answer: After an operation involving the transfer of spent resin (especially highly contaminated resin) it is important to perform post-job surveys. These surveys should focus on suspect areas, such as bends in the transfer line or any low spots where resin may have deposited or where the flow rate may have been low that could have resulted in resin plating out or settling in the header. If there are any valves in the transfer line than radiation surveys should be performed to ensure that there are no localized high radiation levels in the vicinity of the valve. General area radiation surveys should be performed to verify that the area is properly posted.

   In this particular case it has been two days since the resin transfer was completed and radiation levels in the affected areas were not confirmed after the evolution. Therefore a review should be conducted to evaluate any entries made into the area over the interim period. This review should determine if any individual may have accessed the area and received unplanned radiation exposure.

   The lesson-learned for this event is the importance of ensuring that adequate and timely post-job surveys are performed for those activities with the potential to impact radiological conditions in plant areas. Corrective actions should ensure that the requirement to perform a post-job survey be incorporated into the work package job history data base and/or the radiation work permit history files. An extent of condition review may be performed to ensure that other activities with a similar potential also capture the need for a post-job survey.

5. Answer: Measures to control the spread of airborne contamination during machining and welding operations will be necessary to ensure that other activities in the cold machine shop are not impacted by the work. The work area should be enclosed within a temporary structure to confine any possible airborne and loose contamination from spreading to areas outside the immediate work area. This could simply be a tented structure using fire retardant materials (since welding, hot work, is to be performed) attached to a scaffolding-type frame structure or a localized enclosed area equipped with portable filtration units to "capture" any airborne contaminants that are generated during the machining and welding stages of the repair work.

   Arrangements to monitor airborne contamination levels both inside the confined work structure and immediately outside of the structure should be established. Monitoring of airborne concentration levels outside the enclosure is important to confirm that airborne concentration levels in the "uncontrolled" area do not exceed levels that may require controls to be established (e.g., posting or stay-time restrictions or use of RPE). Keep in mind the work area is located outside of the established RCA boundary.

A temporary contamination control zone should also be established around the work area. The use of step-off-pads and provisions to have workers monitor for the presence of contamination upon removal of protective clothing and when exiting the work area should be available. This is to ensure that contamination is not spread beyond the work area to non-RCA areas. Strict control of any radioactive material or radioactive waste generated during the activity must be properly controlled and transported to an approved radioactive material storage area. These type items should not remain outside of the work area without positive controls in place to ensure the proper labeling and handling of the material since the work area is outside of the established RCA.

## Chapter 8

1. Answer: Conductivity is the term used to measure the ability of a solution to conduct a current. When substances are dissolved in a solvent they dissociate to form ions. These ions are capable of conducting an electric current. As the amount of dissolved impurities increase the concentration of ions increases and the ability of a solution to conduct an electric current improve.
   Out-of tolerance conductivity values in various plant systems will lead to increased corrosion rates. As corrosion products are transported through the core they become activated leading to higher source terms of radioactive corrosion products. These activated corrosion products are transported to out-of-core locations and settle out in process piping and various components increasing system radiation levels.

## Chapter 10

1. Answer: Iodine is preferentially taken up (or absorbed by) the thyroid gland located in the neck. Therefore exposure to radio-iodine (primarily I-131) may result in a localized dose to an internal organ.
2. Answer: The ALI values in 10CFR20 are based on an annual exposure limit of 50 mSv/y while ICRP recommendations are based on a 20 mSv/y annual limit. Therefore the current 10CFR20 ALI values are different than those now recommended by the ICRP.

## Chapter 11

1. Ensure that workers have been evacuated from the affected area and establish postings to prevent inadvertent entry to the affected area. If the CAM does not provide read-out or annunciation functions in the main control room then contact operations. If possible enter the area, with appropriate respiratory protection, and observe the CAM reading. Monitor the situation continuously until airborne radioactivity concentrations have returned to normal or acceptable levels. Grab air samples may be necessary to confirm airborne concentration levels in plant areas adjacent to the affected area.

# Appendix E

# Radiation Dose Rates and Activity Conversions

Hopefully this will be one of the last texts written that has the need to show conversions from the outdated "conventional" units to the internationally accepted SI units.


*Radiation dose rates*

| | |
|---|---|
| 1 μSv/h = 0.1 mrem/h | 1 mrem/h = 0.01 mSv/h = 10 μSv/h |
| 1 mSv/h = 100 mrem/h | 1 rem/h = 0.01 Sv/h = 10 mSv/h |
| 1 Sv/h = 100 rem/h | 1 μrem/h = 0.01 μSv/h |

*Activity values*

1 Bq = 1 dps

1 Ci = 3.7E10 dps or 1 Ci = 3.7E10 Bq

$Tera = 10^{12}$    0.037 TBq = 1 Ci

$Giga = 10^{9}$    37.0 GBq = 1 Ci

$Mega = 10^{6}$    37,000 Mbq = 1 Ci

| | |
|---|---|
| 1 TBq = 27 Ci | 1 Ci = 37 GBq |
| 1 GBq = 27 mCi | 1 mCi = 37 MBq |
| 1 MBq = 27 μCi | 1 μCi = 37,000 Bq |
| | 1 pCi = 37 mBq |

R. Prince, *Radiation Protection at Light Water Reactors*,
DOI: 10.1007/978-3-642-28388-8, © Springer-Verlag Berlin Heidelberg 2012

# Appendix F

# Radiological Controls for Steam Generator Work Activities

## *Laboratory Exercise*

### Purpose and Scope

You have been assigned to coordinate the preparation and implementation of radiological control measures for steam generator (SG) work and inspection activities for an upcoming outage. The inspection activities involve eddy current testing of the SG tubes. This requires access to the SG bowl area. The SG manways will be removed and multiple entries (or jumps) into the bowl area to set-up equipment, install nozzle dams, and the performance of visual and eddy current inspections will take place over a several day period. Once the SG eddy current testing equipment is installed the two primary side SG manway covers on the U-tube SG will remain open to allow access of hoses and cables associated with inspection and test equipment. Work activities will be performed over several days.

### Summary of Work Activities

The major work tasks associated with this activity are summarized below:

To access the steam generator manway covers the mirror insulation has to be removed from the covers and adjacent areas of the SG. This activity requires unlatching several pieces of insulation and moving the pieces so access to the work area is not encumbered.

The steam generator manway covers are held in place with about 15 studs that must be detensioned using air operated stud detensioners. The manway cover weighs about 150 kg and is removed utilizing a manway transport device. Removal of the SG manway covers will result in increased radiation levels on the SG work platform due to the shielding afforded by the metal manway covers that

R. Prince, *Radiation Protection at Light Water Reactors*,
DOI: 10.1007/978-3-642-28388-8, © Springer-Verlag Berlin Heidelberg 2012

are about 15 cm in thickness. Once the manway covers are removed an internal diaphragm must be removed. The diaphragms are held in place by several flat-head screws that are flush with the outside surface of the diaphragm. The diaphragms are constructed of a thin piece of stainless steel and do not afford any appreciable shielding benefits. Removal of the diaphragm will expose the internal bowl section of the SG.

Once the SG bowl is open then nozzle dams must be installed in both sides of the bowl (refer to the figure of the steam generator). There is a nozzle leading to the reactor vessel and another nozzle connecting the cross-over leg of the SG to the RCP. These are labeled as the primary coolant inlet and outlet nozzles, respectively. The nozzles are approximately a meter in diameter. To prevent tools or debris from falling into the nozzle and primary system piping, nozzle dams are installed. Nozzle dams consist of several unique inter-locking pieces (depending upon their design) that when installed block access to the primary system piping leading from the SG bowl. An inflatable nozzle dam seal is an integral component of the nozzle dam. The seal is inflated with nitrogen. Nitrogen lines run from the nozzle dam seal to the nitrogen source located nearby. These dams also serve another purpose which allows the reactor cavity to be flooded to support ongoing refueling operations con-currently with SG inspections. Consequently, primary system water will be present on the other side of the nozzle dams, hence the term "nozzle dam".

Installation of the nozzle dams requires 2–4 full-body entries into the SG bowl to perform the installation, inspections to verify proper installation and hook up of nitrogen seal supply.

**Radiological Conditions**
Access to the SG work platform (grating) is via a 3 m vertical ladder. The platform is located just underneath the bottom of the steam generator. The centerline of the SG manways is about 1.2 m above the floor of the platform. Limited work space is available on the platform itself. Based on historical survey records from previous SG inspection activities the following radiological conditions are anticipated:

General area dose rates in the center of the bowl: 10–50 mSv/h (1−5 R/h)
Dose rates on contact to the divider plate: 20–50 mSv/h (2–5 R/h)
Dose rates on contact to the tube sheet: 50–100 mSv/h (5–10 R/h)
Dose rates on the SG platform with manway cover removed: several mSv/h to 12 mSv/h (several hundred mR/h to 1.2 R/h)
Loose surface contamination levels inside the SG bowl: µGy/h (mrad/h) smearable

**Exercise**
Describe the radiological safety measures that should be established to support SG inspection activities. These measures should address the aspects of steam generator work coverage listed below.

1. General area preparation to control airborne contamination and the spread of loose contamination.

2. Posting of the work area including entrance(s) to work area, the steam generator platform and immediate work area(s).
3. Any pertinent access controls during the work activity including required controls when the SG manways are removed?
4. Any engineering controls or special equipment required to minimize radiological concerns.
5. Protective clothing required for general entries to the vicinity of the SG, partial-body entries (upper torso) into the SG manway and full-body entries (jumps) into the SG bowl.
6. Dosimetry requirements associated with the activities noted in item 5.
7. Describe the type and location of air sampling required to support the various tasks associated with SG work.
8. Describe the controls for donning and doffing of protective clothing, including those parameters you would consider when identifying the location where these tasks would be performed.
9. Describe the radiation protection job coverage requirements for the various activities. Explain how teledosimeters could be utilized for these activities and any dose savings that may be achieved by utilizing these devices (hint: how would RP coverage be impacted for entries to HRA and LHRA areas if teledosimetry and/or remote video coverage was available)?
10. Explain how you would control entries to LHRA's and HRA's. During periods when individuals are not present on the SG platform and explain posting and access control requirements based on radiation levels.

Answers:

1. General area preparation to control airborne contamination and the spread of loose contamination.
   Anticipated contamination levels once the SG is opened will be excessive—on the order of $\mu Gy/h$ smearable, perhaps as high as hundreds of $\mu Gy/h$ smearable. Consequently, the student should recognize the need to prepare the work area in anticipation of extremely high contamination levels. The dose associated with contamination control measures should be evaluated to ensure that the overall exposure for the task is not higher than what it would be with less elaborate preparations. In some cases it may be more effective, from a dose reduction perspective, to provide minimal contamination control barriers and simply recognize the need that gross decontamination efforts may be required periodically during the course of the job. Exposure associated with these decontamination efforts may be lower than the exposures received while erecting and removing elaborate contamination control barriers.
   If the decision is made to erect contamination control barriers and enclosures then the following items should be addressed:

   - The enclosures should be equipped with a portable filtration unit to filter air from the enclosures. The filtration units should be equipped with both HEPA

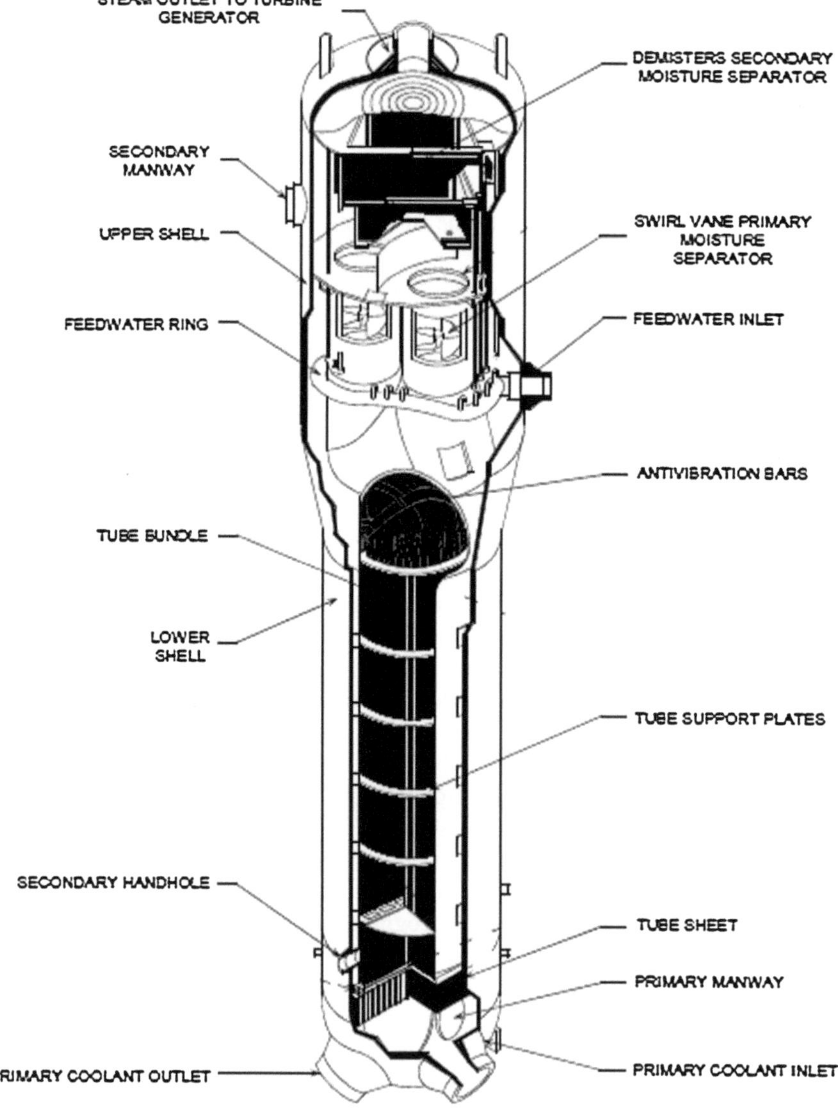


**Fig. F.1** Steam generator and its major components (adopted from www.nrc.gov/reading-rm/basic-ref/teachers)

and charcoal filters to remove airborne particulate and any radio-iodine that may be present.

- Multiple layers of sheeting should be placed on the SG platform. The layers should be designed to facilitate removal as work progresses. As equipment and workers exit the SG over the course of the activity contamination levels on the platform will increase. The ability to quickly remove layers of the contamination control material as contamination levels increase on the platform would minimize contamination control issues and potential airborne contamination concerns.

Note: the student should recognize the need to erect the contamination control barriers and enclosures before the SG manways are physically removed. Once the manway covers are removed area radiation levels will increase.

A portable HEPA filtration unit should be connected to one side of the SG while eddy current inspection activities are performed on the opposite side of the SG bowl. This ensures that air flow will be from the work side of the SG through the SG tubes into the filtration unit. Additionally, as work progresses and the SG tubes continue to dry additional contamination may become airborne from SG internal surfaces. The portable HEPA filtration unit serves to minimize potential airborne concentrations. The exhaust from the HEPA unit should be monitored with a CAM to ensure that the exhaust stream does not pose airborne contamination concerns as the job progresses.

2. Posting of the work area including entrance(s) to work area, the steam generator platform and immediate work area(s).

Postings will be required based on contamination and potential airborne concentrations once the SG is opened. The primary concern dealing with radiological postings will involve the changing radiation levels experienced in the vicinity of the SG manways during various stages of work. Periods when the manway cover is removed will result in greater than 10 mSv/h dose rates on the platform area. Measures must be established to ensure that postings reflect dose rates in the work area. For practical reasons it may be appropriate to treat the SG platform as a locked high radiation area (LHRA) and control access accordingly.

Note: If this approach is taken, access controls and entry requirements must comply with 10CFR20 and plant Technical Specification requirements regarding entry to a locked-high radiation area. Even if dose rates are less than 10 mSv/h, if the area has been posted as a LHRA, appropriate controls must be established and followed. (For USA plants—a licensee could not later claim for example, that the administrative controls required for entry to a LHRA were not implemented, because at the time of entry actual dose rates were less than 10 mSv/h. If the area is posted as a LHRA then access must be controlled accordingly).

Since the SG platform is not enclosed within a well defined room, nor are there solid walls that could be utilized to control access, posting aspects will pose unique challenges. Under these conditions the use of a "flashing" light may be necessary. During plant outages with nearby scaffolds erected individuals could gain access to the SG platform from multiple directions. These factors must be considered to ensure adequate posting from all access routes.

3. Any pertinent access controls during the work activity including required controls when the SG manways are removed?

   Radiation levels streaming directly from the open manway will most likely exceed the criteria for a LHRA. Since the work platform is not a "lockable" location other controls will be required to control access to the LHRA. Utilizing plant Technical Specifications a combination of flashing lights, video surveillance and/or continuous RP coverage will be required during those periods when the manway cover is removed or when the temporary manway shield cover is open (if available).

4. Any engineering controls or special equipment required to minimize radiological concerns.

   As alluded to in the above answers, various engineering controls and equipment may be required to support this activity. The contamination control enclosure (if utilized) may consist of a tented area leading from a SOP positioned on the floor in the vicinity of the vertical ladder leading to the SG platform. The vertical ladder may itself be enclosed in a tent structure to the SG platform, and the platform itself enclosed. This entire enclosure will become highly contaminated during the course of work activities and possibly experience periods of elevated airborne concentrations. The contamination control enclosure should be vented via a portable filtration unit to minimize airborne contamination concerns.

   Once SG eddy current inspection activities commence there will be extended periods of time in which workers will not be present on the platform. The SG inspection team may be located at a remote facility, oftentimes outside of the containment building itself. The use of a temporary SG manway shield cover, designed to allow the necessary cabling, nitrogen airlines to the nozzle dam, and other miscellaneous electric cables to be routed through the cover into the SG bowl area would be beneficial. The manway shield is designed to reduce radiation levels on the platform to levels less than that of a LHRA. This would eliminate the need to control access as a LHRA during those periods when workers are not in the area and while remote inspection activities are in progress. The shielded cover would have to be "lockable" to satisfy access control requirements for a LHRA.

5. Protective clothing required for general entries to the vicinity of the SG, partial-body entries (upper torso) into the SG manway and full-body entries (jumps) into the SG bowl.

   This activity would require the use of a double SOP arrangement. Any activity requiring access to the SG platform, after the SG has been open, would expose workers to high levels of contamination. Controls should be established to minimize the spread of this contamination.

**General entries to the vicinity of the SG**

A double set of protective clothing with the outer coverall consisting of a lightweight fabric, perhaps disposable or one-time use material. If access to the work platform is required then respiratory protection may be necessary, such as a full-face air purifying respirators. The situation is compounded since access to the platform is via a vertical ladder, the rungs of which will become highly contaminated during the course of work activities. The use of a face shield at a minimum may be necessary to protect against facial contamination even if air sample results indicate low airborne radioactivity concentrations.

**Partial-body entries**

In addition to the protective clothing requirements noted for general entries above, respiratory protection will be required. Depending upon airborne concentrations and assuming that any entry into the SG bowl area will be strictly limited based on dose rates; a full-face air purifying respirator should afford the necessary protection. Additional gloves may be prescribed due to the extremely high contamination levels within the bowl area. The outer set of gloves could be removed while still on the platform area after exiting the SG. This would minimize the spread of high-levels of contamination to other areas of the enclosure. A third, outer layer of protective clothing, covering the upper torso could also be prescribed. This in kind could also be removed on the platform after exiting the SG.

**Full-body entries**

Entry into the SG bowl will place an individual in close proximity to extremely high contamination levels. Typically, a direct line of communication is maintained with individuals making full-body entries into a SG. Consequently, respiratory protection may be selected based on the need to provide both radiological protection and the ability to communicate with workers outside the SG. The use of air-supplied hoods (or bubble hoods) offers a cooling component and allows the user to be fitted with a wide option of communication devices. These air hoods also provide a larger field of visibility compared to a full-face respirator. Either two or three coveralls may be used, again depending on heat stress conditions. The outer coverall should be water resistant, offer maximum breathability and provide an effective barrier against the migration of contamination through the outer fabric layers. Extra set of outer gloves and shoe covers may also be required, which could be removed upon exit from the SG. A factor that could influence the type and amount of protective clothing for the three situations noted above would be environmental conditions in the work area. Heat stress considerations could impact the selection of respiratory protection devices or the choice of coverall materials. Based on the unique industrial safety conditions associated with entry into a SG the overall safety of the individual must be considered when prescribing protective clothing and equipment.

Work activities that involve handling eddy current equipment should require the use of an outer glove, most likely a surgeon's type glove to support delicate work activities. These gloves should be changed out frequently for work conducted on

the SG platform and changed immediately after handling EC equipment or tools or items removed from the SG bowl.
6. Dosimetry requirements associated with the activities noted in item 5.

**General entries to the vicinity of the SG**

Dosimetry requirements would include that required for entry into the RCA and containment building during outages, namely a primary dosimeter and an ED. The ED should be equipped with a transmitter unit for entry into the SG work area for monitoring by RP personnel at a remote location.

**Partial-body entries**

In addition to the dosimetry requirements noted above for general entries, extremity monitoring may also be required. Depending upon the design and features of the robotic eddy current equipment utilized a partial-body entry may only involve inserting a forearm into the steam generator bowl for a short period of time. Under these conditions no additional dosimetry may be necessary. If partial-body entries involve the head and upper torso, then these body areas may be located in the highest dose rate areas. Under these conditions the primary dosimeter may have to be relocated to the head region of the individual or a second dosimeter provided. Alternatively a front and back dosimeter may be utilized to evaluate the EDE as noted under full-body entries below. Depending upon the length of exposure extremity dosimeters for the forearms may also be required and perhaps finger rings for the hands, though this eventuality should seldom occur due to the capabilities and design features of eddy current inspection equipment.

**Full-body entries**

The use of two dosimeters (one positioned on the front of the individual and the other on the back) may be utilized to determine EDE. For most situations this methodology should prove effective and results in a less conservative calculation of exposure received by SG jumpers.
A multi-badge pack may be provided to individuals entering a SG if conditions do not lend themselves to using a front and back dosimeter to determine EDE. Due to the high dose gradients that may exist within short distances inside the bowl area, even a momentary movement of an arm or leg could result in a significant exposure to an individual part of the body. For instance, contact to the tube sheet or divider plate with a hand or foot could result in a significant difference between the dose received to the whole-body and that to an extremity. Teledosimeters could be placed on the front and back of the person and even on the top of the head, strapped to the bubble hood, for example. The primary objective is to ensure that sufficient dosimetry is provided to measure the highest dose to the whole-body and the most likely, highest exposed, extremity area.
7. Describe the type and location of air sampling required to support the various tasks associated with SG work.

Assuming that the enclosure arrangement described in the response to questions 1 and 4 above is utilized then air sampling may include continuous air samplers, equipped with both a particulate and charcoal filter head, located on the SG platform and inside the enclosure, in close proximity to the outer SOP where protective clothing is removed. A continuous air sampler should be located immediately outside the enclosure to monitor airborne concentration levels in areas occupied by support personnel. After several days iodine sampling could be eliminated if radio-iodine has not been detected or if concentrations are low (e.g., less than 10% of the applicable DAC). In many cases negative data may prove useful and the decision to eliminate iodine sampling should be well founded.

The use of portable continuous air monitors with live-time readout and alarm functions offer many advantages. If area radiation levels allow and the units can detect the required levels of airborne concentrations then these units should be utilized where possible. The ability to provide live-time read-out and instantaneous alarms could minimize potential radiological incidents. If possible a CAM unit positioned outside the entrance to the tented enclosure, or entry point to the SG work area, set to alarm at a relatively low airborne concentration level would serve to provide early warning of elevated airborne levels in the work area or an indication that contamination control measures have been compromised.

Due to the high contamination levels within the bowl area and limited physical space, the use of lapel air samplers on "jumpers" is usually not warranted due to cross-contamination concerns. Since respiratory protection will be worn by these individuals in any event this should not pose a problem. A good practice is to WBC individuals who enter SGs as a precautionary measure. These WBCs should be performed on the same day as the entry or within a short period thereafter (e.g., within 24–48 h).

8. Describe the controls for the removal of protective clothing, including those parameters you would consider when identifying the location where these tasks would be performed.

Due to the limited space available on the SG platform, at least based on the conditions offered in this exercise, a balance must be achieved between the need to remove highly contaminated articles of protective clothing as soon as possible and the need to move to a lower dose rate area as quickly as possible. An additional concern relates to the need to descend a 3 m ladder. Therefore individuals should not be encumbered by airlines, communication cables, or extraneous protective clothing to the point that industrial safety becomes an overriding concern while descending the ladder.

Provisions to remove outer shoe covers and outer gloves at the base of the ladder once individuals descend from the SG platform should be available. People to assist SG jumpers or platform workers with the removal of any protective clothing should be available. The assistants may also be responsible for securing equipment and breathing lines as the jumpers descend the vertical ladder. Obviously the individuals assisting with these activities will have to be dressed in the appropriate protective clothing and respiratory protection devices.

Once down the ladder an area (i.e., the inner SOP location) to remove respiratory protection equipment and the outer set of protective clothing should be available. This location should be in a relatively low background area and far enough from the SG and other components that may be a source of high radiation, to minimize exposure to workers while removing protective clothing. Since the SGs are located within the biological shield wall, the selection of the location will be a compromise between finding a suitable location to set up the inner SOP, and minimizing the travel path while workers are wearing potentially highly contaminated protective clothing. All things considered an area just outside the loop room or the biological shield wall may be the best compromise. Again, other individuals may be available to assist workers in the removal of these items. Once the required outer protective clothing articles have been removed workers could then proceed to the outer (final) SOP area. The use of temporary shielding racks could be considered to provide shielding at strategic locations to minimize exposure to workers during the exit process.

The outer SOP area may be located in the general walkway area outside the biological shield wall. Alternatively, depending upon past experience and the effectiveness of contamination control measures, the containment building general access SOP area may serve as the "outer" SOP for SG work activities. Workers would remove the final set of protective clothing at this location following standard procedures for removing protective clothing and crossing a SOP area. Controls should be established to ensure that potentially highly contaminated protective clothing or respiratory protection equipment is not handled at the containment building general access SOP area. These items should have been removed, bagged and tagged, or otherwise controlled at the inner SOP area.

9. Describe the radiation protection job coverage requirements for the various activities. Explain how teledosimeters could be utilized for these activities and any dose savings that may be achieved by utilizing these devices (hint: how would RP coverage be impacted for entries to HRA and LHRA areas if teledosimetry and remote video coverage was available)?

   Multiple entry and exit from both high radiation area and locked-high radiation areas will be involved with this activity. RP coverage will be extensive due to controls associated with entry into HRAs and the high levels of contamination. The use of respiratory protection devices will require RP support as well as coordination of the issuance of multi-badges and teledosimetry for SG jumps and other tasks. The strategic use of remote monitoring and teledosimetry systems would minimize exposures to RP personnel for this task.

   Those work activities involving access to locked-HRAs would represent critical RP coverage stages. This could include tasks while workers are on the SG platform as well as for those tasks involving entry pass the plane of the SG manways. The use of remote video coverage along with direct communication capability would allow RP "coverage" to be provided at a remote location. Once workers are equipped with teledosimeters that communicate to a base station, located at the RP remote coverage point, there would be no need for RP technicians to be stationed on the platform or otherwise to remain in HRAs. The details of RP coverage requirements should have been incorporated in pre-job preparations and during mock-up training sessions with the work crew. Entry requirements for entering onto the SG platform based on radiation levels should be discussed during pre-job training sessions.

Vacuum cleaners may be required to remove debris from the EC inspection cables periodically during the course of inspection activities and to clean other miscellaneous tools and equipment. The vacuum cleaner collection drum may approach radiation levels requiring controls as a locked HRA. Arrangements should be established to ensure that the collection drum is surveyed routinely. A teledosimeter placed on the collection drum to provide remote monitoring of radiation levels on the drum would provide warning when dose rates approach 10 mSv/h.

Control of contamination is the other primary area that will involve RP to a large extent. Throughout the task work activities will have the potential to spread high levels of contamination. RP surveillance activities and contamination control measures will have to be adequately implemented to minimize airborne contamination concerns or to prevent the spread of contamination beyond the established boundaries. The top layer of materials placed on the SG platform and other areas within the contamination control zone will have to be removed periodically to maintain contamination within manageable levels. RP coverage activities will involve close monitoring of contamination levels within the work area to ensure that levels do not pose additional radiological concerns.

10. Explain how you would control entries to LHRA's and HRA's. During periods when individuals are not present on the SG platform and explain posting and access control requirements based on radiation levels.

    The use of a SG manway shield cover to reduce dose rates to less than 10 mSv/h outside the SG would eliminate the need to control areas as locked-HRAs, thus, facilitating access control requirements when workers are not present. If a lockable manway shield cover is not available or if dose rates cannot otherwise be reduced to levels less than those requiring controls as a locked HRA, then a flashing light may be provided at each entrance to the SG platform area. Alternatively, or in conjunction with flashing lights, positive administrative controls could be established. Since remote video coverage will most likely be available for SG work activities this function could be performed at the remote monitoring facility.

# Index




R. Prince, *Radiation Protection at Light Water Reactors,*
DOI: 10.1007/978-3-642-28388-8, © Springer-Verlag Berlin Heidelberg 2012

MIX
Papier aus verantwortungsvollen Quellen
Paper from responsible sources
FSC® C105338

FSC
www.fsc.org

If you have any concerns about our products,
you can contact us on
ProductSafety@springernature.com

In case Publisher is established outside the EU,
the EU authorized representative is:
**Springer Nature Customer Service Center GmbH**
**Europaplatz 3, 69115 Heidelberg, Germany**

Printed by Libri Plureos GmbH
in Hamburg, Germany